PRECIOUS STONES

PRECIOUS STONES

A POPULAR ACCOUNT OF THEIR CHARACTERS,
OCCURRENCE AND APPLICATIONS
WITH AN INTRODUCTION TO THEIR DETERMINATION,
FOR MINERALOGISTS, LAPIDARIES, JEWELLERS, ETC.
WITH AN APPENDIX ON PEARLS AND CORAL

BY

Dr. MAX BAUER

Translated from the German with Additions by
L. J. SPENCER, M.A. (Cantab.), F. G. S.

With a New Foreword and Addenda by
DR. EDWARD OLSEN
Curator of Mineralogy
at the Field Museum of Natural History
Chicago

IN TWO VOLUMES
Volume I

New York
DOVER PUBLICATIONS, INC.

International Standard Book Number: 0-486-21910-0
Library of Congress Catalog Card Number: 68-19167

Manufactured in the United States of America
Dover Publications, Inc.
180 Varick Street
New York, N. Y. 10014

FOREWORD TO THE REPRINT EDITION

FROM the very beginning man's survival and well-being has depended on his resourceful utilization of the mineral kingdom. The Stone, Iron, Bronze, Steel, and Atomic Ages have all been built upon a base of mineral technology, from the extremely simple to the very sophisticated. It is also evident that for a very long time man has used minerals for ornamental, religious, and magical purposes. At least one Stone Age site in Europe has revealed the burial of a man accompanied by an assortment of pretty pebbles and mineralized fossils. Among the Stone Age aborigines of North America the Mound Builders of the Ohio Valley are known to have used obsidian and sheets of mica at least partly for ritual purposes. How important such materials were to them may be judged by the fact that the nearest occurrence of mica was over 200 miles away and the nearest occurrence of obsidian over 1500 miles away in the region of the Yellowstone River !

The past several centuries have seen the rise of Western industrial society and with it an educated and affluent middle class. This has created in turn an increased demand for some of the luxuries once enjoyed only by the wealthy or high-born. Precious stones have been among the first such luxuries sought. As a result the gem and jewelry industry today commands an impressive share of the total value of the world mineral market.

With increased interest in precious stones has grown a demand for information concerning them and this has been met over the past century by a number of nontechnical books. Few of these books, however, have been entirely successful. In attempts to avoid being too technical some are rambling, poorly organized, and even misleading. There are, however, several notable exceptions. One of them is this book, *Precious Stones*, by Professor Max Bauer.

Dr. Bauer was a distinguished German mineralogist whose career spanned the turn of the century (1844–1917). For thirty years he was head of the Mineralogical Institute of the University of Marburg. In his many undertakings he epitomized the thoroughness we have come to associate with the German scholar. In his lifetime he published a large number of technical papers and two books, one of them this book on precious stones.

The original German edition, titled *Edelsteinkunde*, was published in 1896. Thirteen years later in 1909 a second German edition appeared, and in 1932 Dr. Karl Schlossmacher, who had studied under Professor Bauer, published a third German edition which was completely revised and largely rewritten by him. Between the first two editions the book enjoyed considerable popularity and in 1904 L. J. Spencer of the British Museum (Natural History) published this English translation of the first German edition with some minor additions and corrections.

Most of the information in this book is as up-to-date today as it was sixty years ago. The latter half of the nineteenth century saw minerals finally classified into a logical and workable system which is still used today—the chemical classification. Thus Bauer was able to group together gem minerals which are related chemically to each other and hence share similar physical properties pertinent to their uses as gems. The book is divided into four main sections, of which the first two comprise the major portion of the text. The first section deals with the general characteristics of gem minerals. The only major addition that might be made to this section today would be a discussion of the vast amount of information which has been accumulated over the past fifty years on the arrangements of atoms within crystals and the direct relationship between the geometry of the many atomic arrays and the external physical properties of minerals. Secondly the present decade has seen the synthesis of precious stones on a scale which makes them competitive with, and virtually indistinguishable from, natural stones.

The second section is a systematic descriptive treatment of precious minerals. In contrast to the majority of books on gems this section is not merely a dry recitation of physical properties and occurrences. Into many of the gem descriptions are woven bits of history, lore, and memorabilia associated with them. For example, in the section on diamonds one encounters such varied items as details of the infamous Arizona diamond swindle of 1870, a discussion of early attempts to make synthetic diamonds, and the quaint ancient test for telling real diamonds from imitations using a hammer, an anvil, and warm goat's blood, the goat having preferably dined on parsley prior to being bled ! Furthermore, gem occurrences as they are described in this section do not consist merely of lists of geographical localities, as is common with many other books. Information is also provided on how the gems occur, in what rock types, in what abundance, and how they are mined. All minerals cannot, of course, be treated with equal weight and some of the less common and little used minerals are passed over in a paragraph or two. There is usually little more to say about them.

All this is not to say that the age of this book does not appear here and there. On page 254 the Excelsior diamond (over 971 carats) is cited as the largest diamond known. Just one year after this book was originally published the Cullinan diamond (3106 carats) was found. Similarly gem-cutting procedures may differ somewhat today because of the introduction of inexpensive synthetic abrasives during the past forty years. Thus his list of recommended abrasives on the bottom of page 83 is of less utility to the present-day gem cutter. Occasionally also Bauer uses a mineral name which is no longer commonly used, and although the chemical compositions of most minerals were accurately known at the time, a few of the compositions given by Bauer have been found to be incorrect. Also during the past sixty years new highly productive gem localities have been discovered and developed.

In going over the book there seemed to be a small enough number of such amendments and corrections that the preparation of a short section of addenda appeared to be feasible. In preparing this the intention was to make only a minimum number of necessary comments designed to make the book more useful to the present-day reader.

In addition to what has been said already it should be noted that *Precious Stones* contains information which is relatively unique in a book of this kind. Professor Bauer has included discussions of a number of topics which are usually only briefly mentioned or left entirely unnoticed in many older and most present-day books on this subject. For example, he discusses in some detail common fraudulent practices in the gem business, the artificial coloring of stones, the making of doublets, the use of glass imitations, and the clever concealment of imperfections in poor quality stones. Such practices are rampant even today. He discusses common flaws in gems and how to detect them. Most unique of all he treats with the many concrete and intrinsic factors that enter into the evaluation of gems. Although he quotes values which have little bearing on today's marketplace the factors contributing to gem evaluation are the same today as they were at the turn of the century.   Professor Chester Slawson in writing the section on gems for the *Encyclopaedia Britannica* has called this book " the most detailed work on gems."

Today in the Americas and Western Europe, in addition to the many gem and jewelry professionals there is an ever-growing number of amateur mineral collectors and hobbyists whose interests range from assembling collections of attractive minerals and rocks to the faceting of gems to standards of extraordinary perfection. It was for these admirers and students of precious stones that Max Bauer wrote this book.

EDWARD OLSEN.

ADDENDA (1968)

p. 7, *line* 16.   The method of x-raying minerals was subsequently discovered in 1917. It is now known that the different geometrical arrangements of carbon atoms in diamond and graphite create the differences in hardness and color which distinguish them.

p. 35, *line* 30.   The x-ray method mentioned here is not the same as *X-Ray Diffraction*, which is used to determine the atomic structures of crystals (cf. above note for p. 7, line 16).

p. 44, *line* 32.   Rutile (*q.v.*) has a dispersive power over six times greater than that of diamond. Today synthetic rutile is commonly faceted into gemstones under the commercial name of " titania."

p. 63, *lines* 47–48 ; *p.* 64, *lines* 1–4.   The small plates referred to here are now known to be finely alternating crystal twins. The labradorescence is due to light interference along the plate-like twin-crystal boundaries.

p. 67, *line* 44.   An electric current can also be produced in a crystal by the application of *pressure*. This is called *piezoelectricity*.

p. 92, *line* 38.   Streaks of " silk " are also called " veils."

pp. 93–94.   Artificial Production.   During the last twenty years the production of large artificial gemstones has made immense strides. Synthetic ruby, sapphire, star sapphire, emerald, spinel, rutile, and other stones are produced in large quantities, and provide high-quality and relatively inexpensive materials for the amateur, professional, and commercial gem cutter.

p. 103, *bottom half of page*.   In 1913 the *metric carat*, equal to 200 milligrams, was adopted in the United States ; in 1932 it became the international standard definition of the *carat*.

p. 126, *line* 9.   The largest Brazilian diamond ever discovered was found in 1938. It weighed 726.6 metric carats and was called the " Presidente Vargas." It was subsequently cut into 29 separate stones, the largest of which is 48.26 carats.

p. 127, *line* 5.   In 1905 the largest diamond discovered to date, the " Cullinan," was found at the Premier Mine, Transvaal, South Africa. It weighed 3106 carats in the rough.

p. 132, *line* 4.   See note for p. 44, line 32.

p. 154, *line* 4.   Dr. Bauer's prediction has been fully realized. Today India's annual diamond production accounts for less than 0.005% of the world total. Also see note below for p. 179, line 44.

p. 159, *line* 17.   See note above for p. 126, line 9.

p. 179, *line* 42.   At the present time Brazilian annual diamond production has become almost negligible, comprising a little less than 1% of the world total.

p. 179, *line* 44.   Over the past fifty years the Congo Republic (former Belgian Congo) has gradually become the major world source of diamonds ; at present it accounts for about 65% of total annual African production and almost 56% of world annual production. South Africa is second, with 18% of African annual production and 12% of world annual production. The Congo diamonds are, however, chiefly industrial grade stones of smaller sizes. South Africa is still the major producer of stones of gem quality. Other large producers in Africa are Ghana, Southwest Africa, Sierra Leone, and Angola.

p. 231, *line* 34.   At the present time the U.S.S.R. produces about 8% of the annual world output of diamonds. Most of its mines are not in the Urals, however, but in new producing districts in Siberia.

p. 233, *line* 14.   There is, at present, a considerable body of evidence indicating that the diamonds found in meteorites were created upon impact with the earth. The tremendous force of the shock is sufficient to convert some small portion of the original graphite, which is very common in meteorites, into minute diamond grains. This point is still in dispute, however, and there are those who argue that the diamonds formed within the meteorites long ago when they were parts of their original planets in space.

pp. 233–238.   In more recent years attempts by others to verify the " successful " experiments of Moissan and Friedländer have been completely unsuccessful. As a result their original experimental results have been seriously questioned.

In 1955 the first known successful synthesis of diamond was achieved in the research laboratories of the General Electric Company at Schenectady, New York. The process requires both high temperature (over 5000°F.) and extremely high pressure (1.5 million lbs. per square inch). These diamonds are all of industrial grade and quite small. It is estimated that such synthetic diamonds comprise between 1% and 2% of the total annual world output of all diamonds.

p. 249, *line* 14.   A *line* as a unit of linear measure (sometimes spelled *ligne*) is equal to 2.26 millimeters, or about 0.09 inches.

p. 252, *line* 23.   See note above for p. 126, line 9.

p. 254, *line* 15.   See note above for p. 127, line 5.

p. 266, *line* 7.   It is now known that even those stones which do lose their coloration on heating are *not* colored by organic compounds.

p. 281, *line* 23.   Synthetic rubies (and sapphires) are presently produced in sizes up to 200 carats by the somewhat more complicated Verneuil process. These stones are competitive with natural stones in price but are not as highly prized.

p. 296, *line* 15.   Spinel, hardness 7.5–8, is actually slightly softer than topaz.

p. 299, *line* 38.   Synthetic gem spinels have been produced since the early 1930's. They are available in a wide variety of colors at very modest prices.

p. 305, *line* 48.   Synthetic chrysoberyl (alexandrite) has never been produced commercially ; there are, however, synthetic spinel and corundum which show, to some degree, a daylight-artificial light color change similar to that of alexandrite. These synthetics are quite inexpensive, whereas genuine faceted alexandrite is very costly.

p. 318, *line* 32.   Today emeralds are produced synthetically by at least two processes and are marketed as gemstones.

p. 324, *line* 13.   In addition, absolutely colorless beryl, known as *goshenite*, is sometimes faceted for gem use. Pale pink to pale peach-colored beryl is known as *morganite*. It is commonly milky and only rarely transparent enough to warrant faceting.

p. 346.   Today six basic species of garnet are recognized :

1. Pyrope	$Mg_3Al_2(SiO_4)_3$.
2. Almandite	$Fe_3Al_2(SiO_4)_3$.
3. Spessartite	$Mn_3Al_2(SiO_4)_3$.
4. Uvarovite	$Ca_3Cr_2(SiO_4)_3$.
5. Andradite	$Ca_3Fe_2(SiO_4)_3$.
6. Grossular	$Ca_3Al_2(SiO_4)_3$.

Gem species are now all considered subvarieties of these basic garnets. Hessonite (sometimes misspelled " essonite ") and rosolite are varieties of grossular. Rhodolite is a variety of pyrope. Topazolite, demantoid, melanite, and schorlomite are varieties of andradite.

p. 364, *line* 31.   Three quite complex generalized species are in use today in descriptions of the chemistry of the tourmalines : dravite, schorl, and elbaite.

p. 374, *line* 19.   Some opal, when kept out in the air over a period of years, as when it is set and worn in jewelry, may gradually lose much of its water content by slow evaporation.   The resulting shrinkage in volume causes severe criss-crossed cracking, sometimes called *checking*, which can ultimately cause the stone to fall apart.   This can be retarded by coating the surface with a little mineral oil periodically, or by storing it, when not in use, in water.

p. 374, *line* 25.   In addition some occurrences of opal consist of opalized fossil bones of extinct reptiles and mammals.

p. 382, *line* 27.   A more recent United States occurrence of opal is at Virgin Valley, Nevada.   Stones from there have a very pronounced tendency to crack unless kept under water.

p. 384, *line* 4.   In 1903 the now famous Australian opal locality of Lightning Ridge, New South Wales, was discovered.   In 1915 precious opal was discovered at Coober Pedy in the Stuarts Range of South Australia, and in 1930 large opal deposits were found at Andamooka, South Australia.

p. 389, *line* 2.   The term *calaite* is no longer used.   The mineralogical name is now *turquois* (spelled *turquoise* in England).

p. 389, *line* 5.   Crystals of turquois were found near Lynch, Virginia, in 1912.

p. 389, *line* 9.   The formula given here is incorrect.   The correct formula is $CuO.3Al_2O_3.2P_2O_5.9H_2O$. In terms of weight percentages this is :

					Per cent
CuO	.	.	.	.	14.58
Al_2O_3	.	.	.	.	56.08
P_2O_5	.	.	.	.	26.03
H_2O	.	.	.	.	3.31
					————
					100.00

p. 389, *Analysis Table*.   The chemical analysis given here reports too little copper (CuO).   The specimen of turquois must have been contaminated by other minerals.

p. 389, *line* 21.   Penfield's formula is incorrect (cf. above note for p. 389, line 9).

p. 402, *line* 10.   " Blue iron-earth," $Fe_3(PO_4)_2.8H_2O$, is now known as vivianite.

p. 403, *line* 7.   Lazulite is now known to be a hydrated iron aluminum phosphate with the formula $FeO.Al_2O_3.P_2O_5.H_2O$.

p. 408, *line* 10   The formula for cordierite is now known to be $2MgO.2Al_2O_3.5SiO_2$ with no constituent water at all.

p. 410, *line* 20.   The formula for idocrase is now known to be $10CaO.2MgO.2Al_2O_3.9SiO_2.2H_2O$.

p. 412, *line* 33.   The formula for axinite is now known to be $4CaO.2FeO.B_2O_3.2Al_2O_3.8SiO_2.H_2O$.

p. 415, *line* 8.   Staurolite is now known to be a hydrated silicate of iron and aluminum, $FeO.2Al_2O_3.2SiO_2.H_2O$.

p. 422, *line* 4.   Chlorastrolite is now known to be a mixture of silicate minerals and not a single mineral.

p. 423, *line* 13.   The name *calamine* is no longer applied to zinc carbonate.   This mineral is now called *smithsonite*.

p. 426, *line* 3.   The color of amazonstone is now known to be due to small amounts of lead in it.

p. 430, *line* 15.   The sheen is now thought to be due to a microscopic interlamination of potash and soda feldspars.

p. 434, *line* 31.   Labradorescence is now thought to be due to optical interference effects caused by the twinning laminations referred to on p. 432.

p. 440, *line* 30.   Natural ultramarine is now called *lazurite*.

p. 450, *line* 2.   Moldavites are now grouped with tektites, which are considered by some to be meteoritic, by others to be terrestrial rock and soil fused by impacts of very large meteorites with the earth.

p. 453, *line* 1.   Bastite is now known to be a variety of serpentine.   It is not a pyroxene mineral.

p. 455, *line* 22.   In addition to the varieties of spodumene described, a lilac to rose-colored variety called *kunzite* is now fashionable as a gem.   It is found principally in San Diego County, California, and in the state of Minas Geraes, Brazil.   Heat treatment of faintly colored crystals is used to produce the desired gem color.

p. 458, *line* 36.   The formula is now known to be $2CaO.5MgO.8SiO_2.H_2O$.

p. 464, *line* 26.   It is difficult to see how Dr. Bauer could ever have concluded that in America jadeite is, or ever was, more common than nephrite.   It is certainly just the reverse.

p. 480, *line* 3.   The coloration of most smoky quartz is now considered due to radiation damage to its electronic structure by slightly radioactive minerals in the rocks in which it is formed. Some cases of coloration by organic inclusions may also occur.

p. 482, *line* 19.   Even today no one is certain what causes the coloration in amethyst.

p. 527.   The term *chessylite* is only rarely used now.   *Azurite* is the official mineralogical name.

p. 531, *line* 4.   Over the past sixty years huge fluorite deposits have been discovered and exploited in and around Hardin County, Illinois.   Magnificent crystal specimens of purple, blue, green, white, and yellow have been found there.   The fluorite is used primarily as a source of industrial fluorine.

p. 534, *line* 39.   See note above for p. 44, line 32.

p. 565, *line* 20.   See note above for p. 35, line 30.

NOTE BY TRANSLATOR

SINCE the publication of Professor Max Bauer's " Edelsteinkunde," in 1896 (first issued in parts during 1895 and 1896), many new facts concerning precious stones have appeared in mineralogical literature. They relate mainly to new localities and to modes of occurrence, but also to the chemical composition of stones, the work, in particular, of Professor S. L. Penfield, of Newhaven, having shown that the generally accepted chemical formulæ of several minerals used as gems required revision.

References to the more important papers published since the appearance of his book have been kindly supplied by Professor Bauer. These and many other memoirs have been consulted, while free use has also been made of the valuable Annual Reports on precious stones, compiled by Dr. George F. Kunz, and published in the volumes of the United States Geological Survey.

The translation has thus been brought up to date, and several additions made to the original, notably under Corundum (ruby and sapphire). Under Diamond a short account of the newly discovered deposits in British Guiana has been given, but very little has been added to the original account of the South African diamond-mines—the most important of all gem-mines—since it was found impossible to incorporate much new matter with this section without rewriting the whole.

Among additions to the work will also be found references to some of the more noteworthy specimens of precious stones in the Mineral Collection of the British Museum (Natural History).

Having these and many other additions of minor importance to incorporate, the translation must necessarily be a somewhat free one, and certain portions of the original have been slightly modified or abridged in deference to the needs of English readers. In the original scarcely enough importance is attached to the optical characters of minerals, to their examination in convergent polarised light and to the measurement of refractive indices, which are of the greatest practical value in the determination of faceted stones. It was felt, however, that the addition of such matter would considerably alter the scope and plan of the work, which it has been the aim of the translator to preserve unaltered throughout.

The text-figures and plates have been reproduced directly from the original with such alterations in the spelling of names on the maps as were necessary. A new figure (Fig. 51) is given of the largest diamond yet found, the " Excelsior," a photograph for this purpose having been kindly supplied by Dr. George F. Kunz.

I could not have undertaken the large amount of work involved in this translation had I not been assured of the assistance of my wife, E. M. Spencer. The actual rendering in English is hers, and she has also carefully revised the whole of the proofs; I feel, therefore, that the work is as much hers as it is mine.

L. J. S.

December 1903.

PREFACE

THE desire of the publishers to present, to the German public, a work on precious stones similar in character to that admirably supplied in American literature by George Frederick Kunz's "Gems and Precious Stones of North America" gave the initiative to the writing of the present book. In this case, however, all precious stones had to be dealt with, and an introduction to the methods employed in their determination had also to be given. For the latter the excellent and exhaustive instructions given by C. Doelter in his "Edelsteinkunde" may serve as a model. These, however, have been somewhat modified and simplified. In particular the examination in convergent polarised light has been dispensed with, since it is unusual for gem-merchants and jewellers to be sufficiently well acquainted with the theory of the subject to make practical use of this method, while information of this description would be superfluous to a trained mineralogist. It has been considered advisable, however, to preface the systematic description of precious stones with a general survey of the related sciences, especially those of physics and mineralogy, in so far as they assist in the understanding of the nature of precious stones.

The reader is assumed to be neither a specialist in science nor wholly without scientific knowledge. It has been sought to treat the subject in such a way that it may be intelligible to any one possessed of a good general education. It is therefore hoped that the book will suffice for those who take a general scientific interest in precious stones, and that it will be specially useful to those engaged in the buying and selling of precious stones and in their application to purposes of ornament, namely, to gem-merchants and jewellers.

It was at first considered that pearls and coral, being not minerals but products of the animal kingdom, could not properly find a place in this work. In deference, however, to the wishes of the readers of the earlier parts of the book, an appendix dealing with these important subjects has been added. In writing the section on pearls, the works of Möbius and von Martens, among others, have been consulted, and for coral, those of Lacaze-Duthiers and of Canestrini.

The author has taken especial pains to treat of the mode of occurrence and the localities of each stone with as much detail as the size of the volume allowed; and the distribution of stones in the most important of the countries in which they are found is graphically shown by small sketch-maps in the text. Many new facts relating to this subject have been communicated by the author's colleagues, but even the latest mineralogical literature shows that inaccuracies still abound. Only those who have themselves gone into this branch of mineralogy and have realised how widely scattered are the accounts which deal with the occurrence of precious stones, and how prevalent errors, uncertainties and mistakes are, can appreciate the difficulties connected with such studies. The difficulty of arriving at the necessary facts in the preparation of the sketch-maps was particularly great, and these are therefore less numerous than was originally intended. Many of my colleagues have helped me in this matter by communicating their personal observations or by sending publications bearing upon the subject ; to each of them I return sincere thanks.

The methods employed in the working of precious stones, and the purposes to which the latter are generally applied, have been gone into in some detail, since these stand in the closest relation to the natural characters of the stone. The general part, therefore, includes sections dealing with the forms of cutting, the process of grinding, &c., the information relating to each particular stone being repeated with the special description of that stone.

In the execution of the work the publishers have as far as possible carried out the wishes of the author. The originals of the coloured plates have been painted by the artistic hand of Herr E. Ohmann of Berlin. Most of the specimens figured are preserved in the mineralogical collection of the Natural History Museum at Berlin. For permission to use these, I return my most grateful thanks to the Director of the collection, Professor C. Klein, Privy Councillor of Mines, as also to the Curator, Professor C. Tenne, for the time and trouble he devoted to work connected with the production of the water-colours. No small part of the success of these coloured plates must be ascribed to his active co-operation in their production. Thanks are also due to Director A. Brezina of Vienna for permission to reproduce the well-known picture preserved in the Mineralogical Department of the Natural History Museum, and representing the Kimberley mine, the richest and most famous of the Cape diamond mines, here published for the first time.

The references to the literature are few, as such references appear out of place in a work primarily intended for general readers. To the narrower circle of mineralogists the author would fain have given more precise and scientific information on innumerable points. For the majority of readers it is desirable that each section should be, as far as possible, complete and independent in itself, so that there is little or no necessity for referring to other parts of the book. This has necessitated the repetition of many statements, but without, it is hoped, the reiteration becoming tiresome.

The alphabetical index has been made as complete as possible, and includes many terms not to be found in the text ; the meaning of each is given together with the page reference.

The author has attained his object if he has succeeded in giving to gem-merchants and jewellers, as well as to admirers of gems, a clear representation of the natural characters and occurrence of precious stones, of the methods according to which they are worked, and of the purposes to which they are applied. If, then, in the pages which follow, the description of certain remarkable minerals should awaken in wider circles a more lively interest in mineralogy as a whole, of which the subject of precious stones is but a part, the author will be gratified, and will consider that he is amply repaid for his trouble.

MAX BAUER.

MINERALOGICAL INSTITUTE OF THE UNIVERSITY, MARBURG,
       Autumn 1896.

CONTENTS

VOLUME I

PAGE

INTRODUCTION 1

FIRST PART

GENERAL CHARACTERS OF PRECIOUS STONES

PAGE

I. NATURAL CHARACTERS AND
OCCURRENCE

A. Chemical Composition 7
B. Crystalline Form 8
C. Physical Characters 11
   (*a*) Specific Gravity 11
   (*b*) Cleavage 26
   (*c*) Hardness 29
   (*d*) Optical Characters 34
     1. Transparency 34
     2. Lustre 35
     3. Refraction of Light . . . 37
     4. Double Refraction of Light . 46
     5. Colour 55
     6. Dichroism 59
     7. Special Optical Appearances and
       Colour Effects 62
   (*e*) Thermal, Electrical, and Magnetic
     Characters 65
     1. Thermal Characters . . . 65
     2. Electrical Characters . . . 66
     3. Magnetism 67

D. Occurrence of Precious Stones . . 68

II. APPLICATIONS OF PRECIOUS
STONES

A. Technical Applications 70
B. Application as Jewels 71
   (*a*) Forms of Cutting 72
   (*b*) Process of Cutting 79
   (*c*) Boring 85
   (*d*) Working on the Lathe . . . 85
   (*e*) Engraving ; Etching . . . 85
   (*f*) Colouring and Burning . . . 87
   (*g*) Mounting and Setting . . . 89
   (*h*) Faults in Precious Stones . . 91
   (*i*) Artificial Production . . . 93
   (*j*) Counterfeiting 94
   (*k*) Value and Price 101

III. CLASSIFICATION OF PRE-
CIOUS STONES 106

SECOND PART

SYSTEMATIC DESCRIPTION OF PRECIOUS STONES

Diamond 113
  A Characters of Diamond . . . 113
    1. Chemical Characters . . . 113
    2. Crystalline Form 119
    3. Specific Gravity 127
    4. Cleavage 128
    5. Hardness 129
    6. Optical Characters 130
    7. Electrical and Thermal Char-
      acters 138

PAGE

B. Occurrence of Diamond . . . 138
   1. India 140
   2. Brazil 155
   3. South Africa . . . 179
   4. Borneo 217
   5. Australia 221
   6. North America . . . 226
   7. British Guiana . . . 228
   8. Urals 230
   9. Lapland 231

PAGE

10. In Meteorites 232
C. Origin and Artificial Production of
   Diamond 233
D. Applications of Diamond . . 238
   1. Application in Jewellery . . 238
   2. Diamond-cutting . . . 242
   3. Technical Applications . . 245
   4. Large and Famous Diamonds . 247
   5. Value of Diamonds . . . 255
   6. Imitation and Counterfeiting . 260

VOLUME II

Corundum 261
   Ruby 265
   Sapphire 282
   Other varieties . . . 292
Spinel 295
Chrysoberyl 300
   Cymophane 302
   Alexandrite 304
Beryl 306
   Emerald 308
   Precious beryl (Aquamarine, &c.) . 318
Euclase 324
Phenakite 326
Topaz 327
Zircon 340
Garnet Group 345
   Hessonite 350
   Spessartite 352
   Almandine 352
   Pyrope (Bohemian garnet, "Cape ruby,"
      Rhodolite) 356
   Demantoid 361
   Grossularite, Melanite, Topazolite . 362
Tourmaline 363
Opal 373
   Precious opal 374
   Fire-opal 384
   Common opal 385
Turquoise 389
Bone-turquoise 402
Lazulite 403
Callainite 403
Olivine 404
Cordierite 408
Idocrase 410
Axinite 412
Kyanite 413
Staurolite 415
Andalusite (Chiastolite) . . . 415
Epidote 417
   Piedmontite 419
Dioptase 419

Chrysocolla 420
Garnierite 420
Sphene 421
Prehnite (Chlorastrolite, Zonochlorite) . 421
Thomsonite (Lintonite) . . . 422
Natrolite 422
Hemimorphite 423
Calamine 423
Felspar Group 424
   Amazon-stone 425
   Sun-stone 426
   Moon-stone 429
   Labradorescent Felspar . . 431
   Labradorite 432
Elæolite 436
Cancrinite 437
Lapis-lazuli 438
Haüynite 445
Sodalite 446
Obsidian 446
Moldavite 449
Pyroxene and Amphibole Groups . 451
   Hypersthene 451
   Bronzite 452
   Bastite 453
   Diallage 453
   Diopside 453
   Spodumene 455
   Rhodonite (and Lepidolite) . . 456
   Nephrite 457
   Jadeite (Chloromelanite) . . 465

Quartz 471
A. Crystallised Quartz . . . 474
   Rock-crystal . . . 474
   Smoky-quartz 479
   Amethyst 481
   Citrine 486
   Rose-quartz 488
   Prase 488
   Sapphire-quartz . . . 488
   Quartz with enclosures . . 489

	PAGE		PAGE
Cat's-eye	491	Agate (Onyx)	511
Tiger-eye	493	Malachite	524
B. Compact Quartz . . .	496	Chessylite	527
Hornstone	496	Satin-spar (Calcite, Aragonite, and Gypsum)	527
Wood-stone	496	Fluor-spar	528
Chrysoprase . . .	497	Apatite	531
Jasper	499	Iron-pyrites	532
Avanturine	502		
C. Chalcedony	504	Hæmatite	533
Common Chalcedony . .	506	Ilmenite	534
Carnelian	508	Rutile	534
Plasma	510	Amber	535
Heliotrope	511	Jet	556

THIRD PART

DETERMINATION AND DISTINGUISHING OF PRECIOUS STONES

	PAGE		PAGE
General Methods	561	Yellowish-green stones . .	574
A. Transparent stones . . .	566	Green stones	575
Colourless stones	566	B. Translucent and Opaque stones .	576
Greenish-blue stones . . .	567	White, faintly-coloured, and grey	
Pale blue stones	568	stones	577
Blue stones	568	Blue stones	577
Violet stones	569	Green stones	578
Lilac and rose-coloured stones .	570	Black stones	579
Red stones	570	Yellow and brown stones . .	579
Reddish-brown stones . . .	571	Rose-red, red, and lilac stones .	580
Smoke-grey and clove-brown stones .	572	Stones with more than one colour .	580
Reddish-yellow stones . . .	572	Stones with metallic lustre . .	581
Yellowish-brown stones . .	573	C. Stones with special optical effects .	581
Yellow stones	573		

APPENDIX

	PAGE		PAGE
Pearls		Coral	601
Nature and Formation . .	585	The Coral Skeleton . . .	601
Application	592	The Living Coral . . .	605
Pearl-fishing	594	Distribution : Coral-banks . .	608
Imitation	600	Coral-Fishing : Application : Trade .	611

EXPLANATION OF PLATES

Plates I, XII-XVI, XVIII, and XX, all in color, are in volume II following page 376.

PAGE

PLATE I. (Following p. 376.)

DIAMOND, CORUNDUM, SPINEL, AND ZIRCON.

Fig. 1. Diamond, crystal in matrix : Brazil . 162
,, 2. ,, ,, ,, S. Africa 190
,, 3. ,, sphere of Bort . . 125
,, 4. ,, Carbonado . 125, 173
,, 5. Ruby, crystal 265
,, 6. ,, cut 265
,, 7. Sapphire, crystal 282
,, 8. ,, cut 283
,, 9. Spinel (Balas-ruby), crystal . 297
,, 10. ,, (Ruby-spinel), crystal . 297
,, 11. Zircon (Hyacinth), crystal . . 341
,, 12. ,, ,, in basalt . 343
,, 13. ,, cut 341

PLATE II. (Facing p. 74.)

FORMS OF CUTTING.

Fig. 1a, b, c. Brilliant, double-cut . . 74
,, 2a, b, c. ,, English double-cut
(double–cut brilliant with star) 74
,, 3a, b, c. Brilliant, triple-cut, old form . 74
,, 4a, b, c. ,, ,, new form,
round 74
,, 5a, b, c. Brilliant, the same, oval . . 74
,, 6b, c. ,, the same, pear-
shaped 74
,, 7a, b, c. Brilliant, the same, triangular 74
,, 8a. Half-brilliant 75

PLATE III. (Facing p. 76.)

FORMS OF CUTTING.

Fig. 1a, b, c. Star-cut (of Caire) . . . 75
,, 2a, b. Step-cut, four-sided . . . 76
,, 3b. ,, six-sided . . 76
,, 4b, c. ,, eight-sided . . . 76
,, 5a, b. Mixed-cut 76
,, 6a, b. Cut with double facets . . 76
,, 7a, b, c. Cut with elongated brilliant
facets 76
,, 8a, b, c. Maltese cross 77

PLATE IV. (Facing p. 78.)

FORMS OF CUTTING.

Fig. 1b. Rosette (Rose-cut), round . . 77
,, 2b. ,, pear-shaped . . 78
,, 3a. ,, Dutch Rose . . . 78

PAGE

Fig. 4a. Rosette Brabant Rose . . . 78
,, 5a, 6a. ,, Other forms . . . 78
,, 7a, b. ,, Rose recoupée . . 78
,, 8a, b. ,, Cross-rose . . . 78
,, 9a. Double Rosette (Pendeloque) . 78
,, 10. Briolette 78
,, 11a, b. Table-stone 77
,, 12a, 13b. Thin-stone 77
,, 14a, b. Table-stone, with brilliant form
above 77
,, 15a, b, 16b. Thick-stone . . . 77
,, 17b. Cabochon, simple (hollowed) . . 79
,, 18b. ,, ,, with facets . 79
,, 19b. ,, double 79

PLATE V. (Facing p. 152.)

DIAMOND MINE AT PANNA, INDIA . . . 151

PLATE VI. (Facing p. 168.)

DIAMOND-WASHING IN BRAZIL . . . 168

PLATE VII. (Facing p. 198.)

KIMBERLEY DIAMOND MINE IN 1872 . . 197

PLATE VIII. (Facing p. 202.)

KIMBERLEY DIAMOND MINE IN 1874 . . 199
,, ,, ,, (West side) in 1885 200

PLATE IX. (Facing p. 240.)

ACTUAL SIZES OF BRILLIANTS (DIAMONDS)
of $\frac{1}{4}$ to 100 carats 241

PLATE X. (Facing p. 248.)

FAMOUS DIAMONDS (Natural Size).

Fig. 1a, b, c. Orloff 249
,, 2. Great Mogul 248
,, 3a, b. Shah 249
,, 4a, b. Koh-i-noor, Indian cut . 248
,, 5a, b, c. ,, new form . . 248
,, 6. Stewart (from South Africa) . 253
,, 7b, c. Mr. Dresden's (from Brazil) . 252

PLATE XI. (Facing p. 252.)

FAMOUS DIAMONDS (Natural Size).

Fig. 8a, b, c. Regent 250
,, 9a, b, c. Star of the South (from Brazil) 252
,, 10a, b. Florentine 250
,, 11a, b. Sancy 250
,, 12. Pasha of Egypt 251
,, 13a, b, c. Nassak 251
,, 14. Star of South Africa . . . 253
,, 15. Polar Star 249

PAGE

PLATE XII. (Following p. 376.)

BERYL AND CHRYSOBERYL

Fig. 1. Emerald, crystal in calcite : Colombia
308, 314
„ 2. „ crystal in mica-schist : Salzburg 308, 316
„ 3. Emerald, cut 308, 310
„ 4. Beryl (golden beryl), crystal . . 324
„ 5. Aquamarine, crystal : Siberia . . 322
„ 6, 7. „ cut 319
„ 8. Chrysoberyl (Alexandrite), crystal :
Urals 305
„ 9a. Alexandrite, cut-stone by daylight 305
„ 9b. „ „ by candlelight 305
„ 10. Chrysoberyl, crystal : Brazil . . 300
„ 11. „ (Cymophane), cut . 302

PLATE XIII. (Following p. 376.)

TOPAZ AND EUCLASE

Fig. 1. Topaz, blue crystal : Urals . 330, 338
„ 1a. „ the same, cut . . 330, 338
„ 2. „ dark yellow crystal : Brazil
331, 336
„ 2a. „ the same, cut . . 331, 336
„ 3. „ pale yellow crystals : Saxony
331, 333
„ 3a. „ the same, cut . . 331, 333
„ 4. „ rose-red crystal : Brazil 332, 336
„ 4a. „ the same, cut . . 332, 336
„ 5. Euclase, crystal : Brazil . . . 325

PLATE XIV. (Following p. 376.)

EPIDOTE, GARNET, AND OLIVINE.

Fig. 1. Epidote, crystals : Salzburg . . 418
„ 2. „ cut 418
„ 3. Almandine, crystal in mica-schist 352, 353
„ 4. „ cut. [See also Pl. XVIII.,
Fig. 7] 352, 353
„ 5. Pyrope (Bohemian garnet) in matrix
356, 357
„ 6. „ (" Cape ruby "), cut . 356, 360
„ 7. Hessonite, crystals with diopside :
Piedmont . . . 350, 352, 454
„ 8. Hessonite, cut : Ceylon . . . 350
„ 9. Demantoid, rough : Urals . . 361
„ 10. „ the same, cut . . 361
„ 11. Olivine (Chrysolite), crystal . . 405
„ 12. „ „ cut . . 406

PLATE XV. (Following p. 376.)

IDOCRASE, DIOPTASE, AND TOURMALINE.

Fig. 1. Idocrase, crystals : Piedmont . . 411
„ 2. „ the same, cut . . 411
„ 3. „ cut : Vesuvius . . 412
„ 4. Dioptase, crystals : Siberia . . 420
„ 5. Tourmaline, rose - red and green
crystal : Elba 366
„ 6. Tourmaline, red crystal : Siberia . 370

PAGE

Fig. 7. Tourmaline, green crystal : Brazil . 371
„ 8, 9. „ red and green crystal :
Massachusetts . . . 366, 371
„ 10. Tourmaline, brown, cut : Ceylon . 372
„ 11. „ blue, cut : Brazil . 372

PLATE XVI. (Following p. 376.)

FELSPAR AND OPAL

Fig. 1. Amazon-stone, crystals . . . 426
„ 2. Labradorite, polished . . . 433
„ 3. Labradorescent Felspar, polished . 432
„ 4. Moon-stone, rough . . . 429
„ 5. „ cut 429
„ 6. Precious opal, rough : Australia 375, 382
„ 7. „ the same, cut . 375, 382
„ 8. „ rough : Hungary 375, 379
„ 9. „ the same, cut . 375, 379
„ 10. Fire-opal, rough 384
„ 11. „ cut 384

PLATE XVII. (Facing p. 472.)

GROUP OF QUARTZ CRYSTALS (ROCK-CRYSTAL):
DAUPHINÉ. 472, 474, 476

PLATE XVIII. (Following p. 376.)

QUARTZ (AND GARNET).

Fig. 1a. Amethyst, crystals 482
„ 1b. „ cut 486
„ 2. Rock-crystal with enclosures (" needle-
stone ") 475, 489
„ 3a. Smoky quartz, crystals . . . 480
„ 3b, c. „ „ cut . . 480, 481
„ 4a, b. Cat's-eye, green and brown, cut
491, 492
„ 5. Tiger-eye, polished 493
„ 6. Heliotrope, polished . . . 511
„ 7. Almandine, cut. [This Fig. should have
been placed with the other garnets
on Plate XIV.] . . . 352, 353

PLATE XIX. (Facing p. 512.)

AGATE.

Fig. a. Fortification-agate : Oberstein 512, 513
„ b. Onyx-agate : Brazil . . . 512, 513

PLATE XX. (Following p. 376.)

LAPIS-LAZULI, TURQUOISE, MALACHITE,
CHALCEDONY, AND AMBER.

Fig. 1. Lapis-lazuli, polished . . . 438
„ 2. Turquoise, blue, cut . . . 392
„ 3. „ green, in matrix . 392
„ 4a. Malachite, rough 525
„ 4b. „ polished . . . 525
„ 5a, b. Onyx, cut 523
„ 6. Carnelian (Intaglio) . . . 521
„ 7. Carnelian-onyx (Cameo) . . . 521
„ 8. Chrysoprase, cut 497
„ 9. Amber, polished in part . . 537, 540

LIST OF TEXT FIGURES

VOLUME I

Fig. 1. Pycnometer 12
„ 2. Basket of platinum wire for holding stone in determination of specific gravity 13
„ 3. Ordinary balance, with arrangement for hydrostatic weighing . . 14
„ 4. Bench for use in hydrostatic weighing with an ordinary balance . 15
„ 5. Westphal's balance for determining the specific gravity of solids 17
„ 6. Jolly's spring-balance for determining specific gravity . . . 19
„ 7. Westphal's balance for determining the specific gravity of liquids . 21
„ 8. Cleavage of calcite 27
„ 9. Refraction of light on passing into a precious stone 37
„ 10. Refraction of light on passing out of a precious stone 39
„ 11. Total reflection 40
„ 12. Total reflection in diamond when surrounded by air 41
„ 13. Total reflection in diamond when surrounded by methylene iodide . 41
„ 14. Path of a ray of light through a plate with parallel sides . . . 42
„ 15. Path of a ray of light through a prism 42
„ 16. Dispersion of light 42
„ 17. Dispersion of light by a plate with parallel sides 43
„ 18. Dispersion of light by a prism. Formation of the spectrum of white light 43
„ 19. Path of the rays of light through a prism 44
„ 20. Path of a ray of light in a brilliant . 45
„ 21. Double refraction of a ray of light . 46
„ 22. Path of light through a doubly refracting plate . . . 47
„ 23. Double refraction of calcite or doubly refracting spar 48

Fig. 24. Path of light through a doubly refracting prism 48
„ 25. Path of light through a doubly refracting prism . . . 49
„ 26*a*. Images of a flame observed through a doubly refracting stone . . 49
„ 26*b*. Images of a flame observed through a singly refracting stone . . 49
„ 27. Polariscope for observation in parallel light 50
„ 28. The dichroscope 60
„ 29. Brilliant (triple-cut) 74
„ 30. Rosette 77
„ 31*a—s*. Crystalline forms of diamond 120, 121
„ 32. Actual sizes of octahedral crystals of diamond of 1 to 1000 carats . 126
„ 33. Diamond-fields of India . . . 144
„ 34. Diamond-fields of Brazil . . . 156
„ 35. Diamond-fields of the Diamantina district, Brazil 157
„ 36. Occurrence of diamond in the Serra da Cincorá, Bahia 172
„ 37. Occurrence of diamond in South Africa 181
„ 38. Diamond mines at Kimberley . . 183
„ 39. Diagrammatic section through the Kimberley mine 189
„ 40. Section through the Kimberley mine 200
„ 41. Section through the De Beer's mine 201
„ 42. Diamond-fields of the Island of Borneo 217
„ 43. Diamond-fields of New South Wales 222
„ 44. Actual size of rose diamonds of 1 to 50 carats 241
„ 45. Directions of least hardness on the facets of a brilliant . . . 244
„ 46. First stage in the development of a brilliant from an octahedron . 244
„ 47. The "Great Table," a large Indian diamond mentioned by Tavernier 250

PAGE

Fig. 48. "Star of the South." Two views of the rough stone 252

„ 49. "Victoria" diamond of 457½ carats from South Africa . . . 253

„ 50. Outline of a diamond of 428¼ carats from South Africa . . . 253

PAGE

Fig. 51. Largest known diamond, the "Excelsior," weight 971¾ carats. From the Jagersfontein mine, South Africa 254

„ 52. The "Tiffany Brilliant," 125½ carats 254

VOLUME II

Fig. 53. Crystalline forms of corundum (a—d, ruby ; e—i, sapphire) . . . 262

„ 54. Occurrence of ruby and sapphire in Burma and Siam 269

„ 55. Ruby-fields of Burma . . . 269

„ 56. Ruby and sapphire mines of Muang Klung, Siam 276

„ 57. Ruby mines in Badakshan on the Upper Oxus 278

„ 58. Crystal of artificially prepared ruby 280

„ 59. Occurrence of sapphire in Ceylon . 287

„ 60a—d. Crystalline forms of spinel . 295

„ 61a—c. Crystalline forms of chryso-beryl 301

„ 62a—e. Crystalline forms of beryl (emerald and aquamarine) . . 307

„ 63. Crystal of emerald belonging to the Duke of Devonshire . . . 310

„ 63a. Occurrence of beryl near Mursinka, Urals 321

„ 64. Crystalline form of euclase . . 325

„ 65a—c. Crystalline forms of phenakite 326

„ 66a—d. Crystalline forms of topaz . 328

„ 67. Occurrence of yellow topaz near Ouro Preto, Brazil . 335

„ 68a—d. Crystalline forms of zircon . 341

„ 69a d. Crystalline forms of garnet . 347

„ 70a—c. Crystalline forms of tourmaline 365

„ 71. Crystalline form of olivine (chryso-lite) 405

Fig. 72. Crystalline form of cordierite . . 408

„ 73a, b. Crystalline forms of idocrase . 411

„ 74. Crystalline form of axinite . . 412

„ 75. Crystalline form of kyanite . . 414

„ 76. Crystalline form of andalusite . . 415

„ 77. Chiastolite 416

„ 78a—c. Crystalline forms of epidote . 417

„ 79. Crystalline form of dioptase . . 419

„ 80a—c. Crystalline forms of felspar . 424

„ 81. Crystalline form of amazon-stone . 426

„ 82. Occurrence of lapis-lazuli in Badak-shan 441

„ 83. Occurrence of lapis-lazuli in the neigh-bourhood of Lake Baikal . . 443

„ 84. Occurrence of lapis-lazuli on the Talaya River 443

„ 85a—d. Crystalline forms of quartz . 471

„ 86. Sceptre-quartz 483

„ 87. Occurrence of amethyst near Mursinka in the Urals 485

„ 88. "Flèches d'amour" from Wolf's Island, Lake Onega, Russia . . 489

„ 89. Mocha-stone 507

„ 90. Agate-grinding and polishing work-shop at Oberstein 519

„ 91. Agate-grinding workshop of Herr August Wintermantel at Waldkirch (Baden) 521

„ 92. Antique intaglio 523

„ 93. Antique cameo 523

„ 94. Antique cameo 523

INTRODUCTION

Certain minerals occurring in the earth's crust are distinguished by special beauty, and have therefore been used since the earliest times for personal ornament and for decorative purposes generally. The beauty of such minerals depends upon their transparency, lustre, or colour, or in some cases upon a play of colour, due to the modification of rays of light reflected from the surface or transmitted through the stone. This beauty is not manifested to its fullest extent until the stone is cut and polished, when these features become more conspicuous. In some cases, for example in the rarely occurring diamonds of a beautiful red or blue colour, all these features are present in the same stone; in others, as in the ruby, the play of colour is absent, and the effect of the stone depends upon its transparency, lustre, and colour. In a stone such as precious opal, the beauty of the mineral depends solely upon a play of colour, which is independent of the colour of the stone itself. Opaque minerals, like turquoise, with but little lustre, owe their beauty to their fine colour; finally, colour may be completely absent, and the beauty of the stone due to its transparency, lustre, and play of colour, as in the purest colourless diamonds.

Those minerals which, through the possession of some or all of the features enumerated above, lay claim to beauty of appearance are not all equally suitable for gems. Besides the possession of undeniable beauty, which for its use as a gem-stone is naturally a *sine quâ non*, a mineral must also possess a certain degree of hardness, for otherwise a very small amount of wear will suffice to dim its beauty. Even should it successfully resist the effect of contact with the moisture of the skin, a comparatively soft stone will succumb to the action of grit and dust, which for the most part consist of particles of the mineral quartz. It is desirable, therefore, that all minerals used for personal ornament should possess at least the hardness of quartz, and a still greater hardness, the so-called gem-hardness, is an advantage. Some discretion should be exercised in the setting of stones of different degrees of hardness; thus a comparatively soft stone may be quite suitable for the ornamentation of a brooch, but should not be set in a ring, where it is likely to get much hard wear. In the same way minerals, which, however beautiful in the fresh condition, lose their beauty on exposure to the atmosphere are unsuitable for use as gems.

We have now shown that only those minerals, which combine beauty of appearance with considerable hardness, and a power of resisting external influences, are suitable for use as ornaments. The mineral substances so distinguished are known as precious stones or gems.

All the essential characters of a gem occur together in but few minerals, so that the number of precious stones is small compared with that of all the mineral species known. Moreover, the minerals distinguished by the possession of the characters of a gem occur for the most part, at least in pieces of good size and quality, in very sparing amount, so that besides their intrinsic beauty they possess the added charm of rarity. This latter character is perhaps the one to which is most largely due the costliness of these objects, for the possession of something rare, something that it is impossible for every other person to possess, has ever exercised a fascination upon human nature.

The minerals which combine the highest degrees of beauty, hardness, durability, and rarity—diamond, ruby, sapphire, and emerald, for example—are by common consent placed in the foremost rank of gems; those in which these characters, especially that of hardness, are less conspicuous, are less highly esteemed. The former may be grouped together as precious stones, while the latter, characterised sometimes by greater beauty, but by lower hardness and more common occurrence, are known as semi-precious stones.

It is impossible to draw a hard and fast line between precious and semi-precious stones; a stone which one person might regard as precious, another would place with semi-precious stones. In deciding the point, not one character but all must be taken into consideration, and the rarity or commonness of the stone set against its other characters. Thus the emerald, though comparatively soft, is one of the costliest of stones owing to its magnificent colour, and the rarity of faultless specimens. Again, precious opal and turquoise, though opaque and comparatively soft, are more valuable than is amethyst, which, though transparent and harder, is yet, on account of its commonness, regarded as only a semi-precious stone.

Just as the essential characters of a gem are developed in different minerals to different degrees, so do these characters differ in degree in different specimens of the same mineral species. The hardness is in all cases the same in the same species, but the transparency and colour may be very different in different specimens, so that while some furnish us with the finest examples of precious stones, others may be entirely unsuitable for ornamental purposes, and the remainder furnish stones of inferior quality. Thus the mineral species beryl includes not only the costly emerald, but also the far less beautiful and less valuable golden beryl and pale greenish-blue aquamarine, and the cloudy and opaque common beryl, which is destitute of the essentials of a gem, and is therefore never used as such. The transparent varieties of minerals used as precious stones are distinguished by the prefix " precious " from the opaque or " common " varieties.

The value of a mineral as a gem does not depend solely upon its natural characters, but is influenced to a very large extent by the fashion of the day. A stone which at one time commands a high price, at another, for no apparent reason, will scarcely find a purchaser; while minerals, at first comparatively unknown to the jeweller, will leap suddenly into popular favour. At one time diamonds are most in favour, at another the so-called fancy stones. At the present time the latter is the case; scarcely more than a dozen years ago the jeweller's stock consisted of diamonds, rubies, sapphires, emeralds, and garnets, with an occasional topaz or aquamarine. Now almost all the stones described in this book, a list of which will be found in the table of contents, and most of which are coloured, are commonly bought and sold in the precious stone market. Many of the minerals now cut as gems are not suitable in every respect for the purpose, and this is specially so in the case of those minerals which are cut and worn more for the association of the wearer with the place of their occurrence than for their intrinsic beauty or suitability.

The minerals which must be reckoned as precious stones are by no means fixed in number, nor is it a fact that a mineral once used as an ornamental stone is ever after used as such. New minerals come into favour and old ones fall out of use, but a certain number, the richest and most beautiful of gems, survive all the changes and chances of fashion, and stand now, where they did ages ago, highest of all in popular esteem.

Of the natural substances used for ornamental purposes other than precious stones, pearls and coral are the most important. These, however, are not members of the mineral kingdom, being products of animal life, and are therefore excluded in the consideration of precious stones. Amber, on the other hand, though not a true mineral substance, finds a place here. It is the fossilised resin of extinct trees, but being a constituent of the earth's crust, as are precious stones, it is customary to regard it and similar substances as within the domain of mineralogy.

Since precious stones are minerals, their study must be considered as a branch of mineralogy. It includes the investigation of their natural characters, such as chemical composition, crystalline form, specific gravity, hardness, cleavage, and their behaviour towards light, as well as an inquiry into their occurrence in the earth's crust, their mode of origin there, and the localities where they are to be found. But since the application of precious stones is a matter of practical importance, a thorough knowledge of the subject must include an acquaintance with the mining and working of precious stones, their use in jewellery and ornaments, and other subjects of a somewhat technical nature.

Another important branch of the same subject is the determination of gems and the recognition of the features which distinguish them from deceptive imitations in glass, or from less valuable minerals of similar appearance. Long familiarity with the appearance of different stones will, in many cases, enable a dealer or amateur, even though destitute of mineralogical knowledge, to identify any given stone at a glance. In other cases, however, a person who depends in this way upon his memory and sense of recognition, is very likely to fall into error, which a scientific mineralogist would avoid by the use of exact methods of determination. It is therefore greatly to the advantage of persons who buy and sell these costly objects to make themselves acquainted with the principles and methods of mineralogy here laid down. Not only does a sufficient practical knowledge of this subject help in avoiding errors of determination, but it also enables one to gather from rough stones valuable indications of the purposes to which they are most suited, and the methods of working most advantageously employed.

The first part of the present work will be devoted to a consideration of the mineral characteristics which are of importance to the specialist in gems, a general consideration of the mode of occurrence of precious stones, and, finally, certain matters relating to the application and working of these stones. The second part will contain a particular and detailed account of every mineral which has been used for ornamental purposes, with special reference to precious stones; while the third part epitomises the characters to be relied on in determining precious stones and for distinguishing them from other precious stones and from imitations.

FIRST PART

GENERAL CHARACTERS OF PRECIOUS STONES

I. NATURAL CHARACTERS AND OCCURRENCE.

A. CHEMICAL COMPOSITION.

PRECIOUS stones in their general chemical relations do not differ essentially from other minerals. They are composed of the same chemical elements, which are combined together according to the same laws. At one time, however, it was believed that, since the characters of precious stones were so remarkable, their chemical composition must also be unique. Hence it was assumed that all precious stones contained a rare and precious earth as a fundamental constituent. More exact chemical investigations have shown, however, that the constituents of the rarest of precious stones are frequently very common substances, such as carbon or alumina. The precious metals—gold, platinum, &c.—never enter into the composition of gems, and the rare elements very exceptionally so. As examples of such occurrence may be mentioned the element zirconium, present in zircon, and the element beryllium, present in emerald, aquamarine, and a few other rarer stones.

The chemical composition of different stones varies considerably in complexity. While it is very simple in some, in others it is complicated by the presence of numerous constituents. In the case of the diamond, the chemical composition is very simple; this, the most important of gems, consists solely of the common and widely distributed element carbon. The carbon of diamond, however, is endowed with special properties, and differs very widely from graphite, the other crystallised modification of carbon, and from coal, which consists largely of carbon. Among gems, the diamond stands alone in the simplicity of its chemical composition.

At least two, and in the majority of cases a number of, elements enter into the composition of all other precious stones. The rarest and most costly of all stones, the red ruby, contains only two elements; and the blue sapphire is identical with the ruby in chemical composition, differing from it only in colour. The two elements of which the ruby and sapphire consist are aluminium and oxygen. The former, an important constituent of clays and other widely distributed minerals, is a metal which in recent years has become of great importance in the arts and manufactures; the latter is an important constituent of atmospheric air. The combination of aluminium and oxygen, known as oxide of aluminium, or alumina, is an essential constituent of many other valuable gems. Rock-crystal, amethyst, agate, opal, and other stones also consist of a simple oxide, the oxide of silicon. This oxide, which is known as silica, is the most important constituent of the earth's crust. Zircon, spinel, and chrysoberyl furnish examples of slightly more complex oxides.

While the group containing the oxides furnishes so many important gems, there are other groups of minerals which are unimportant from this point of view.   Such groups are those of the metallic sulphides (compounds of metals with sulphur), the haloid compounds (combinations of metals with chlorine, bromine, iodine, and fluorine), and the sulphates (compounds of sulphuric acid).   Although these three groups may include minerals which are occasionally used as ornamental stones, we find none possessing the essentials of a gem to any marked degree.

The group containing the silicates is again an important one, for it embraces the emerald, garnet, chrysolite, topaz, and many other precious stones.   Tourmaline may be mentioned as an example of the few gems belonging to this group, the chemical composition of which is specially complex.

Of the other divisions of the mineral kingdom there remains only to be mentioned that in which the phosphates are placed.   This division contains only one gem, the turquoise. This important and valuable stone, which is composed of phosphoric acid combined with alumina and water, is remarkable, inasmuch as it is the only costly stone which contains any considerable amount of water as an essential constituent.

The ornamental stone malachite may be mentioned here as being the representative of the carbonates or compounds of carbon dioxide, and at the same time as containing a considerable amount of water as an essential constituent.

To identify any given stone and to determine the mineral species to which it belongs, a chemical analysis is often desirable and, in some cases, essential.   Since this method involves the complete destruction of the substance experimented on as such, it is obviously of very limited application in the determination of precious stones of great intrinsic value. In the case of uncut stones a chemical analysis may be made of detached fragments.   But with cut and polished stones, not only is a complete chemical analysis impossible, but the mere testing with acid must be avoided.

B. CRYSTALLINE FORM.

Most chemical compounds, including the majority of minerals, frequently occur as solid bodies bounded by plane faces.   These shapes have been assumed on the solidification of the substance, and are due to the internal molecular forces exercised by the substance, and not to any external influence.   Such definite shapes are known as crystalline forms, and substances occurring in this condition are said to be crystallised.   With very few exceptions all precious stones are crystallised.   Diamond, ruby, sapphire, emerald, topaz, &c., occur naturally as crystals of the finest development.   Only a few, of which the most important is opal, are not bounded by the plane faces characteristic of crystallised substances, but occur only as irregularly-shaped masses.   Such substances without definite external form are said to be amorphous.

Crystallised bodies, therefore, differ from amorphous bodies in that on solidification they assume a regular form bounded by plane faces, which is the outward expression of internal molecular forces.   Certain peculiarities in the physical characters of crystallised bodies, which are absent in amorphous substances, are also due to these internal molecular forces. Thus it is still possible to distinguish a crystallised from an amorphous body, even though the characteristic regular boundaries of the former should happen to be absent.

The absence of the regular boundaries of a crystallised substance may be due to one or more of a variety of causes.   Their free development may have been hindered by external conditions ; as, for instance, when a substance crystallises in a confined space where free development in all directions is impossible. Or, again, as often happens, in extricating a crystal from the matrix, a blow from the hammer may destroy some of the plane faces of the specimen. Moreover, in the process of cutting and polishing precious stones, the natural plane faces are always destroyed. In all these cases, however, the substance still possesses the internal structure characteristic of a crystallised body. The essential difference between a crystallised and an amorphous body lies in their internal structure, on which depends the character of the substance. The presence of plane faces in the crystallised substance is merely the outward expression of its internal structure.

A crystallised body which shows no regular boundaries is said to be *crystalline* or *massive*. When these boundaries are present the body is termed a *crystal*. Portions of crystalline, massive material cannot be distinguished in their external form from an amorphous substance, but their internal structure shows a very essential difference, which will be described later. A crystal, however, on account of its regular boundaries can never be confused with an amorphous body.

The knowledge of crystals and the laws governing the relations between their faces belongs to the special science of *crystallography*. A knowledge of the subject is essential to the correct understanding of the natural relations of minerals, including also precious stones.

It has been established that each crystallised substance, including precious stones, having a definite chemical composition, has also a crystalline form which is characteristic of the substance, or to be more correct, it may exhibit a series of crystallised forms related in such a way that each may be derived from another. Moreover, bodies of different chemical composition will in general be characterised by different crystalline forms, having, as a rule, no mutual relations.

Hence it is possible to distinguish bodies not only by their chemical composition but also by their crystalline form, and this applies equally to precious stones. It is thus obvious that a knowledge of the crystallographic relations of precious stones is not only of theoretical importance, but also of the highest practical importance, for it would enable a buyer of rough stones to distinguish a genuine from a false stone by the form alone, thus avoiding injury to the stone. This method of identification, however, is applicable only when the specimen is crystallised. In the case of massive or crystalline material, the data for its scientific determination can only be obtained from the physical characters of the specimen.

The science of crystallography is not one of which the general principles can be conveyed in a few words. Generally speaking, a complete and thorough study of the subject is necessary to obtain a knowledge of practical value. Since a detailed account of the science of crystallography is quite outside the scope of the present work, the reader must be referred to special works on the subject and to the various text-books of mineralogy, which usually contain a section devoted to crystallography. It will, therefore, be assumed in what follows that the reader possesses a knowledge of at least the elements of the subject, and is further acquainted with the elements of those sciences, such as chemistry, physics, geology, the aid of which is necessary in the study of minerals and precious stones.

It may be stated briefly here that all crystals with few exceptions can be cut by a plane into two equal parts, having the same relation between them as exists between an object and its image in a mirror. Such a plane is known as a plane of symmetry, and crystals of different substances possess different numbers of these planes. The greater the number of planes of symmetry possessed by a crystal the higher its degree of symmetry. Those crystal forms which may be cut in the same manner by the same number of planes possess the same degree of symmetry, and are grouped together into the same *crystal-system*. There are six of these systems, to one or other of which every mineral and every crystallised precious stone must of necessity belong. The names of the different crystal-systems with the number of planes of symmetry characteristic of each are given below:

1. The Cubic System   with 9 planes of symmetry.
2. The Hexagonal System ,, 7   ,,   ,,
3. The Tetragonal System ,, 5   ,,   ,,
4. The Rhombic System   ,, 3   ,,   ,,
5. The Monoclinic System ,, 1   ,,   ,,
6. The Triclinic System   ,, 0   ,,   ,,

Sometimes the symmetry exhibited by a crystal is such that only half the typical number of faces are developed. These derived forms are known as *hemihedral*, or half-faced forms. These forms must be distinguished from those possessing the full number of faces, which are known as *holohedral*, or full-faced forms. Again, from hemihedral forms may be derived, by a development of only half the faces, another group, the members of which are known as *tetartohedral*, or quarter-faced forms. The hemihedral and tetartohedral classes of the different systems receive special names, which, however, need not be mentioned here. All the holohedral and several of the hemihedral and tetartohedral classes are represented among precious stones.

All precious stones of the same kind, *i.e.*, all diamonds, all emeralds, &c., exhibit forms belonging to the same crystal system; they all possess the same degree of symmetry, and all show the same hemihedral or tetartohedral development if such is present.

It not unfrequently happens that two similarly developed crystals of one and the same mineral are so grown together as to be symmetrical with respect to each other about a certain plane, one crystal being a reflection of the other in this plane, as, for example, is shown for spinel in Fig. 60d. A regular grouping of two crystals in this way is known as a *twin*. Twins may generally be recognised by the presence of re-entrant angles between the faces at the edge of the plane of junction of the two crystals. Simple crystal individuals do not show such re-entrant angles. Sometimes on the second crystal of a twin a third individual may be grown in the same manner, thus giving rise to a triplet. Similarly four crystals grown together in a certain regular manner give rise to a quartet. Such regular growths are often very complex, and it is then no easy matter to discover the mutual relations of the several simple crystals.

The principal crystalline forms are too important to be ignored; they will be described and figured below with the description of the various precious stones. To those who possess even a small acquaintance with the laws and terms of crystallography, the description of the different forms and their mutual relations will be easily intelligible; to others, however, it may present some difficulty. But as all the crystallographic details are collected together in a small space, it is open to such persons to omit them. Though their conception of a precious stone in its natural condition will, in such case, suffer, yet a fairly correct idea of the aspect and crystalline form of uncut crystallised precious stones may be obtained from an inspection of the figures.

Amorphous substances, such as opal, which are incapable of assuming a crystalline form, usually occur in irregular masses of indefinite shape, but rounded, spherical, botryoidal, reniform, or nodular masses are also found.

Crystallised bodies, and, consequently, many precious stones, are frequently not of uniform structure throughout; they do not consist of a single crystal individual, but of several irregularly grown together. The compact mass which results from such a collection of crystalline individuals is known as a crystalline or massive *aggregate*. The constituent particles of such an aggregate may be of various shapes; they may be developed fairly equally in all directions, or they may be considerably elongated or shortened in one or more directions. Thus arise granular, columnar, fibrous, shelly, scaly, or other kinds of aggregates. A granular aggregate is coarsely or finely granular according to the size of the constituent particles.

Sometimes the particles are so fine that they cannot be distinguished with the naked eye, nor even with the help of a simple lens, and the mass then appears to be perfectly homogeneous. The microscope, however, reveals the fact that it is in reality an aggregate of minute grains, fibres, or scales. A truly homogeneous body appears homogeneous even under the highest powers of the microscope. A mass built up of minute particles, but with an external appearance of apparent homogeneity, is known as a *compact* aggregate. It often shows the rounded exterior of an amorphous body, and while its constituent particles may show regular crystal-faces the aggregate as a whole never does.

Specimens of such compact aggregates are frequently opaque, and their microscopic examination necessitates the preparation of a slice sufficiently thin to be transparent. A plate with parallel sides is cut and one side is polished. The plate is then fixed to a slip of glass with Canada-balsam, the unpolished surface being uppermost. The plate is then ground down till it is so thin as to be transparent, when the upper surface is polished. To preserve the section a glass cover-slip is cemented over it with Canada-balsam. Many important and interesting facts respecting the character of minerals and precious stones have been learnt from the microscopic examination of such *thin sections*, as they are called. The method has been specially useful in the examination of turquoise, chalcedony, and agate, where special difficulties lie in the way of other methods.

C. PHYSICAL CHARACTERS.

A. SPECIFIC GRAVITY.

One of the most important characters of a precious stone is its density. On this quality depends the weight of a stone of any given size. Thus of two bodies of equal size but of different material, the one having the greater density will exceed the other in weight. To give a concrete example, a cubic inch of iron weighs rather more than a quarter of a pound, while a cubic inch of oak weighs half an ounce. The cube of iron is, therefore, eight times heavier than the cube of wood.

Instead of measuring the density of a substance it is more convenient to compare the weight of any given volume with the weight of an equal volume of some standard substance. The substance usually selected as the standard is water at a temperature of 4° C. The ratio of the weight of any volume of a substance to the weight of an equal volume of water at the above temperature is known as the *specific gravity* of that substance. The specific gravity of a body is found, therefore, by dividing its weight by the weight of an equal volume of water. To calculate how many times one substance is denser than another the specific gravity of the former must be divided by that of the latter.

Experience has shown that each chemical substance, each mineral, and also each precious stone, has a definite specific gravity, which in most cases differentiates it from all other substances. This character furnishes an important means of identification, which is specially valuable in the case of precious stones, for by this method the most costly cut stone can be identified and suffer no injury in the process. With the exception, perhaps, of the optical characters, no other feature is as important as a means of determination.

Every dealer in jewels should therefore be able to make a rapid determination of the specific gravity of any given gem with sufficient accuracy for practical purposes. The time expended in acquiring the necessary manipulation and the cost of the apparatus will be amply rewarded. It is proposed to give here a detailed account of the methods in use for the determination of specific gravity, not only for those conspicuous for the accuracy of the results they furnish, but also such as would be chosen when rapidity and ease of manipulation rather than extreme accuracy are needed.

It will be noticed that the value of the specific gravity of any given mineral, as quoted in text-books or works of reference, varies between certain narrow limits. This is due, firstly, to small errors in determination ; and, secondly, to the impossibility of obtaining two absolutely identical specimens of the same mineral, since there may be, in certain cases, small or even considerable variations in chemical composition (isomorphous replacement); and, again, crystals frequently contain as enclosures various amounts of different impurities.

Methods for the Determination of Specific Gravity.

1. *Method with the Pycnometer.*—This is perhaps the most accurate of the several methods available for the determination of specific gravity. The pycnometer or specific gravity bottle (Fig. 1) is a small vessel of thin glass with a wide neck, into which fits a ground-glass stopper. The stopper is perforated longitudinally by a very fine canal.

FIG. 1. Pycnometer

To find the specific gravity of a stone its weight g is first observed. Next the weight p of the specific gravity bottle, filled with distilled water, is found ; this observation can be made once for all (for the same flask), and need not be repeated at each new determination. Care must be taken that the water rises to the top of the perforation in the stopper ; this is usually effected by filling the flask quite full and then inserting the stopper. Before weighing, the flask must be well dried on the outside. The stone is now placed in the flask. It displaces an amount of water equal in volume to its own bulk. The stopper is replaced so that water rises to the top of the perforation. The flask must be again dried and weighed. Let this weight be denoted by q. The weight of the flask, full of water, together with the weight of the stone is $g + p$. The weight of water displaced by the stone in the second part of the operation is consequently $g + p - q$, and this represents an amount of water equal in volume to the volume of the stone. Since then the weight of the stone is g, and the weight of an equal volume of water is $g + p - q$, the specific gravity, d, is therefore given by

$$d = \frac{g}{g + p - q}.$$

To take a numerical example ; let the weight of a stone be $g = 4\cdot382$ in grams or whatever unit is used ; the weight of the bottle filled with water be $p = 15\cdot543$; the weight of the bottle containing both the stone and the water $q = 18\cdot680$. Then the weight of the displaced water is $g + p - q = 4\cdot382 + 15\cdot543 - 18\cdot680 = 1\cdot245$ and

$$d = \frac{g}{g + p - q} = \frac{4\cdot382}{1\cdot245} = 3\cdot52.$$

This is the specific gravity of the stone, in this case a topaz.

To avoid serious error in determinations of specific gravity the presence of bubbles of air must be carefully guarded against. These often cling with great pertinacity to the stone, but may usually be dislodged with the help of a clean platinum wire or by gently warming the flask. Should any collect under the stopper, the flask must be emptied and refilled with distilled water.

When all the precautions mentioned above are carefully attended to, the accuracy of the determination varies with the delicacy of the balance. With a good balance and a little practice the pycnometer method is capable of giving results accurate to the third place of decimals. It has the further advantage of enabling the operator to determine the specific gravity of several small stones or fragments taken together. The application of this method of determination is limited in two directions: it is useless, on the one hand, for determining the specific gravity of stones too large to pass through the neck of the bottle, and, on the other hand, is not sufficiently accurate when only very small quantities of material are available.

2. *Method with the Hydrostatic Balance.*—This method is more frequently used, and with careful manipulation is perhaps quite as accurate as the method just described. It depends on the fact that a body when immersed in water weighs less than when in air. According to the well-known principle of Archimedes, the loss in weight is equal to the weight of the water displaced, that is, of a volume of water equal to the volume of the body.

The stone is first weighed in air; let this weight be g. It is then suspended from the arm of the balance by a hair, fine wire, or thread, and weighed in water; let this weight be f. It is obvious that the difference in weight, $g - f$, is the weight of the water displaced, that is, of a volume of water equal to the volume of the stone. The specific gravity, d, of the stone is therefore given by $d = \dfrac{g}{g - f}$.

The hydrostatic balance used in the above operation differs in no essential respect from an ordinary balance. It is merely arranged so that the right-hand scale-pan hangs from a much shorter support than the other. Frequently the right-hand scale-pan of an ordinary balance can be replaced by one with shorter supports and a small hook on the under side. When the stone is immersed in water, the thread or wire which carries it is attached to this hook. For the purpose of conveniently holding the stone, the lower end of the thread or wire may carry a small clip, as shown in Fig. 5, or the wire may be simply wound round the stone so that it rests in a stirrup or in a small spiral. A basket of this description is shown in Fig. 2. The water is placed in a glass vessel under the right-hand scale-pan with the short support, and the wire carrying the clip or other arrangement is immersed in water during the whole operation, so that it is weighed in water even before the stone has been fixed in position for the second weighing. It is then unnecessary to specially determine the weight of the wire with its attachment and the loss of weight of this in water, as the effect in one weighing is cancelled by that in the other.

The use of a balance with specially arranged scale-pans, as described above, is by no means essential, however, for the determination of specific gravity. Every jeweller has a good balance with scale-pans having supports of equal length, and this can easily be used as a hydrostatic balance, as illustrated in Fig. 3. For this purpose, the vessel containing water is placed on a small table or bench, such as is shown in Figs. 3 and 4, which stands over

FIG. 2. Basket of platinum wire for holding stone in determination of specific gravity.

the scale-pan of the balance, leaving the latter quite free to move.   The thread or wire with the attachment for carrying the stone is fastened to the hook on the upper part of the right-hand scale-pan.   Obviously the water-containing vessel must not be so large as to interfere with the movement of the balance, and the length of the wire must be such that

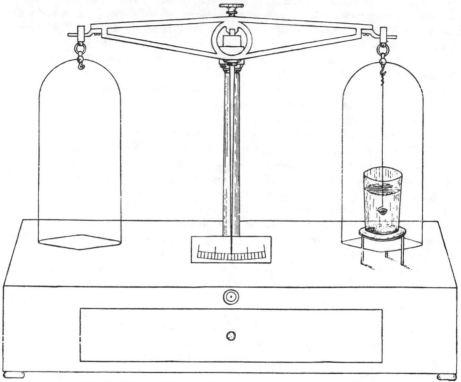

FIG. 3.   Ordinary balance with arrangement for hydrostatic weighing.

the stone it carries will not touch the bottom of the vessel nor rise partly out of the water when the balance is swinging.

In the determination of a specific gravity with such a balance the best mode of procedure is as follows: A fragment of a mineral or metal or any object, the weight of which is greater than that of the stone whose specific gravity is to be determined, is placed in the left-hand scale-pan.   This acts as a counterpoise, and must remain throughout the whole operation.   It has been pointed out above, that the wire with its carrying attachment must remain immersed in water till the completion of the operation.   Weights are then placed in the right-hand scale-pan so as to balance the counterpoise; let the weight required be m. These weights are then replaced by the stone; equilibrium is restored by adding a weight, l, in the right-hand scale-pan.   The weight of the stone will therefore be $m - l = g$ say.   The stone is then attached to the end of the wire and immersed in water, care being taken that it hangs freely without touching the sides and bottom of the vessel.   By adding more weights to the right-hand scale-pan the balance is again brought into equilibrium.   If the total weight is now t, then, the loss in weight of the stone, when weighed in water, is $t - l$, and this is also the weight of the water displaced, that is, the weight of a volume of water equal to the volume of the stone. Hence the specific gravity of the stone is, as before,

$$d = \frac{g}{t - l}.$$

To take a numerical example. If the weight first added to balance the counterpoise be $m = 10\cdot784$ grams (or other unit), and the weights which, together with the stone, balance the counterpoise be $l = 4\cdot803$ grams, then the weight of the stone is $g = m - l = 10\cdot784 - 4\cdot803 = 5\cdot981$ grams. Let the weight added when the stone is immersed in the water be $t = 7\cdot060$. Then the loss of weight of the stone will be $t - l = 7\cdot060 - 4\cdot803 = 2\cdot257$ grams, and $d = \frac{g}{t - l} = \frac{5\cdot891}{2\cdot257} = 2\cdot65$. This is the specific gravity of rock-crystal (quartz).

Other conditions being equal, the more delicate and sensitive the balance the more accurate will be the specific gravity determination. Under favourable conditions and with careful weighing the determination should be correct to the third place of decimals.

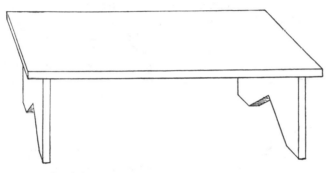

FIG. 4. Bench for use in hydrostatic weighing with an ordinary balance.

Certain precautions, however, must be attended to. In the first place, all parts immersed in water must be quite free from adhering air-bubbles; these may often be dislodged with the help of a clean platinum wire, but frequently it is advisable to bring the water almost to the boiling-point while the stone and wire are immersed in it. For this purpose the wire must be detached from the balance; but before replacing the vessel of water on the balance it should be allowed to cool, which process may be hastened by immersing the vessel, still of course containing the wire and stone, in a bath of cold water. Should the stone and wire be at all greasy through being handled, they will not be properly wetted by the water, and should then be washed in alcohol, ether, benzene, or a solution of soda; in the latter case they must be afterwards rinsed in clean water. Finally it must be borne in mind that the percentage of error in the determination of a very small stone is greater than in that of a larger one, since small errors in weighing influence results more in the former case than in the latter.

When the specific gravity of a number of stones is to be taken, a counterpoise is chosen which exceeds the heaviest of them in weight. This may then remain throughout the whole series of operations and the value of m be determined once for all. Two weighings only are then necessary to find the value of l and t for each stone. From l and t and the constant value m the specific gravity can be calculated in each case.

The specific gravity of several fragments or small stones may be determined by placing them together in the wire basket, shown in Fig. 2, and weighing them as a single stone.

The use of the counterpoise, as described above, has the effect of neutralising any error in the zero point of the balance.   Unless special accuracy is required the counterpoise may be dispensed with, especially when a good balance with the zero accurately adjusted is available.   The stone is then weighed in the usual way by placing it in one scale-pan and the necessary weights in the other.   Let its weight thus determined be p, and its weight when immersed in water be q; then the loss of weight will be $p - q$, and the specific gravity is $d = \dfrac{p}{p - q}$.

In the above determination two weighings only are necessary, whereas, when a counterpoise is used, three must be made.   When, however, a series of determinations are made with one counterpoise two weighings suffice, and either method is equally expeditious.   For the purpose of determining a stone or distinguishing it from another, the value of its specific gravity, obtained without the use of a counterpoise, is generally sufficiently accurate.   The following is a numerical example of this method.   A garnet (cinnamon-stone) weighed in air 4·375 grams (p), and in water, 3·168 grams (q); the loss in weight is therefore $p - q = 4·375 - 3·168 = 1·207$ grams.   Hence the specific gravity is

$$d = \frac{p}{p - q} = \frac{4·375}{1·207} = 3·63.$$

Since determinations of specific gravity are not, under ordinary conditions, made with water at the temperature of 4° C., it is necessary, when accuracy is required, to reduce the results of calculation to this standard.   For the practical purpose of the jeweller, however, the direct observation is quite sufficient.

3. *Method with Westphal's Balance.*—While methods 1 and 2, described above, are susceptible, as has been shown, of extreme accuracy, they have a disadvantage in the amount of time which must be expended on careful weighing.   The method now under consideration combines a degree of accuracy sufficient for all practical purposes with a considerable economy in time; it is therefore valuable to the practical jeweller.   The balance used in this method is named after its maker, the mechanician Westphal, of Celle, in Hanover.   It enables a specific gravity to be readily and quickly determined, correct to the second place of decimals, or under favourable conditions to the third place.   It has the advantage of cheapness over the hydrostatic balance, and, moreover, can be used for other purposes, as will be shown later.

Westphal's balance with accessories for determining the specific gravity of solids, and therefore of precious stones, is illustrated in Fig. 5.   In principle it is really a simplification of the hydrostatic balance with a counterpoise, the left-hand scale-pan and counterpoise of the latter being replaced in Westphal's balance by a counterpoise fixed to the beam.   It consists of a beam, *abc*, to which is fixed a knife-edge of hardened steel at *b*.   This knife-edge, which is directed downwards, is supported by, and turns on, a grooved steel plate fixed to the curved brass piece *de*, itself supported by the brass column *f*.   The latter slides in the tube *hh*, and can be fixed at any convenient height by means of the screw *g*.   Passing through the disc *l*, attached to the foot *k*, of the instrument, is a levelling screw *m*, directly beneath the end of the balance beam; by means of this the column *f* can be adjusted so as to be accurately vertical.

The beam carries at one end (the left in Fig. 5), a heavy weight of brass, which takes the place of the counterpoise and left-hand scale-pan in the ordinary hydrostatic balance.   At the same end, *a*, of the beam is a pointer which, when the beam is horizontal, indicates zero on the scale attached to the piece *de*.   At the opposite end, *c*, of the beam (on the right in Fig. 5) is fixed a knife-edge with its edge directed upwards.   On this rests a hook, from which is suspended, by short platinum wires, the scale-pan *n*. On the under side of this pan is a hook for attaching the fine platinum wire and clip *p*, which holds the stone to be tested. This portion hanging from *c* corresponds to the right-hand scale-pan with short supports of the hydrostatic balance.

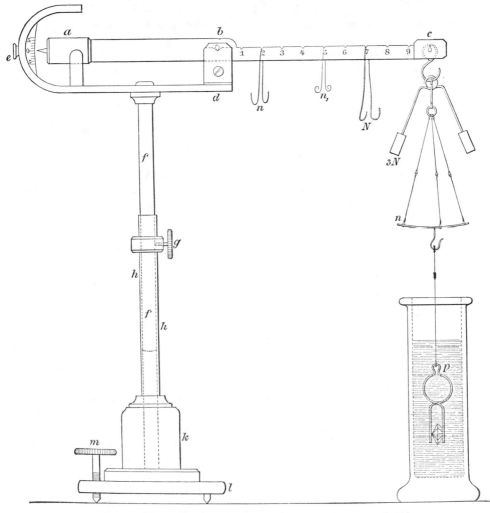

FIG. 5. Westphal's balance for determining the specific gravity of solids.

Between the knife-edge *b* and the end *c* the beam is divided into ten equal parts, the equally distant lines of division being numbered consecutively from *b* towards *c*. The upper side of the beam is notched at these lines of division. The weights are of special construction; some being for the purpose of hanging on the hook under *c*, and others, known as riders, rest when in use in the notches on the beam itself, as shown in Figs. 5 and 7. These weights are quite arbitrary, and need not necessarily be multiples or sub-multiples of any recognised unit; they are related to each other, however, in a certain definite manner suitable to the purpose of the instrument. The normal weight, *N*, when placed on the hook at c, corresponds to the unit weight, but when placed in the notches of the beam numbered 1, 2, 3, &c., it has the value $\frac{1}{10}N$, $\frac{2}{10}N$, $\frac{3}{10}N$, &c., that is, $\frac{1}{10}$, $\frac{2}{10}$, $\frac{3}{10}$, &c., of the unit. A second weight n has one-tenth the value of N, or $n = \frac{1}{10}N$. When placed on the hook at c its value is $\frac{1}{10}N$, and when placed in the notches 1, 2, 3, &c., it has the values $\frac{1}{10}n$, $\frac{2}{10}n$, $\frac{3}{10}n$, &c., or $\frac{1}{100}N$, $\frac{2}{100}N$, $\frac{3}{100}N$, &c. Lastly, there is a third weight $n_1 = \frac{1}{10}n = \frac{1}{100}N$, which is its value when suspended from c, but when placed in the notches 1, 2, 3, &c., it has the values $\frac{1}{10}n_1$, $\frac{2}{10}n_1$, $\frac{3}{10}n_1$, &c., or $\frac{1}{100}n$, $\frac{2}{100}n$, $\frac{3}{100}n$, &c., or $\frac{1}{1000}N$, $\frac{2}{1000}N$, $\frac{3}{1000}N$, &c. The weights n and n_1 are used only as riders, and of the normal weights, N, one will be required for use as a rider and the others for suspending from the hook under c. For convenience, larger weights, multiples of the normal weight ($2N$, $3N$, &c.), are also supplied for suspending at c.

The operation of weighing is performed in the following manner: When the beam is horizontal and the pointer indicates zero on the scale at e, let there be a certain number of the normal weights N on the hook c, one of the normal weights in one of the notches on the beam, and weights n and n_1 in certain other notches. The number of N weights on the hook c will give the whole number of the required reading, and the riders N, n and n_1 on the beam give the first, second, and third decimal places of the same unit. For example, there may be at c three of the weights N, and the riders N, n and n_1 at the divisions 7, 2, and 9 respectively; the reading will then be 3·729 of the unit N, or 3·729N. If there is no rider N and riders n, n_1 at divisions 3 and 5 respectively, then the reading is 3·035. The arrangement of the weights shown in Fig. 5, corresponds to the reading 3·725N, and in Fig. 7 to 2·707N.

Throughout the whole series of weighings for the determination of a specific gravity the clip for holding the stone and part of the suspending wire must remain immersed in water. The support f is adjusted in the tube h so that, during the swinging of the beam, the clip shall neither touch the bottom of the vessel nor rise out of the surface of the water, but remain completely immersed in approximately the central part of the vessel. The determination of specific gravity according to this method involves three separate weighings, as described below. One of these may, however, be reduced to a constant and used throughout a series of determinations.

(1) Sufficient weights are placed on the beam to bring the pointer to zero.

The following example illustrates the method of manipulating the weights: The normal weights N, $2N$, $3N$, $4N$, successively hung on the hook c were each found to be insufficient to move the beam, but $5N$ brought it past the horizontal position. One N was therefore removed and $4N$ left on the hook. The rider N was then placed on the beam at the divisions 9, 8, 7, &c., successively; at the third division the beam was still tilted down to the right, but when the rider was at the second division the beam was tilted to the left; the rider was therefore left at the second division. In the same way the rider n finds a place at 5, and n_1 at the same division, in which case it may hang from the rider n, as shown in Fig. 7, where n_1 hangs from N. The counterpoise of the balance therefore corresponds to the weight 4·255N. A fourth decimal place in the reading may be obtained by placing the rider n_1 between the two divisions 5 and 6. If, for instance, it is placed midway between divisions 5 and 6, the reading would be 4·2555N; if nearer to 5 than to 6, then it would read 4·2553N.

(2) The stone of which the specific gravity is to be determined is now placed in the scale-pan and the weights readjusted until the beam is again horizontal. In the case quoted, the weight required was 3·812N. Hence the real weight of the stone was $(4·255 - 3·812)N = 0·443N$.

(3) Finally the stone is fixed in the clip p and immersed in the water; the weight now required to restore equilibrium was $3\cdot9785N$. The loss of weight of the stone was therefore $(3\cdot9785 - 3\cdot812)N = 0\cdot1665N$, and the specific gravity

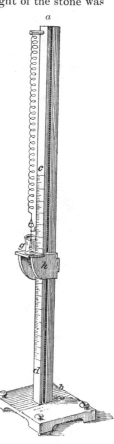

$$d = \frac{0\cdot443}{0\cdot1665} = 2\cdot66.$$ This is again the specific gravity of rock-crystal (quartz), and the stone may therefore be identified as quartz.

This method of determination enables the specific gravity of a stone to be found correctly to two places of decimals with very little trouble, provided that the precautions already mentioned have been observed : namely, that the stone is free from air-bubbles, that it does not come in contact with the sides of the vessel nor rise out of the water, and that it is not too small. The specific gravity of a stone weighing half a carat, that is about one-tenth of a gram, can be determined accurately to the second place of decimals ; though the figure in the second place will be uncertain when the stone only weighs $\frac{1}{4}$ or $\frac{1}{5}$ of a carat, the determination is still useful for practical purposes. With stones smaller than this, however, the results are not sufficiently reliable. If supported in the basket shown in Fig. 2, or in a net made of platinum gauze fixed to the clip, a number of such small stones may be weighed together.

In this method, as previously, the first weighing, giving the constant for the counterpoise of the instrument and the particular wire and clip to be used, may be performed once for all, so that afterwards only two weighings are necessary, and this effects a considerable saving in time. Should the stone, whose specific gravity is to be determined, be so heavy that it raises the counterpoise by its own weight, then suitable additions must be made to the weight of the latter, but in this case the balance will lose in sensitiveness. Usually, however, the weight of the counterpoise is sufficient for most of the purposes for which the instrument is intended.

(4) *Method with Jolly's Spring-balance.*—The spring-balance invented by, and named after, Jolly, formerly a physicist in Munich, possesses considerable advantages, for by its means the specific gravity of stones of fair size can be determined with sufficient accuracy and very simple manipulation, no weights being required. The construction of the instrument is shown in Fig. 6. A vertical rectangular support, acd, about a yard and a half in length, stands on a base b furnished with levelling screws. The vertical support carries on one face a strip of plane mirror on which a scale is engraved. From a horizontal projection at the upper end, a, of the vertical support hangs a spiral of fine steel wire. This carries at its lower end a fine

FIG. 6. Jolly's Spring-balance for determining specific gravity.

platinum wire, to which are attached, one above the other, two small cups, m and m' (Fig. 6, A), of glass or platinum gauze. The length of wire lying between the two cups also bears two reference marks at o and o'. The lower cup m' is immersed in water contained in the vessel g which rests on the stand h. This stand, h, can be moved up and down the vertical support and fixed in any desired position.

In using the instrument, which for the first reading must remain unloaded, the lower cup m' is immersed in the water in the vessel g until the reference mark at o' is exactly in the surface of the water ; the position of the reference mark at o on the vertical scale is then read.   To effect this, the eye of the observer is so placed that the reference mark o, which is an acute triangle, exactly covers its reflection in the mirror, when the position of the upper angle on the scale is read; by this means error due to parallax is avoided.   As an example, let us suppose that o stands at division 45 on the vertical scale.   The stone to be determined is then placed in the upper cup m ; this will stretch the spiral spring, and the stand h bearing the vessel of water must be lowered until o' again stands in the level of the water.   The new position of o in the mirror scale is now read ; let it be at division 75. Then the weight of the stone in air corresponds to $75 - 45 = 30$ scale divisions.   The stone is now placed in the lower cup m', which always remains immersed in water ; since the weight is diminished the spiral will shorten.   The stand h will now require to be raised until o' is again in the surface of the water ; suppose the reading on the scale is now 65. The loss in weight of the stone in water is represented by $75 - 65 = 10$ scale divisions and the specific gravity is, therefore $\frac{30}{10} = 3\cdot0$, corresponding to colourless, transparent tourmaline.

(5) *Method with heavy liquids.*—Within recent years this method for the determination of specific gravity has become of considerable importance.   It depends on the fact that a body placed in a liquid will sink or float according as the density of the liquid is less or greater than that of the body.   A stone placed in a liquid heavier than itself will rise to the surface ; if placed in a liquid lighter than itself it will sink to the bottom ; while if the stone and liquid are of equal density the former will remain stationary at any point within the latter.   The movement upwards or downwards of a stone placed in liquid is quicker or slower according as the difference in density between the stone and the liquid is greater or less.

The liquid for use in the application of this method must fulfil a variety of conditions. First, it must be as heavy as possible, so that its use may be extended not only to the lighter gems, but also to as many as possible of the denser ones, which, if placed in a light liquid, would merely sink to the bottom and render their discrimination impossible.   Secondly, the liquid should be clear, transparent, and colourless, so that there is no difficulty in observing the movement of the immersed stone.   Thirdly, it must not be thick or viscous, otherwise the free movement of the stone would be impeded.   Finally, it must mix readily and perfectly in all proportions with a second lighter liquid, so that its density may be easily varied.

All these conditions are fulfilled by methylene iodide, a compound of carbon, hydrogen, and iodine having the chemical formula CH_2I_2.   It is one of the heaviest liquids known, having at ordinary temperatures a specific gravity of about $3\cdot3$.   Owing to its high coefficient of expansion, its density varies, however, very considerably with the temperature ; at $10°$ C. it is $3\cdot3375$; at $15°$ C. it is $3\cdot3265$; and at $20°$ C. it is $3\cdot3155$.   Methylene iodide is, further, perfectly transparent, very mobile, and of a pale yellow colour.   It is readily miscible in all proportions with benzene, which is a light liquid of specific gravity $0\cdot88$.   Thus by mixing methylene iodide and benzene together a series of liquids is obtained varying in density between $0\cdot88$ and $3\cdot3$, the lower value being less than the density of water, and the higher being three and a third times as great.

To determine the specific gravity of a stone by this method, it is first placed in pure methylene iodide contained in a tall cylindrical vessel, such as is shown in Figs. 5 and 7. If the stone sinks, we know it is heavier than the liquid, that is, it has a specific gravity greater than $3\cdot3$, but how much greater cannot be determined.   If it remains suspended in the liquid, neither rising nor falling, even when moved about with a glass rod, we know that its specific gravity is identical with that of the liquid, which is about 3·3 at ordinary temperatures.   In the third case, if the stone should rise to the surface, even after being pressed down with a glass rod, we know that its specific gravity is less than that of the liquid. When this happens, benzene is added drop by drop and the mixture well stirred.   The addition of benzene is cautiously continued until a mixture of such a density is arrived at that

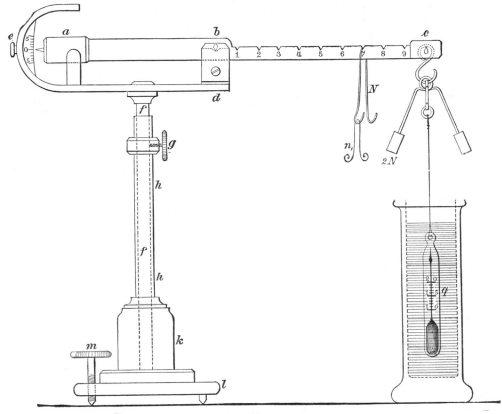

Fig. 7. Westphal's balance for determining the specific gravity of liquids.

the stone no longer rises to the surface, but remains suspended in the liquid, moving neither up nor down.   We then know that the mixture has the same specific gravity as the stone.

The value of the specific gravity of the mixture of methylene iodide and benzene, and consequently that of the stone, has now to be determined.   This may be done by means of the pycnometer, which is weighed first full of water, and then full of the mixture. Deducting from each of these weights the weight of the pycnometer itself, the weights of equal volumes of water and of the mixture are known ; and dividing the latter by the former the specific gravity of the mixture is found.

The specific gravity of a liquid is determined more conveniently, and with sufficient accuracy for practical purposes, by means of Westphal's balance.   This instrument, together with the accessories necessary for determining the specific gravity of a liquid, is shown in Fig. 7.   The small scale-pan previously used (Fig. 5) is now replaced by a glass float, q, containing a thermometer. This hangs by a fine wire from the hook c, and is sufficiently heavy to balance the counterpoise and bring the pointer to zero. If, however, the pointer should not stand exactly at zero, it may be brought to this position by turning the levelling screw m. When the float is immersed in a vessel containing distilled water, its loss in weight is equal to the weight of the displaced water. The pointer is brought to zero again by placing a unit weight, N, on the hook, c, the sizes of the float and the weights having been so arranged that this shall be the case. The loss of weight in the float is, therefore, represented by $1N$. Any other liquid which requires a unit weight, N, to bring the beam again into the horizontal position will have the same specific gravity as water, that is, a specific gravity of 1. Should twice or thrice the unit weight, $2N$ or $3N$, be required, then the specific gravity of the liquid is 2 or 3, that is, it is two or three times as heavy as water. For intermediate values, the riders must be used in the manner explained previously. Thus, for example, let us suppose the float immersed in some liquid, which to bring the beam into the horizontal position requires $3N$ at c, N at the second division, n ($= \frac{1}{10}N$) at the fifth division, and n_1 ($= \frac{1}{100}N$) at the ninth division. Then the specific gravity of the liquid is given directly by the reading 3·259. The specific gravity of the liquid contained in the vessel in Fig. 7 is, in the same way, 2·707. Hence the specific gravity is given directly by the readings, and there is no necessity for the smallest calculation.

With this instrument it is possible with a little practice to determine the specific gravity of a liquid correct to the second place of decimals, and the whole operation can be performed in a few minutes. All that is necessary is to immerse the float in the narrow glass cylinder containing the liquid, and by the addition of weights to bring the beam into the horizontal position.

The use of **indicators** enables the determination of the specific gravity of a liquid to be performed with still greater rapidity ; though the values obtained by this method are only approximate, they are sufficient for all practical purposes connected with the determination of precious stones. There are used as indicators either small, differently weighted glass bulbs, known as specific gravity beads, or small mineral fragments of different specific gravities, ranging by small amounts from the specific gravity of the lightest to that of the heaviest of precious stones. The following minerals, among many others, may be selected for such a series of indicators: chalcedony (sp. gr. = 2·560), microcline (sp. gr. = 2·591), petalite (sp. gr. = 2·648), labradorite (sp. gr. = 2·686), calcite (sp. gr. = 2·728), &c., this being a portion of a series of indicators suitable for the use under discussion.

Such a series would be used in the following manner: When the liquid has been so diluted that its specific gravity is identical with that of the stone to be determined, the lightest of the series of indicators, chalcedony, say, is put into it. Should it float it is taken out (and should be washed with benzene), and each of the series tried in succession until one is found which sinks to the bottom of the vessel. To take a special case, let us suppose that petalite floats while labradorite sinks. We should then know that the specific gravity of the liquid, and therefore of the stone under examination, lies between 2·648 and 2·686. There would then be a probability, or at least a possibility, that the stone is quartz (rock-crystal, amethyst, &c.), which has a specific gravity of 2·65. In practice, a set of mineral fragments, the specific gravity of each of which has been accurately determined, should be kept solely for use as indicators, and it should be possible to readily distinguish one fragment from another.

The method of determining specific gravity by the aid of heavy liquids, and specially by that of methylene iodide, has the advantage of giving results which are quickly and easily arrived at and sufficiently accurate for practical purposes; moreover, it is equally useful in the case of fragments or even splinters of material so small that with other methods the result would be unreliable. Unfortunately, however, the application of this particular liquid is limited to the determination of stones whose specific gravity does not exceed 3·3. It is true that other liquids are known having a greater specific gravity than that of methylene iodide, but each has some disadvantage and fails to fulfil all the necessary conditions mentioned above.

The density of methylene iodide may be increased to 3·6 by dissolving in it iodine and iodoform to the point of saturation. Stones with a specific gravity of 3·6 will then remain suspended at any point in the liquid, while those that float can be made to remain suspended by diluting the solution with benzene or pure methylene iodide. As before, the specific gravity of the liquid can be determined by the help of a series of indicators, or by Westphal's balance. The denser liquid obtained in the way just mentioned has the disadvantage of being deeply coloured and almost opaque; it is therefore difficult to observe the movements of a stone immersed in it, and to determine at a glance its approximate specific gravity. Nevertheless, in the absence of a better substitute it must be made use of, and is valuable in certain cases.

Recently, however, a liquid has been discovered even heavier than methylene iodide saturated with iodine and iodoform, dense enough, in fact, for the determination of the heaviest of precious stones. This is a double nitrate of silver and thallium, with the chemical formula $AgTl(NO_3)_2$, which, although solid at the ordinary temperatures, melts at 75° C. to a perfectly clear and transparent liquid with a mobility equal to that of water. It has a specific gravity of about 4·8, and in it the heaviest of transparent precious stones, namely, zircon, will float. It is miscible in all proportions with water, and hence liquids of any required density can be obtained by simple dilution. When the liquid has been so diluted that its specific gravity is identical with that of the stone under examination, its value can be found by means of Westphal's balance, or by the aid of a series of indicators, the latter being the more convenient with this liquid.

Such a determination presents rather more difficulty than one performed with methylene iodide, inasmuch as the liquid must be kept at a certain temperature. For this purpose the solid silver thallium nitrate is placed in a thin glass beaker of about the size and shape of the vessel shown in Figs. 5 and 7, and heated on a water-bath or over a small spirit or gas flame until it fuses. A little water is then added; this not only lowers the specific gravity but also the temperature at which fusion takes place, the melting-point sinking to 60° or even to 50° C., a fact which adds considerably to the value of the liquid for practical use. In diluting the liquid with water, great care must be taken to avoid adding too much, since a small amount of water will make a considerable difference in the density of the mixture. This mistake can be avoided by adding a little too much water at first, and then driving off the excess by evaporation, constantly stirring and watching the behaviour of the stone, which will show when the density of the liquid becomes identical with its own by remaining suspended at any point. Since the specific gravity of the liquid varies not inconsiderably with the temperature, it is important that it should be determined either by indicators or by Westphal's balance at the temperature the liquid had when the stone was observed to remain suspended in it. By means of this heavy liquid, then, the specific gravity of the heaviest of precious stones may be determined with a sufficient degree of accuracy, the only exceptions being those with metallic lustre, namely, iron-pyrites and hæmatite, which are sometimes used for ornamental purposes.

For the practical worker in precious stones, the determination of specific gravity is simply a means to an end, namely, the discrimination of stones similar in appearance and the identification of others. Hence it is often only necessary to ascertain whether the specific gravity of any particular stone exceeds a certain value. As an example, let us suppose that a doubt exists as to whether a colourless stone is rock-crystal (quartz, sp. gr. = 2·65) or topaz (sp. gr. = 3·5); by simply placing it in methylene iodide (sp. gr. = 3·3) the doubt is settled at once, for if rock-crystal it will float, but if topaz it will sink.

Methods such as this can be very advantageously used for the rapid discrimination of precious stones. It is necessary, however, to be provided with a series of liquids of various known densities in which the stone under examination may be dropped. The approximate value of its specific gravity can be learnt by simply observing its movement upwards or downwards in these liquids. As has been pointed out before, a slow movement up or down indicates that the difference in specific gravity between the stone and the liquid is small ; a quick movement indicates a greater difference ; while an absence of movement shows that the specific gravity of the stone is identical with that of the liquid.

In practice a good series is furnished by the following four liquids :

No. 1. Methylene iodide saturated with iodine and iodoform. Specific gravity = 3·6.
No. 2. Pure methylene iodide. Specific gravity = 3·3.
No. 3. Methylene iodide diluted with benzene. Specific gravity = 3·0.
No. 4. Methylene iodide further diluted with benzene. Specific gravity = 2·65.

The liquids of this series may be numbered consecutively from the heaviest to the lightest, and will be frequently referred to below as liquids No. 1, No. 2, No. 3, and No. 4.

With such a series of liquids it is possible to make an approximate determination of the specific gravity of almost any precious stone with great ease and rapidity, and this will be of considerable aid in its recognition. The stone under observation should be first placed in liquid No. 1 ; should it sink, which can be easily seen in spite of the deep colour of the liquid, its specific gravity is greater than 3·6. If, on the other hand, it rises to the surface, it must be removed with a pair of forceps, wiped with a cloth moistened with benzene, and placed in liquid No. 2. Should it sink in this liquid, we then know that its specific gravity lies between 3·6 and 3·3 ; if it remains suspended then its specific gravity is exactly 3·3. If, on the contrary, it floats to the surface, it must be taken out and placed successively in liquids No. 3 and No. 4, and similar observations made. When a stone slowly sinks or slowly rises in either of the liquids we know that its specific gravity is slightly greater or slightly less than that of the liquid. A stone of specific gravity 3·02, for example, would slowly sink in liquid No. 3.

The series of liquids for such determinations should be kept ready to hand in wide-mouthed glass bottles properly labelled and closed with ground-glass stoppers. The latter precaution is to avoid evaporation of the methylene iodide, which is somewhat expensive (four shillings per ounce), and of the very volatile benzene, and thus prevent alterations in the specific gravity of the liquids. When not in use the bottles should be kept in a dark place, since exposure to light causes methylene iodide to slowly decompose with separation of iodine which darkens the liquid. When after long use, methylene iodide has become dark in colour it may be decolourised by shaking it up with a dilute solution of caustic potash, which must afterwards be poured off or removed by means of a separating funnel.

The specific gravity of the four standard liquids is very liable to change, partly on account of evaporation and partly on account of the small quantities of liquid introduced with the stones if they are not well dried after immersion. Hence the specific gravity of each of the four liquids must be frequently checked by means of Westphal's balance or, better still, by the use of indicators. These indicators may conveniently be kept in the bottles ; each should be chosen so that its specific gravity is near that of the liquid in which it is to be kept. When the indicators show that the specific gravity of any one liquid has altered, this may usually be corrected by simply adding more benzene.

A crystal or fragment of quartz (sp. gr. = 2·65) is a good indicator for liquid No. 4, and should always remain suspended in the liquid. Liquid No. 3 may contain as indicators a phenakite (sp. gr. = 2·95) and a rose-red tourmaline (sp. gr. = 3·02); the former will float and the latter slowly sink if the liquid is of the correct specific gravity. In liquid No. 2 dioptase (sp. gr. = 3·29) should float and olivine (sp. gr. = 3·33) should slowly sink. Finally in liquid No. 1 topaz (sp. gr. = 3·55) must float and hessonite (sp. gr. = 3·65) must sink.

These four liquids enable us to classify for purposes of identification all precious stones according to their specific gravity into five groups, namely :

I. Stones with a specific gravity greater than 3·6.
II.   „   „   „   „   between 3·3 and 3·6.
III.   „   „   „   „   „   3·0 „ 3·3.
IV.   „   „   „   „   „   2.65 „ 3·0.
V.   „   „   „   „   of   2·65 „ less.

The stones of group I. will sink in each of the four liquids, while those of other groups will float in one or other of the liquids.

As an example of the help such a series of liquids can give in the identification of a precious stone, let us suppose that a doubt exists as to whether a colourless, transparent stone is rock-crystal (sp. gr. = 2·65), phenakite (sp. gr. = 2·95), or colourless tourmaline (sp. gr. = 3·02). If it remains suspended when placed in liquid No. 1 it must be quartz. If it sinks in liquid No. 4 but floats in liquid No. 3 it will be phenakite ; if, on the other hand, it sinks in No. 3 it must be tourmaline. Supposing the unknown, colourless stone should be either diamond (sp. gr. = 3·5) or zircon (sp. gr. = 4·65), then in the former case it will float in liquid No. 4, and in the latter case it will sink.

It is important that while observations are being made the liquids should be at the ordinary temperature (15 – 20° C. = 59 – 68° F.), and should not be subjected to any great variations in this condition, since, as was noted above, the specific gravity of methylene iodide is considerably altered by changes of temperature.

In the third section of this book, which deals with the determination of precious stones, full use will be made of these four liquids, and of the convenient classification of precious stones into five groups, based on their differences in specific gravity. At this point it will be useful to give a tabular list of the more important precious stones, arranged according to their specific gravity from the heaviest to the lightest, and divided into the five groups as determined by their behaviour in the four standard liquids. As already mentioned, the specific gravity of any one precious stone shows small variations, which are indicated in the table.

Precious Stones arranged according to Specific Gravity.

GROUP I. (sp. gr = 3·6 or more).

Zircon	.	.	.	.	4·6—4·7
Almandine	.	.	.	.	4·11—4·23
Ruby	.	.	.	.	4·08
Sapphire .	.	.	.	.	4·06
"Cape Ruby"	.	.	.	3·86
Demantoid	.	.	.	.	3·83
Staurolite	.	.	.	.	3·73—3·74
Pyrope	.	.	.	.	3·69—3·78
Chrysoberyl	.	.	.	.	3·68—3·78
Kyanite .	.	.	.	.	3·60—3·70
Hessonite	.	.	.	.	3·60—3·65
Spinel	.	.	.	.	3·60—3·63

GROUP II. (sp. gr. = 3·3—3·6).

Topaz	.	.	.	.	3·50—3·56
Diamond .	.	.	.	.	3 50—3·52
Epidote .	.	.	.	.	3·35—3·50
Idocrase .	.	.	.	.	3·35—3·45
Sphene	.	.	.	.	3·35—3·45
Olivine	.	.	.	.	3·33—3·37

GROUP III. (sp. gr. = 3·0—3·3).

Jadeite .	.	.	.	.	3·3
Axinite .	.	.	.	.	3·29—3·30
Diopside .	.	.	.	.	3·2—3·3
Dioptase .	.	.	.	.	3·29
Andalusite	.	.	.	.	3·17—3·19
Apatite	.	.	.	.	3·16—3·22
Spodumene	.	.	.	.	3·15—3·20
Tourmaline (green)	.	.	.	3·1
,, (blue)	.	.	.	3·1
,, (red)	.	.	.	3·08

Euclase . . . . . 3·05
Fluor-spar . . . . 3·02—3·19
Tourmaline (rose-red) . . . 3·02
,, (colourless) . . . 3·02

GROUP IV. (sp. gr. = 2·65—3·0).

Nephrite .	.	.	.	.	3·0
Phenakite	.	.	.	.	2·95
Turquoise	.	.	.	2·6—2·8
Labradorite	.	.	.	.	2·70
Aquamarine	.	.	.	2·68—2·75
Beryl	.	.	.	.	2·68—2·75
Emerald .	.	.	.	.	2·67

GROUP V. (sp. gr. = 2·65 and less).

Quartz . . . . . ⎫
Smoky-quartz . . . . . ⎬ 2·65
Amethyst . . . . ⎪
Citrine . . . . . ⎭

Jasper . . . . . ⎫
Hornstone . . . . . ⎬ 2·65
Chrysoprase . . . . ⎭
Cordierite . . . . 2·60—2·65
Chalcedony . . . . ⎫ 2·60
Agate, &c. . . . . ⎭
Obsidian . . . . . 2·5—2·6
Adularia . . . . . 2·55
Haüynite . . . . 2·4—2·5
Lapis-lazuli . . . . 2·4
Moldavite . . . . . 2·36
Opal . . . . . 2·19—2·2
Jet . . . . . . —1·35
Amber . . . . . 1·0—1·1

B. CLEAVAGE.

It has previously been briefly pointed out that crystallised bodies are distinguished rom amorphous bodies by certain peculiarities of internal structure. Whereas the substance of an amorphous body possesses identical properties in every direction through it, the substance of a crystallised body possesses different properties in different directions. An important character of crystallised bodies, and one which shows considerable variation in different directions, is the degree of cohesion existing between the ultimate crystalline particles of which the mass is built up. In many such bodies the cohesion in certain directions is so feeble that a slight blow is sufficient to cause them to break into fragments. On examination, these fragments will be found to possess perfectly plane surfaces, which are the surfaces of minimum cohesion in the substance. This phenomenon is shown to perfection by calcite, a crystal of which, if allowed to fall, will break into fragments bounded by perfectly bright and even surfaces.

The best way to produce this separation or splitting along plane surfaces is to place a chisel or knife edge on the crystal in the proper position, and to drive it in with a single sharp blow from a hammer.

The direction within a crystal, along which there is minimum cohesion, is known as the *direction of cleavage*, and the plane surface of separation is known as the *cleavage face*. The cleavage directions and faces are always identical both in number and direction in all specimens of the same mineral species. The ease with which cleavage takes place, and the perfection of the resulting cleavage faces, are also constant in all specimens of the same species, but different in different species. Some minerals exhibit little or no cleavage, while in others, notably calcite, cleavage is produced very easily, and the surfaces of separation are perfectly bright, smooth, and even.

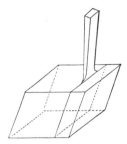

FIG. 8.   Cleavage of calcite.

Among precious stones possessing the property of cleavage to a high degree may be mentioned topaz, which cleaves in one direction, and diamond which cleaves in four directions. In others, as, for example, emerald, cleavage takes place with difficulty, and the cleavage faces are uneven and frequently interrupted by irregular areas. Quartz, garnet, tourmaline, &c., are other examples of precious stones possessing no distinct cleavage ; the difference of cohesion in different directions being so small that the stones will not split along plane surfaces.

In amorphous bodies, as, for instance, opal, the degree of cohesion between the constituent particles, like all other physical characters, is identical in every direction, so that here plane cleavage faces are impossible, and as a fact never occur. When a body shows sure indications that it possesses the property of cleavage, we are safe in inferring from that fact alone that the material of which the body is composed is crystallised and not amorphous. Hence it is sometimes possible to distinguish between a genuine crystallised precious stone and a glass imitation, since glass, being amorphous, can have no cleavage.

If in the same crystal there are three or more directions of cleavage, it will then be possible to develop out of it by cleavage a body bounded entirely by cleavage faces ; such a body is known as the *cleavage form* of that particular crystal. Calcite, for example, cleaves with equal facility in three directions, inclined to one another at equal oblique angles. It is therefore possible to obtain from any crystal of calcite a cleavage fragment having the form of a rhombohedron, a solid figure which resembles a cube with two of the opposite corners pressed together. In the same way, the four cleavages of diamond will give a cleavage form identical with the regular octahedron.

Such cleavage forms resemble natural crystals in possessing plane regular faces, but whereas in crystals these faces are the result of natural growth, in cleavage forms they have been produced by artificial means. In connection with the cutting of precious stones and the purposes to which they are to be applied, a knowledge of the cleavage possessed by different stones is most desirable. The property of cleavage considered from this point of view will be treated in detail under the descriptions of individual stones.

Cleavage frequently affords a simple means by which a stone in the rough condition may be identified or distinguished from others of similar appearance. As we have already seen, the cleavage directions and faces are always identical in number, direction, and quality in all specimens of the same mineral species, and in general differ from those of other species. The cleavage of a stone is thus one of its characteristic and distinguishing features. As an example of the use which can be made of this character, let us suppose a case in which it might be very difficult to decide which of two stones is an aquamarine, and which a certain colour-variety of topaz, both being of a sea-green colour and very similar in general appearance. Aquamarine has a very imperfect cleavage in one direction, while topaz has a perfect cleavage also in one direction.   Should one of the stones show a distinct cleavage, there can then be no doubt that it is topaz and not aquamarine.   If, however, no distinct cleavage can be made out, the evidence must be considered as negative, since a cleavage face need not necessarily be developed or outwardly visible even on a mineral which cleaves with great facility.

The cleavage of a mineral is not always expressed as a cleavage surface forming one of the external boundaries to the stone.   Fairly perfect cleavage is often indicated by the presence of plane cracks running in a certain direction inside the crystal itself.   Frequently such crevices give rise to the brilliant rainbow colours of thin films ; the film here being air or simply a vacuous space.   The cleavage of the stone is thus manifested by these iridescent colours in a very beautiful manner.   On a surface parallel to which there is a very perfect cleavage there is often to be seen a peculiar lustre resembling that of mother-of-pearl ; this is limited to crystals possessing a perfect cleavage, and hence its occurrence may be taken as an indication of the presence of such.   Even in faceted stones cleavage may be sometimes recognised by the iridescent colours and pearly lustre due to internal plane fissures.

From the æsthetic and commercial points of view, however, the presence of such cleavage fractures in a cut stone is far from desirable, since they give rise to irregularities in the reflection and refraction of light which seriously diminish the beauty and consequently the value of the stone.   The presence of cleavage cracks or " feathers " is a very bad fault in a transparent precious stone, for a small and scarcely noticeable crack may in course of time extend and cause the stone to break into fragments.   Rough stones showing any marked cleavage cracks are useless for cutting, since in the process they will probably break.

Stones which cleave with great facility should be treated with special care when mounted as gems, for a fall or blow, or a sudden rise in temperature (arising perhaps from immersion in hot water), may be sufficient to give rise to, or further develop, a cleavage crack, which may result in the complete fragmentation of the stone.

Although the property of cleavage in some cases leads to undesirable results, yet considerable advantage may be derived from it in others.   A stone with a distinct cleavage, such as topaz, too large for a single gem, may be easily reduced by cleaving to any desired size with no loss of material ; a stone not possessing this property must be sliced to the required size, a process involving the expenditure of much time and trouble.   Again, by cleavage, portions of rough stones may be easily and quickly removed which otherwise would have to be got rid of by grinding, a laborious and costly operation.   Moreover, the cleavage fragments can be utilised in the fashioning of smaller gems, and waste of material thus avoided.

The facile cleavage of the diamond is utilised very largely in the production of cut stones.   As we have already seen, the cleavage form of the diamond is an octahedron, and this approximates to the shape of a brilliant, which is the form of cutting usually adopted for the diamond.   The first stage in the transformation of a rough diamond into the cut stone is therefore the development of the octahedral cleavage form, and this is quickly and easily performed owing to the ready cleavage of the mineral.   This property of the diamond then obviates both the necessity for the laborious and expensive process of grinding in the production of the cut stone, and also the waste of material consequent on this process.

Fracture.—The fractured surface of a mineral not possessing the property of cleavage is not plane, but uneven and irregular.   The particular character of the fractured surface, or *fracture* as it is called for brevity, is different in different minerals, and, as it is more or less characteristic, it must be taken into account in the identification of rough stones.

The fractured surface frequently has the rounded form of a molluscan shell, and the two surfaces, respectively convex and concave, which fit together, both exhibit regular, circular ridges and grooves concentric about the point where the specimen received the blow which caused the fracture. Since these circular markings resemble the lines of growth on a molluscan shell, this type of fracture is known as a conchoidal fracture. Perfect conchoidal fracture is shown by artificial glass and also by the natural volcanic glass, obsidian. The surfaces of the conchoidal fracture may vary in extent, and also in the degree of curvature, being at times almost plane, hence such fractures may be distinguished as flat- or deep-conchoidal, and large- or small-conchoidal. The last, or sub-conchoidal fracture, merges into what is known as uneven fracture. As mentioned above, the fractured surface may approximate to a plane surface, never, however, being truly plane; this is known as even fracture, and is well shown by jasper: it merges into the flat- and large-conchoidal types.

Sometimes the fragment separated from a specimen by a blow from a hammer shows on the fractured surface loosely attached splinters, lighter in colour than the main mass of the stone. This class of fracture is said to be splintery, and is excellently illustrated by chryso-prase. Naturally there may exist every gradation between a typical splintery fracture and a smooth fracture.

Precious stones are frequently penetrated by cracks which are the forerunners of fractures. Such cracks considerably lessen the transparency and beauty, and consequently, the value of a stone; specimens showing such flaws are avoided by gem-dealers. Though of rare occurrence in many stones, in others, e.g., emerald, they are often present in great numbers.

These internal cracks due to fracture resemble the cracks shown by stones possessing a cleavage in that they often exhibit iridescent colours; as, for example, is sometimes seen in rock-crystal. Here, however, the cracks, and the bands of colour to which they give rise, are more or less markedly curved, and are thus quite distinct from the plane surfaces of cleavage cracks.

C. HARDNESS.

For a mineral which is to be used as a gem, an important and, indeed, indispensable property is that of hardness. By the hardness of a mineral is understood the resistance which it offers to being scratched or marked by another body. The greater the resistance, the harder the mineral. Only the harder stones, when used as gems, are capable of preserving unimpaired their transparency, lustre, and play of colours. Softer stones when newly cut may display all these qualities, but in use they soon become scratched on the surface, which detracts considerably from their beauty; a single scratch on the back of a transparent stone, that is on the side away from the observer, is many times reflected, and thus the bad effect is multiplied. The beauty of opaque stones is also greatly marred by scratches, but those on the front side of the stone only will be observable in this case.

The degree of hardness is of considerable importance in identifying and distinguishing precious stones, and is a character of which the dealer in gems should make frequent use. Hence the necessity for acquiring a knowledge of the different degrees of hardness possessed by different stones.

The method of testing the relative hardness of two stones is simplicity itself. A sharp corner of the one is rubbed with a certain pressure across a smooth surface of the other;

the stone which is scratched is the softer of the two, and if neither is scratched the stones are of equal hardness.

In this way it can be shown that all specimens of the same mineral species have the same degree of hardness, and that this quality will differ more or less considerably in different minerals. The hardness of a mineral is, therefore, one of its characteristic features, and it affords a means whereby precious stones may be identified or distinguished.

From observations on the relative hardness of minerals made in the way just described, all minerals can be arranged according to the degrees of hardness they possess in a series ranging from the softest to the hardest. From such a series, the late Viennese mineralogist, Mohs, selected the hardest, the softest, and eight minerals of intermediate hardness, and with them constructed a table for use as a **scale of hardness.** The ten selected minerals were numbered consecutively from 1 to 10, No. 1 being the softest and No. 10 the hardest, The scale is given below :

1. Talc.	6. Felspar.
2. Gypsum.	7. Quartz.
3. Calcite.	8. Topaz.
4. Fluor-spar.	9. Corundum.
5. Apatite.	10. Diamond.

It must be borne in mind that the difference in hardness between any two consecutive members of this series is by no means identical. Thus, for example, the difference in hardness between diamond (10) and corundum (9) is vastly greater than between corundum (9) and topaz (8); greater indeed than the difference which exists between corundum (9) and talc (1). The different minerals in this scale of hardness are chosen solely with a view to practical convenience in mineralogy, namely to afford a means whereby the relative hardness of any mineral may be expressed with clearness and brevity by a number.

To express the hardness of any given stone in this way we must first ascertain to which member of the scale it corresponds in hardness ; the number of this member then expresses the hardness of the stone, which may be written $H = 8$, for example. By this it is understood that the hardness of the stone corresponds to No. 8 on Mohs' scale, that is, it has the same hardness as topaz. In the same way, if the hardness of the stone lies between that of quartz and of topaz, $H = 7—8$; should it be nearer quartz, then $H = 7\frac{1}{4}$; or nearer topaz, then $H = 7\frac{3}{4}$; while if it lies apparently midway between them, $H = 7\frac{1}{2}$. In the last case the stone would scratch quartz with the same ease as it is itself scratched by topaz. $H = 7\frac{1}{4}$ means that topaz scratches the stone more easily than the stone itself scratches quartz ; while $H = 7\frac{3}{4}$ means that the stone easily scratches quartz but is not so easily scratched by topaz. Beyond this these numbers have no exact meaning.

Specimens of the ten minerals which form the scale of hardness should be kept specially for the purpose of determining hardness. They should be crystals with sharp edges and smooth faces, and of a convenient size. In testing any given stone a sharp edge of the scale mineral is rubbed over a smooth face of the stone, the scale minerals being used in consecutive order from the softest to the hardest. It is important to distinguish carefully between a scratch on the surface of the stone undergoing examination and a streak of powder which may arise from abrasion of the corner of the scale mineral. To avoid this mistake the surface should be wiped and then examined with a lens.

To take an example, let us suppose that the stone to be tested is not scratched by any member of the scale until topaz is tried ; its hardness will then lie between 7 and 8. If it is not capable of itself scratching quartz its hardness is exactly that of quartz, namely, 7. Should the quartz, however, be scratched by the stone, its hardness is then $H = 7—8$, or, as explained above, it may be fixed at $H = 7\frac{1}{4}$, $7\frac{1}{2}$, or $7\frac{3}{4}$ according as its hardness approximates more nearly to that of quartz, lies midway between that of quartz and topaz, or approaches more nearly to that of topaz.

It is sometimes sufficient to determine the hardness of a stone approximately, and, in such cases, when the exact degree of hardness is not required, the scale of hardness need not be used. The softest mineral, talc (No. 1 of the scale), is greasy to the touch. No. 2 on the scale, namely gypsum, can be easily scratched with the finger-nail, but this is impossible in the case of calcite (No. 3). A knife scratches calcite easily, fluor-spar (No. 4) less easily, apatite (No. 5) still less easily, and felspar (No. 6) only with difficulty, while quartz (No. 7) cannot be scratched at all with a knife. Quartz and members higher in the scale will strike fire with steel more or less readily, while felspar (No. 6) only does this with difficulty and to a small extent, and lower members of the scale not at all. Minerals harder than apatite (No. 5) are capable of scratching ordinary window-glass, and this substance may be fixed approximately at No. 5 in the scale of hardness. The more a mineral exceeds apatite in hardness the more easily will it scratch glass.

Minerals used as precious stones have the highest degree of hardness. The hardness of the most valuable corresponds to Nos. 10, 9, and 8 on the scale; those of less value have a hardness denoted by 7, rarely lower than this. Hardness above that of quartz is therefore known as gem-hardness; a mineral below this standard has little application as a gem, since it will be readily scratched even by dust. Among other constituents, dust contains minute mineral particles especially of quartz, and in cleaning a stone of hardness less than 7 by rubbing it with a cloth it will be scratched by these particles of quartz, and in course of time lose its beauty, becoming dull, rough, and lustreless. Stones of gem-hardness are not so scratched and damaged. Moreover, if stones of different hardness are allowed to rub against each other, as may easily happen if several mounted gems are kept together loosely in a box, then the harder stones will scratch and damage the softer ones. Since diamonds are usually represented in collections of jewels, all other stones, including ruby and sapphire, being softer, are liable to suffer if due care in this respect be not taken.

With few exceptions, therefore, precious stones possess a high degree of hardness, and very nearly all are capable of scratching glass, a substance usually conveniently at hand as window-glass or as a watch-glass. Since glass naturally will not scratch glass, a genuine precious stone may be easily distinguished from its imitation in glass by the test of hardness. As an aid to the identification of a stone, its position in the scale of hardness should be determined as described above. In the case of a cut stone, however, the process must be reversed to avoid damage to the stone; that is, the scratching power of the cut stone must be tried upon the scale minerals, commencing with the lower members, until the hardest the stone is capable of scratching is found. In such cases only an approximate value of the hardness can be arrived at, but it will usually be sufficient.

For such purposes there is no need to use the complete scale of hardness; the softest as well as the hardest members may be omitted. As a standard for the fifth degree of hardness a small plate of glass serves excellently, and is more easily obtainable than a good piece of apatite. There will be required in addition pieces of felspar, quartz, and topaz. The quartz should be a colourless transparent crystal (rock-crystal); the topaz should have a smooth, cleavage surface, so that the slightest scratch may be easily and surely recognised whenever the surface is examined with a lens. For the practical determination of gems the use of any other than these four members of the scale of hardness is superfluous. The softer precious stones, with a hardness less than that of apatite, will be recognised by their incapability of scratching glass; combined with other easily observed characters, this will usually be sufficient to determine the identity of the stone. Stones which are harder than topaz are but few in number; they are corundum (including ruby and sapphire), chrysoberyl, and diamond, the hardest of all stones. These stones stand alone in their power of scratching topaz; they may be readily distinguished from each other by the specific gravity or other determinable characters, as in the case of the few stones which do not scratch glass.

Determinations of hardness must be performed on cut stones with the greatest care, for it is possible that corners of the stone may chip off even when being pressed against a softer stone, and especially so when the cut stone possesses a good cleavage as in topaz or diamond. The loss of a corner would not be serious in a rough stone, since, in the process of cutting, broken edges are removed; but it would be fatal to the perfection and beauty of a cut stone. This test of hardness then, though useful in the case of rough stones, must only be used with caution in the case of valuable cut stones.

In place of the scale of hardness, the use of which has been just explained, the dealer in precious stones more frequently uses other instruments, and specially a hard steel file. This easily scratches minerals with a hardness of 5, and only slightly those with a hardness of 6, giving more or less powder according as the hardness of the stone is less or greater. Quartz is of about the same hardness as hardened steel of good quality, of which the file should be made. Stones with a hardness of 7 are therefore only with difficulty marked by the file, while harder stones will rub and polish the file, which will leave a shining, metallic mark on the stone. An approximate idea of the hardness of a stone may be obtained from the pitch of the sound emitted when the file is rubbed on the stone. Provided that stones similar in size are used for testing, then the harder the stone the higher the note emitted.

In the case of cut stones the file is too clumsy an instrument, and the practical jeweller uses in its place a pencil of very hard steel furnished with a sharp point. This pencil scratches felspar easily and glass still more easily; it scarcely touches quartz, however, and has no effect on harder stones. The girdle of a cut stone is a suitable part on which a trial of its hardness may be made; it being by the girdle that the stone is fixed in its setting, a small scratch in this region is unnoticed. The steel pencil is especially useful in distinguishing genuine precious stones from their softer glass imitations, since the former cannot be marked by it, while the latter are scratched with ease. As before mentioned, however, the greatest care is needed in testing cut stones, especially the transparent kinds, so that even this more refined method has certain limitations.

The hardness of a stone is naturally a question of great importance in the process of cutting; the material of the grinding disc and the grinding powder, to be described later, must be chosen according to the degree of hardness of the stone. When worked under similar conditions, with the same kind of abrasive material, the harder the stone the longer and more difficult will be the process of grinding. As a rule also, the harder the stone the sharper will be its edges and corners when cut, and the more susceptible will be its faces of a brilliant polish. The edges and corners of softer stones are much less sharp, and consequently these stones have a less pleasing appearance. A high degree of hardness is thus not only essential for the preservation of the beauty of a stone, but is also one of the properties on which its beauty depends.

From the time required for grinding a facet on a stone, it is possible to form an estimate of the hardness of the stone on this facet. It not infrequently happens that a stone can be more easily and quickly cut in certain directions than in certain others; moreover, not only do the different natural faces of a crystal vary in hardness, but the hardness on any face is not identical in all directions. The hardness of a crystal, therefore, like the other physical characters, varies with the direction.

These differences in hardness, however, are usually very small, and require for their detection an instrument of special construction capable of precise measurement; such an instrument is known as a sclerometer. The somewhat rough method of scratching, described above is useless for the detection of such small differences in hardness; it is applicable, however, in the case of kyanite. The hardness in different parts of a crystal of this mineral varies between that of apatite (5) and that of quartz (7). No other mineral used as a gem shows variation in hardness between such wide limits. Certain differences in hardness, shown by different specimens of the same mineral species, may be attributed in part to the fact that the hardness has been determined in different directions; this difference in crystallised precious stones, however, is so small as to be of little importance. In amorphous stones, such as opal, and in glasses, the hardness, like all other physical characters, is the same in all directions.

Finally, it must be noted that the hardness of a mineral is distinct from its **frangibility,** the quality on which depends the ease or difficulty with which a stone is broken by a blow from a hammer. The frangibility of a stone depends not only on its hardness, but also, and to a great extent, on the quality of cleavage it possesses. Contrary to popular opinion, the diamond, in spite of its enormous hardness, is very brittle and can be easily broken to pieces. Certain peculiarities of structure greatly diminish the frangibility of some minerals; this is especially the case in those of which the structure is that of a matted aggregate of very fine fibres or needle-shaped crystals of microscopic size. An example of such a mineral is furnished by nephrite (jade), which, although its hardness is scarcely equal to that of felspar, offers a very great resistance to the hammer, and can only be broken with considerable difficulty. Such substances are described as being tough, while those which are easily frangible would be described as brittle. A high degree of brittleness is not a desirable quality in a precious stone, since the stone is liable to be broken in use unless special care is taken.

In the following table are given all the more important minerals, which may be used as gems or as ornamental stones, arranged in order of hardness, from the softest to the hardest. The numbers refer to their degrees of hardness on Mohs' scale:

Precious Stones arranged according to Hardness.

Amber	$2\frac{1}{2}$	Demantoid	$6\frac{1}{2}$
Jet	$3\frac{1}{2}$	Idocrase	$6\frac{1}{2}$
Malachite	$3\frac{1}{2}$	Olivine	$6\frac{1}{2}$
Fluor-spar	4	Chalcedony (agate, carnelian, &c.)	$6\frac{1}{2}$
Dioptase	5	Axinite	$6\frac{3}{4}$
Kyanite	5—7	Jadeite	$6\frac{3}{4}$
Haüynite	$5\frac{1}{2}$	Quartz (rock-crystal, amethyst, citrine,	
Lapis-lazuli	$5\frac{1}{2}$	jasper, chrysoprase, &c.)	7
Sphene	$5\frac{1}{2}$	Tourmaline	$7\frac{1}{4}$
Hæmatite	$5\frac{1}{2}$	Cordierite	$7\frac{1}{4}$
Obsidian	$5\frac{1}{2}$	Garnet (red)	$7\frac{1}{4}$
Moldavite	$5\frac{1}{2}$	Andalusite	$7\frac{1}{2}$
Opal	$5\frac{1}{2}$—$6\frac{1}{2}$	Staurolite	$7\frac{1}{2}$
Nephrite	$5\frac{3}{4}$	Euclase	$7\frac{1}{2}$
Diopside	6	Zircon	$7\frac{1}{2}$
Turquoise	6	Beryl (emerald, aquamarine)	$7\frac{3}{4}$
Adularia	6	Phenakite	$7\frac{3}{4}$
Amazon-stone	6	Spinel	8
Labradorite	6	Topaz	8
Iron-pyrites	6	Chrysoberyl	$8\frac{1}{2}$
Prehnite	$6\frac{1}{2}$	Corundum (ruby, sapphire, &c.)	9
Epidote	$6\frac{1}{2}$	Diamond	10

D. OPTICAL CHARACTERS.

Those qualities of precious stones which depend on their behaviour towards light are known as optical characters, and are of special interest and importance. The transparency and lustre, the colour and play of colours of a stone, depend largely on its optical characters. Moreover these qualities furnish a means whereby the stone may be easily determined, and with no risk of injury such as accompanies the testing of its hardness. It is important, therefore, to be acquainted, at least to a certain extent, with the laws of optics, and with some of the instruments used in the investigation of optical phenomena ; these are dealt with below as fully as space allows.

1. Transparency.

The majority of precious stones are transparent, but in the uncut condition the free passage of light through the stone is often obstructed by rough and uneven faces. When such faces are removed by cutting, and the cut surfaces polished, an apparently cloudy specimen often becomes beautifully clear and transparent. The transparency of such costly stones as diamond, ruby, and sapphire, is often very perfect, while the same property exists in less costly stones, such as rock-crystal and amethyst. The greater the transparency of a specimen of any given precious stone the more highly is it prized. The only jewels of the first rank which are not transparent are the noble opal and the turquoise ; in stones of less value, such as agate, chrysoprase, malachite, and others, this is more frequently the case.

Transparent bodies allow the free passage of light through their substance ; an object viewed through a perfectly transparent substance will have no blurred edges, but will present a clear and sharp outline. A stone which combines complete absence of colour with perfect transparency, as in diamond and rock-crystal, would be described as water-clear or limpid. A perfectly water-clear stone with the highest degree of transparency, and free from any trace of colour, is known to jewellers as a stone of the first or purest *water*, and stones of this high quality are especially prized. Should the stone show a very slight cloudiness or tinge of colour, scarcely noticeable perhaps to the unpractised eye, it is known as a stone of the second water ; similarly a stone which shows a further departure from the standard of perfection in these qualities is known as a stone of the third water. This subject will be again reverted to under the special description of diamond.

A substance which in a mass of some thickness allows a large proportion, but not all, of the light emanating from any source to pass through it, is known as a semi-transparent substance. Any object, for example a flame, viewed through such a substance will not be seen distinctly, but will be blurred in outline. A substance through which some of the light of a flame can pass, but through which it is impossible to see even a blurred outline of the flame, is described as being *translucent*. As an example of a semi-transparent stone chalcedony may be mentioned, while opal is an example of a translucent stone. In some cases light can only pass through a very thin splinter or a sharp edge of a broken stone ; such stones are translucent only at their edges, being quite opaque in mass. A chrysoprase held in front of a light will show a dark centre surrounded by a lighter border. Opaque stones, even when thin, completely cut off all light, hence when held before a light they present a uniformly dark outline with no lighter border. Opaque stones then, for example hæmatite, owe their beauty not to their transparency, but to the fineness of their lustre and colour.

Different specimens of the same kind of stone vary greatly in transparency, and consequently in value ; while one specimen may be perfectly transparent, another may be so cloudy and opaque as to be useless as a gem. The cloudiness in such a case is due to the presence of numerous cracks and fissures in the stone, or to foreign matter included in its substance, either of which obstructs the free passage of light through the stone, scattering it at the surface.

Cracks and fissures are specially frequent in stones, such as topaz, which possess a good cleavage. They are not confined to such stones however, but are frequently found in those which possess no distinct cleavage, such, for instance, as the green emerald, which is almost always penetrated by numerous cracks. Their presence naturally reduces the transparency of the stone ; perfectly faultless specimens of emerald are of the greatest rarity. Enclosures of foreign matter are of not uncommon occurrence in crystallised minerals. Thus black and other coloured grains are sometimes found in the substance of diamond, and numerous scales of mica in that of emerald. These enclosures may be so small as to be only visible under a high power of the microscope ; when distributed evenly throughout the substance of the stone, as they usually are, their effect is to make the whole stone cloudy. On the other hand a few enclosures of larger size will leave portions of the stone clear.

The substance of some precious stones contains enormous numbers of extremely minute cavities often arranged in strings, and giving rise to a silky or cloudy glimmering or sheen, which greatly impairs the transparency and beauty of the stone in which they occur. This kind of cloudiness, present as a fault in precious stones which would otherwise be transparent, is known to jewellers as " silk."

The transparency of a mineral is largely dependent upon its structure. While crystals, at least as far as precious stones are concerned, are usually transparent, an aggregate of small crystals of the same kind, that is a compact crystalline aggregate, is usually opaque, or at most translucent. The reason for this is clear ; at the boundaries of each of the minute constituent grains, fibres or scales of a crystalline aggregate, a certain amount of light will be scattered and lost, and thus never reach the eye. For this reason chalcedony, chrysoprase, &c., are not transparent, although they are built up of minute transparent crystalline grains of quartz, a mineral which in its most perfectly crystallised and transparent condition is known as rock-crystal.

The varying degrees of transparency possessed by different precious stones for Röntgen (X) rays has an important application in their determination ; this will be dealt with in the third part of the present work.

2. LUSTRE.

When light falls upon a body a portion of it is thrown back or reflected from the surface, while another portion enters its substance. It is the portion of light reflected from the surface on which depends the lustre of the body.

The lustre of a body varies with the proportion of light reflected at its surface ; hence different **degrees of intensity of lustre**, distinguished as splendent, shining, glistening, glimmering, and dull, exist in different stones. The lustre of a perfectly smooth surface, reflecting a sharp image of an object, is splendent ; a surface which reflects a less sharp image is shining ; the image reflected from a surface of glistening lustre is still less sharp ; a surface giving only a feeble reflection has glimmering lustre ; lastly, a dull surface reflects no light.

The lustre of most precious stones, especially the more valuable, is splendent, as is often seen on the natural faces of a crystal, but more frequently when the stone has been cut and polished. A high degree of lustre adds very considerably to the beauty of a stone, and the object of polishing is to render this quality as perfect as possible. What is known as the " fire " of a precious stone is connected with its lustre, and this quality only exists with a specially high degree of lustre ; the term " fire " has, however, sometimes another meaning, as we shall see further on under Optical Dispersion.   Very few of the more valuable precious stones are devoid, in the cut condition, of a brilliant lustre ; of these the most important is turquoise, which, even after being polished to the fullest possible extent, shows a certain dulness of surface.   This is due to a certain extent to the softer character of the stone ; for the harder precious stones, such as diamond and ruby, are susceptible of a higher degree of polish than the softer stones, such as turquoise.

Each kind of stone is capable of receiving a certain degree of polish depending on the physical characters of its substance.   And while there is for every stone a certain maximum lustre, which cannot be exceeded, even with the most persistent polishing, this lustre may often, from a variety of causes, fall below the maximum.

Different precious stones are not only characterised by different degrees of intensity of lustre, but also by different **kinds of lustre ;** and this in many cases enables one to distinguish stones which are otherwise similar in appearance.   Thus it would be possible for the least practised eye to distinguish at a glance a genuine diamond from its imitation in rock-crystal, simply by the difference in lustre.

Degree of lustre and kind of lustre are both loosely referred to as lustre.   Though it has been attempted to give some idea of the different degrees of lustre existing in different stones, yet no adequate conception of the different kinds of lustre can be derived from a mere verbal description.   It is far preferable to acquaint oneself with the different kinds of lustre by actually comparing substances showing the different kinds.   Thus a piece of burnished metal, a sheet of glass, a polished diamond, the mother-of-pearl lining of a shell, a layer of greasy oil, or a piece of satin, may each possess a high degree of lustre, the quality or kind of which is however different in every case.   The different kinds of lustre exhibited by the substances just mentioned are taken as types ; and the different kinds of lustre existing in minerals can be referred to one or other of these types.   Thus we have metallic lustre, glassy or vitreous lustre, adamantine (diamond) lustre, pearly lustre, greasy lustre, and silky or satiny lustre ; while more minutely descriptive terms can be derived from these, as, for example, metallic-adamantine, metallic-pearly, &c.

In the description of a mineral it is important to be able to recognise and name its particular kind of lustre.   As explained above, each kind differs in intensity, so that we may have strong or feeble vitreous lustre, &c.   Both degree and kind of lustre depend on the properties of the particular substance and are therefore, under certain conditions, characteristic of that substance.   It is, however, to be noted that the kind of lustre may be considerably modified by certain peculiarities of structure shown by a mineral, so that it sɪ not an invariable character of a mineral species.   The kind of lustre depends not only on light reflected from the surface of a stone, but also, in part, on light which has penetrated into its substance and suffered some modification before again passing out.   As examples of this influence of structure on lustre may be cited the silky lustre of satin-spar, a finely fibrous variety of calcite, and the greasy lustre of elæolite, a variety of nepheline containing vast numbers of microscopic enclosures ; contrasted with these is the vitreous lustre shown by the more usual varieties of the minerals calcite and nepheline.

Metallic lustre, being exhibited exclusively by perfectly opaque substances, is found in only a few of the less important precious stones, as, for example, hæmatite.

Vitreous lustre, on the other hand, is best shown by perfectly transparent minerals, and is of very frequent occurrence ; it is present to a more or less marked degree in the majority of transparent precious stones, such as rock-crystal, topaz, ruby, sapphire, emerald, and others. It is, however, liable to be modified in substances possessing certain properties; thus, should a substance possess a high refractive index and also a high dispersion, the vitreous lustre will pass into **adamantine lustre**, which, again, may show an approach to metallic lustre. Characteristic adamantine lustre is possessed only by diamond, but an approach to it is shown by zircon, especially when colourless.

Silky or **satiny lustre** is exhibited by minerals possessing a finely fibrous structure, as, for example, the ornamental stone satin-spar, the fine green malachite, and the golden tiger-eye.

Pearly lustre is exhibited exclusively in those faces of crystals parallel to which there is a direction of perfect cleavage, and then only when cleavage cracks have been developed in the interior of the crystal. Topaz, felspar (the moonstone variety), and other stones, sometimes show pearly lustre, but only on faces parallel to the perfect cleavage; all other faces have the usual vitreous lustre.

Greasy lustre appears to be associated with the presence of numerous microscopically small enclosures, which in many minerals are of constant occurrence. Elæolite, which is sometimes cut for ornamental purposes, shows a typical greasy lustre; other minerals, such as olivine, of which the lustre is usually vitreous, may show an approach to greasy lustre.

The lustre of some minerals, as, for example, turquoise, resembles that of wax, and is described as **waxy lustre**. Others, again, have the lustre of resin: **resinous lustre** is shown by many garnets, such, for instance, as hessonite, which, when massive, sometimes closely resembles resin in appearance.

3. REFRACTION OF LIGHT.

The refraction of light, and the phenomena connected with it, have an important bearing on the study of precious stones.

We have already noticed that of the light which falls upon a transparent body, such as a precious stone, a portion is reflected at the surface, while another portion enters its substance and is propagated in straight lines through it. When the incident ray of light strikes the bounding surface of the transparent body perpendicularly, the light which passes into the interior of the body is propagated in the same direction as that of the incident ray. If, however, the incident ray strikes the surface obliquely, the path taken by the light through the substance of the body will not coincide in direction with the incident ray, but will be in a new direction; the ray may then be said to be bent or refracted.

In Fig. 9 let MN be the surface of separation between the transparent body (precious stone), S, and the air, L. A single ray of light travelling in air in the direction AC, and striking the separating surface at C, will not be propagated

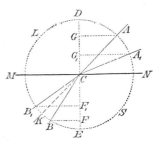

FIG. 9. Refraction of light on passing into a precious stone.

in the stone in the same straight line, namely, along CK, but will be bent or refracted into the direction CB. CB is then the refracted ray corresponding to the incident ray AC. The directions AC and CB, and also DE, the normal to the surface at C, all lie in the same plane, which is perpendicular to MN, and in Fig. 9 is the plane of the paper. This plane is known as the plane of incidence. Whenever light passes from air into a stone, the refracted ray CB is always nearer the normal DE than is the incident ray AC; that is, the light is bent *towards* the normal. This may be expressed otherwise by stating that the angle of incidence, ACD, is greater than the angle of refraction BCE.

When the angle of incidence is greater than ACD, the angle of refraction will be greater than BCE. In Fig. 9, let A_1C represent another ray of light in the same plane of incidence, such that the angle of incidence A_1CD is greater than ACD, then the corresponding refracted ray is represented by CB_1. It will be seen from the figure that the angle of refraction B_1CE is greater than BCE, which is the angle of refraction corresponding to the angle of incidence ACD. Similarly in every case the angle of refraction varies with the angle of incidence, and this variation is governed by a definite law, namely, the law of refraction.

In the plane of incidence and with C as centre (Fig. 9), let a circle of any convenient radius be drawn. Let the circle cut the incident rays AC and A_1C at the points A and A_1, and the refracted rays BC and B_1C at the points B and B_1. From these points drop perpendiculars, AG, A_1G_1, BF, B_1F_1 upon the normal DE. Then the ratio of the perpendicular AG to the corresponding perpendicular BF is the same as the ratio A_1G_1 to B_1F_1, and will always be constant for the same substance whatever may be the angle of incidence. Hence, when light passes from air into any particular stone, the following relation holds for all rays:

$$\frac{AG}{BF} = \frac{A_1G_1}{B_1F_1} = \ldots = n,$$

where n is a constant for that particular substance, but is different for different substances. This constant n is known as the refractive co-efficient, the refractive index, or the **index of refraction** of the substance. As just pointed out, the refractive index is independent of the angle of incidence; it has a certain definite value in all specimens of the same kind of precious stone, and will be different in different stones.

Since, in the passage of light from the air into a stone, the angle of incidence is always greater than the angle of refraction, it is easily seen from Fig. 9 that the perpendiculars AG and A_1G_1 will always be greater than the perpendiculars BF and B_1F_1; the index of refraction of all precious stones is therefore, when compared with that of air, always greater than unity.

The index of refraction of a stone can be determined accurately to several places of decimals by various methods. For the purpose of identifying precious stones, however, such a degree of accuracy is unnecessary, and refractive indices will be given here, as a rule, to only two decimal places. The following values may be given as examples, that of air, of course, being 1:

Water	$n = 1{\cdot}33$
Fluor-spar	$n = 1{\cdot}44$
Spinel	$n = 1{\cdot}71$
Garnet	$n = 1{\cdot}77$
Diamond	$n = 2{\cdot}43$

In passing from air into any of these substances, the bending of the rays of light is greater the greater the refractive index of the substance, and conversely. The value of the refractive index is in many precious stones very high, but is far higher in diamond than in any other gem. The values for other stones will be given under the special description of each. In comparing two substances with different refractive indices, the one with the higher refractive index is known as the "optically denser" substance, while the other would be described as the "optically rarer"; thus precious stones are "optically denser" than water or air.

A ray of light in passing from air into a stone immersed in liquid will be twice bent; once at the surface of separation between the air and the liquid, and again at the surface of separation between the stone and the liquid. The amount of bending depends in each case upon the difference in the refractive indices of the two media through which the ray passes. Thus, if the refractive index of the liquid is much greater than that of air, the ray of light will be much bent in passing from air into the liquid. Similarly, if the refractive index of the stone is not much greater than that of the liquid, the ray will experience but little bending when entering the stone ; while if the index of refraction of the stone and that of the liquid are identical, there will be no bending of the ray of light, and it will travel through both in the same straight line.

The use of methylene iodide in determining specific gravities has already been described. Another of its convenient properties is a very high index of refraction, the value of which, moreover, can be diminished by diluting the liquid with benzene. If a stone be immersed in methylene iodide so diluted with benzene that the refractive index of the liquid is the same as that of the stone, there will be no bending of the rays of light, and they will pass in straight lines through the liquid and the stone. Provided that the liquid and the stone are of the same colour, the result will be that the latter becomes invisible and cannot be detected. If the index of refraction of the liquid be changed by the addition of benzene or of methylene iodide, the boundaries of the stone will become visible ; its outlines will grow sharper and more distinct as the difference between its refractive index and that of the liquid is increased by the further addition of either one or other of the liquids.

The phenomenon just described is sometimes made use of for the purpose of discovering hidden cracks, enclosures, and other flaws in precious stones. The stone is immersed in a strongly refracting liquid such as methylene iodide ; its external boundaries will then become less distinct or, if the stone has the same refractive index as that of methylene iodide, invisible. Any flaws in the interior of the stone will thus be rendered prominent and can be easily seen.

Light is refracted not only when passing from an optically rarer into an optically denser medium, as, for instance, from air into precious stone, but also in the reverse case, as, for example, when a ray of light in a stone passes out into the air. In the passage of light from a denser to a rarer medium, the law of refraction still holds good. We shall see from Fig. 10, however, that the refracted ray is in this case bent *away* from the normal, or, in other words, the angle of incidence is less than the angle of refraction ; while in the previous case the refracted ray was bent *towards* the normal, and consequently the angle of incidence was greater than the angle of refraction.

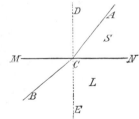

FIG. 10. Refraction of light on passing out of a precious stone.

In Fig. 10, let MN be the surface of separation between the stone S and the air L. It will be seen that the angle of incidence ACD of the ray AC in the stone is less than the angle of refraction BCE of the refracted ray; also that the refracted ray BC is bent away from the normal. In this case also, the bending of the ray is greater the greater the index of refraction of the stone, but the amount of bending is the same whether the light passes from stone to air or *vice versâ*. In one case the light travels in the direction ACB, and in the other in the direction BCA.

In the case also of the passage of light from a denser to a rarer medium, the angle of refraction increases with the angle of incidence. In Fig. 11, where MN is the surface of separation between the precious stone S and the air L, the ray AC, incident upon the surface MN at C is bent into the direction CB, A_1C into the direction CB_1, and so on. As the angle of incidence ACD becomes greater and greater so the angle of refraction BCE also becomes greater and greater. When the angle of incidence reaches a certain value, represented by A_2CD, the corresponding angle of refraction B_2CE will be a right-angle; the refracted ray will then emerge from the stone in a direction parallel to the bounding surface MN.

Obviously at 90° the angle of refraction has reached its maximum value and no further increase is possible. Should the angle of incidence now be increased, even by a small amount, it will then be impossible for the ray of light to leave the stone, and it will be refracted no longer, but simply reflected by the bounding surface back into the stone. In Fig. 11, the incident ray A_3C is reflected from the surface MN, along the line CB_3 inside the stone. This takes place according to the usual laws of reflection, the angle of incidence A_3CD being equal to the angle of reflection B_3CD. In the same way, every ray incident upon the surface MN at a greater angle than A_2CD, will be unable to pass out of the stone, and will be reflected back again by the surface MN; A_4CB_4, for example, is the path of such a ray and its reflection.

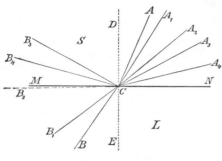

FIG. 11. Total reflection.

When light, travelling in one medium, as, for example, air, strikes the surface of a denser medium, such as a precious stone, a portion of it enters the stone and is refracted as described above, while the remaining portion is reflected from the surface. This takes place invariably, whatever may be the angle at which the incident light strikes the surface of the denser medium. In the reverse case, when light travelling in one medium, for example a precious stone, strikes the surface of a rarer medium, for instance air, the same thing may happen, that is, the light may be partly reflected and partly refracted, but this does not happen invariably as in the former case.

It was seen from Fig. 11, that when the angle of incidence exceeds a certain fixed value (A_2CD) the light is not refracted at all, but is reflected from the bounding surface back into the stone. In all other cases, as has been shown, light incident upon the surface of separation of two media is divided into a refracted portion and a reflected portion. Since in this particular case the light is not so divided, but the whole of it is reflected, this kind of reflection is known as internal total reflection, or, briefly, as **total reflection.**

Total reflection takes place at the surface of separation of two media only when the light travelling in the denser medium strikes the surface at an angle exceeding a certain degree of obliquity. Total reflection never takes place when light passes from a rarer to a denser medium. In this case there will always be refraction, for when the incident angle reaches a maximum of 90°, since the refracted ray is bent *towards* the normal, the angle of refraction will be less than 90°, and the light will pass out of the rarer into the denser medium. It is always possible then for light to pass from air into a precious stone, but it cannot pass out again unless it strikes the surface of the stone at an angle not exceeding a certain degree of obliquity.

The limiting angle A_2CD in Fig. 11 is known as the *critical angle* or the *angle of total reflection.* Its value depends upon the refractive indices of the two substances at the boundary of which reflection and refraction takes place. The greater the difference in the refractive indices the smaller will be the angle of total reflection, A_2CD. If the difference is very small the incident ray will make a large angle with the normal before total reflection takes place.

In diamond, which has a very high refractive index relative to air, the angle of total reflection is small, namely 24° 24', which is represented by the angle A_1CD in Fig. 12. A ray of light inclined to the normal at an angle slightly less than A_1CD will be refracted and pass out into air in the direction CB_1 while one inclined at a slightly greater angle will be totally reflected in the stone in the direction CB'_1. The ray A_3C, making a still larger angle with the normal, will be totally reflected along CB_3; while the ray A_2C will pass out of the stone along CB_2, not undergoing total reflection.

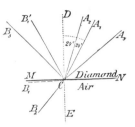

FIG. 12. Total reflection in diamond when surrounded by air.

If the optically denser body is, instead of diamond, glass, having, say, a refractive index of 1·538, then the angle of total reflection will no longer be 24° 24' but 40° 30', the body, as before, being surrounded by air. In this case, only those rays which are incident at an angle greater than 40° 30' will be totally reflected.

Since the angle of total reflection increases when the difference between the refractive indices of the two media decreases, it follows that the angle of total reflection will be greater if the stone is surrounded by water instead of air. The angle of total reflection for a diamond placed in methylene iodide, the refractive index of which is 1·75, will be 46° 19', the angle A_1CD in Fig. 13. Some of the rays, the obliquity of which causes them to be totally reflected when the diamond is surrounded by air, will be refracted when the surrounding medium is methylene iodide; thus fewer rays will in this case be totally reflected. The use made of this fact will be mentioned later.

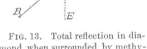

FIG. 13. Total reflection in diamond when surrounded by methylene iodide.

Total reflection has a considerable influence on the path taken by the rays of light in a transparent cut stone. The beauty of transparent cut stones largely depends on the fact that the light which falls on the front of the stone is totally reflected from the facets at the back and passes out again from the front to the eye of the observer. If the light were allowed to pass out at the back of the stone, the latter would lose much of its brilliancy; only when there is total reflection at the back of the stone does it appear, as it were, to be filled with light. The greater the proportion of light thus reflected from the back of a stone, the more brilliant will be its appearance. But to enable us to trace out the exact path of a ray of light in a cut stone, we must first consider some of the phenomena of refraction rather more closely.

Up to the present we have considered only the behaviour of a ray of light at the boundary of different bodies, namely, in passing from air into a liquid or into a precious stone, and *vice versâ*, in passing from a precious stone into air or liquid. By combining these observations, the complete path of a ray of light passing through a precious stone is easily arrived at.

In Fig. 14, let MN, PQ, be parallel sides of a transparent body, and let AB be a ray of light from a source, such as a small bright flame, falling obliquely upon MN. On passing into the plate, the ray is bent towards the normal, DE, and takes the path BC. This portion of the ray meets the second surface PQ at C, the angle of incidence, BCD_1, being equal to the angle of refraction, since the normals are parallel. On passing out into the air, the ray is again refracted, this time away from the normal, and takes the path CF. From the geometry of Fig. 14, it is easily seen that the second angle of refraction, FCE_1, is equal to the first angle of incidence, ABD; and that the paths of the ray outside the plate, namely AB and CF, are parallel. The direction of the ray on emerging from the plate is therefore the same as the original direction, but its path has been shifted a small distance, represented in Fig. 14 by $B'F$. On observing a small object A through a transparent body with parallel sides, it will be seen in very nearly the position it really occupies; this will not be the case however when the bounding surfaces, MN, NP (Fig. 15), are not parallel.

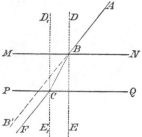

FIG. 14.   Path of a ray of light through a plate with parallel sides.

Let the bounding surfaces, MN, NP, of the transparent body in Fig. 15 be inclined to each other at an angle MNP. We are then dealing with the path of a ray of light through a prism. As before, let the path of the ray of light incident upon MN be AB; on entering the solid it will be bent towards the normal GH, and will take the path BD. On emerging into air, the ray will be bent away from the normal KL and will take the path DE. The angle between the original path of the ray in air and its final path is ACF, and this measures the total amount of bending it has undergone. The source of light A, if observed through the prism, will appear not in the position it actually occupies, but at some point on the line ECF which makes with the direction AB an angle ACF. This angle varies under different conditions. It will be greater the greater the refracting angle MNP of the prism, and the greater the refractive index of the substance of the prism; it depends also upon the angle of incidence, ABG. But it has a certain minimum value which cannot be diminished by either increase or decrease of the angle of incidence; this minimum value of the angle ACF is known as the angle of minimum deviation.

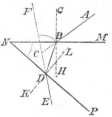

FIG. 15.   Path of a ray of light through a prism.

In passing through a prism, the differently coloured constituents of white light are separated, and we have the phenomenon known as **dispersion**. The beautiful appearance of many precious stones, and specially of diamond, is due to their dispersion of light. The coloured constituents of white light, from a source such as the sun or a lamp, differ not only in colour but also in refrangibility or capacity for being refracted. Thus, the refrangibility of red light is the smallest, and that of violet light the greatest; yellow, green, and blue light occupy in this respect intermediate positions in the order in which they stand.

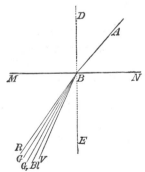

FIG. 16.   Dispersion of light.

It follows, then, that though we have hitherto spoken of a substance as having a single refractive index, this is only strictly true for monochromatic light, such as that given out by a Bunsen flame, or a spirit-lamp flame coloured by the vapour of either of the metals lithium, sodium, thallium, and indium. If white light be used, the refractive index of the substance will be different for each constituent of the light, that for red light being the least and that for violet light the greatest.

When white light passes through a prism, then, the red rays will be deviated or

bent out of their course least, the violet rays most, and the other rays will fall in their
proper order between the two extremes.   In Fig. 16 the ray of white light AB falls upon
the surface MN of a refracting substance.   Owing to the different refrangibilities of the
constituents of the ray, these latter are separated and we get the original single ray of white
light split up into red (R), yellow (G), green (Gr), blue (Bl), and violet (V), rays, deviating
from each other slightly in direction.   Between the rays of the colours just mentioned lie
rays of intermediate tints.   The decomposition of white light into its coloured constituents, or the dispersion of light, varies according to the
dispersive power of the refracting substance.   It is the
more distinct the greater the angle between the extreme
red ray and the extreme violet ray.

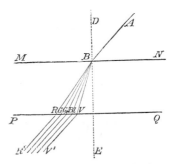

We have now to consider the dispersion of the light
which passes through a precious stone.   We will take first
the case in which the stone has the form of a plate with
parallel sides, as in Fig. 17, and afterwards the case in
which these bounding surfaces are inclined to each other
and so form a prism.

FIG. 17.   Dispersion of light by a
plate with parallel sides.

The ray of white light, AB, falling obliquely on the
surface MN of the precious stone, is split up into the
differently coloured rays lettered BR, BG, BGr, BBl, BV.
These rays pass out of the precious stone at the surface
PQ in the directions RR', VV', &c., all being parallel to the original white ray AB, as
was explained before in connection with Fig. 14.   The eye placed at $R'V'$ will receive all
these differently coloured rays at the same time and in the same direction ; the effect of
this will be to produce in the eye the sensation of white
light just as if the ray of light from A had not passed
through the plate.   With such a parallel-sided plate, then,
a decomposition of white light into its coloured constituents
takes place, but is not observable, since the effect produced
by the first surface is neutralised by the parallelism imparted to the rays at the second surface.

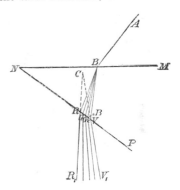

The dispersion of light produced by a prism, on the
other hand, is very noticeable, and is illustrated in Fig. 18.
A ray of white light, AB, falls upon the surface MN of the
prism, and is separated into its variously coloured constituents.   Between the extreme red ray, BR, and the
extreme violet ray, BV, lie the yellow, green, blue, and
rays of intermediate colours.   On passing again into air
at the second surface, NP, of the prism, these rays are
again refracted, and emerge still more widely separated.

FIG. 18.   Dispersion of light by a
prism.   Formation of the spectrum of
white light.

The angle between the extreme red ray RR_1 and the extreme violet VV_1 is R_1CV_1 ; and, as
before, this measures the amount of the dispersion, and varies with the substance of which
the prism is made.   An eye placed at R_1V_1 will receive this bundle of coloured rays,
diverging apparently from C, the various colours being perfectly distinct and brilliant.
The ray of white light thus gives rise to an elongated band of colour which is known as a
spectrum.   The red end of the spectrum lies nearest to the refracting edge, N, of the
prism, and the violet end furthest away from it ; the other colours lying between these
two, and following each other with no break or interruption in the same order as the

colours of the rainbow, namely, red, orange, yellow, green, blue, indigo, violet. The spectrum may be conveniently shown to a number of persons at once by placing a white screen in the path of the coloured rays.

Fig. 19 gives a perspective view of the path of rays of light from a candle-flame, A, through the prism $MNPM'N'P'$. The ray of light AB falls on the face $MNM'N'$ of the prism, and is resolved into the prismatic colours; the red ray travelling along BR, the violet ray along BV, and rays of other colours between. These rays inside the prism meet the face $NPN'P'$, and on passing into the air are further refracted and separated, the red ray taking the path RR', the violet VV', and so on. Since rays of light emanate from every luminous part of the candle flame, a complete image of it will be seen at A' by an eye placed at $R'V'$. In the direction $V'V$ the image will be coloured violet, that is, the side v of the image nearer the refracting edge NN' of the prism will be violet, while the margin r of the image lying on the line $R'R$ will be coloured red. To an eye placed at $R'V'$ the image will be seen to the left of the actual position of the object, as is shown in Fig. 19; the image is thus nearer the refracting edge of the prism than is the object.

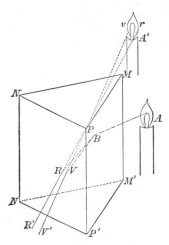

FIG. 19. Path of the rays of light through a prism. (Perspective view.)

The length of the spectrum formed by a prism depends upon a variety of conditions. It is longer the greater the angle between the rays RR^1 and VV^1; this, in its turn, depends upon the dispersive power of the substance of the prism, for the spectra given under similar conditions by two similar prisms, but constructed of different substances, will differ in length. The amount of dispersion produced by a prism will obviously vary with the difference between the degree of refraction of the red rays and of the violet rays; the difference between the refractive indices of a substance for red light and for violet light is indeed frequently regarded as a measure of the dispersive power of the substance.

Amongst precious stones, and indeed the majority of known substances, diamond has by far the greatest dispersive power. The differences in the refractive indices of diamond and of window-glass for red and for violet light, that is the dispersive power of these substances, are given below :

> Red light, $n = 2\cdot407$
> Violet light, $n = 2\cdot465$
> Dispersive power of diamond $= 2\cdot465 - 2\cdot407 = 0\cdot058$.

For window-glass :

> Red light, $n = 1\cdot524$
> Violet light, $n = 1\cdot545$
> Dispersive power of glass $= 1\cdot545 - 1\cdot524 = 0\cdot021$.

The dispersion produced by diamond is therefore more than double that produced by window-glass ; as a result of this, the spectrum given by a prism of diamond will be more than twice the length of that given by a prism of glass having the same refracting angle. The prismatic colours are transmitted to the eye by diamond widely separated from each other, and the stone owes much of its beauty to this fact ; in glass and other substances of less dispersive power, more or less overlapping of the prismatic colours takes place, and this renders them less perceptible to the eye as separate sensations.

The beautiful play of prismatic colours, shown by many precious stones, and especially by diamond, is quite independent of the colour of the stone itself, but is due to the decomposition of white light into its coloured constituents by refraction within the stone. The greater the dispersive power of a stone the more marked will be this play of prismatic colours ; on account of the specially high power of dispersion of diamond, the play of colours exhibited by this gem is far in advance of any other precious stone.

This play of prismatic colours is sometimes, especially by English jewellers, referred to as the " fire " of a stone. The same term, " fire," is, however, also used to denote the brilliancy of lustre of a stone ; it was used in this sense above when dealing with the quality of lustre.

Any two facets of a cut stone which are not parallel may constitute a prism and thus give rise to the decomposition of white light into its coloured constituents. The facets at the back and front of a cut stone should be so related as to give the maximum decomposition of white light. Further, the faces at the back of the stone must be steeply inclined, so that light, entering the stone from the front and being resolved into its component colours, will strike the back faces of the stone at such angles that it is totally reflected by them and passes out again at the front of the stone.

The more perfectly the form of cutting fulfils these conditions, namely, the greatest possible decomposition of white light into its coloured components, and the greatest possible internal reflection of this light from the back facets, the more beautiful will be the cut stone. The form of cutting most suitable for bringing out the beauty of the diamond is that known as the brilliant. This form is shown from different points of view in Figs. 29 and 52 among others, and in section in Fig. 20.

The form of a brilliant will be discussed in detail later ; here it need only be mentioned that its numerous facets give it approximately the shape of a double four-sided pyramid, of which one apex is truncated by a large plane, the table, and the other by a smaller plane. A brilliant is placed in its setting so that the table lm (Fig. 20) is at the front towards the observer, while the small truncating plane hi is turned to the back away from the observer.

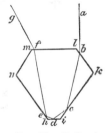

FIG. 20. Path of a ray of light in a brilliant.

The path of a ray of light inside a stone cut as a brilliant is shown in Fig. 20. Let us suppose a ray of light ab to fall on the oblique facet kl, and to be refracted within the stone in the direction bc. The refracted ray bc falls very obliquely on the facet ki, and forms with the normal to this facet an angle greater than the critical angle of the substance ; it will therefore be totally reflected in the direction cd, and cd and cb will be equally inclined to ki. In the same way the ray travelling along dc is again totally reflected from the surface hi in the direction de, and is then reflected from the surface hn in the direction ef. The ray travelling along ef strikes the facet lm at a high angle, that is, at an angle less than the critical angle of the substance ; it is therefore possible for it to pass out of the stone into air along the path fg. This direction, fg, will not, as a rule, coincide with the original direction of the ray ab, since in its journey through the stone it has undergone two refractions and three internal reflections. Moreover, as a consequence of the two refractions undergone by the original ray of white light, ab, it will be split up into its component colours, and, on emerging from the stone, will present to the observer a beautiful play of prismatic colours. To avoid obscuring the diagram, the different paths of differently coloured rays are not shown in Fig. 20, as they are in Figs. 17 and 18 ; the path, as shown in Fig. 20, may be regarded as the mean path for the several colours, or, more correctly, as the path which would be taken by monochromatic light within the stone. Other rays of light entering the front and side facets of the stone will be refracted and totally reflected in the same manner, and will therefore follow a path very similar in direction to the one shown in Fig. 20. The whole stone will therefore appear to be full of light, and will emit flashes of rainbow colours.

The many beauties of the diamond can be traced back to the optical characters of the stone; its high index of refraction causes a large proportion of the light which enters the front facets of a suitably cut stone to be totally reflected from the back faces, while from its high dispersive power results a wide separation of the rays of differently-coloured light, and, in consequence, a fine play of prismatic colours. These features of a cut diamond are specially noticeable when the stone is contrasted with another colourless stone, cut in the same manner, for example, rock-crystal. The latter appears in comparison dull and dead, owing to the fact that it possesses neither the high index of refraction nor the great dispersive power of the diamond. The highly refractive and dispersive glass called strass, when cut in the form of a brilliant, may, however, closely resemble a diamond in these characters.

From what has been said above, it is easy to see that the cutting of a stone is a very important factor in developing the potential beauty with which its optical characters endow it. A diamond cut in the form of a good brilliant, far exceeds in play of colour and general brilliance a similar stone cut in any other form, such, for instance, as a rosette (rose-cut), which does not fully utilise the optical characters of the stone.

4. DOUBLE REFRACTION OF LIGHT.

Hitherto we have considered only those substances in which a single refracted ray corresponds to a single incident ray. There are, however, many bodies, including many

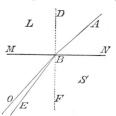

precious stones, which have the property of splitting a single incident ray of light into two refracted rays which are propagated in their substance along paths differing slightly in direction.

In Fig. 21 the ray AB, travelling in air L, is incident upon the surface MN of the stone S at B, where it is split into the two refracted rays BO and BE, inclined to one another at a very small angle OBE, which never exceeds a few degrees.

FIG. 21. Double refraction of a ray of light.

Bodies which behave towards light in this way are described as being **doubly refracting** or birefringent, in contradistinction to the **singly refracting** bodies hitherto considered. Substances exhibiting the phenomenon of double refraction may also be described as **optically anisotropic,** while those which exhibit single refraction are described as being **optically isotropic.**

As far as regards the transparency, lustre, colour, and play of colour of a stone—those characters in short which affect the beauty of the stone—it is unimportant whether the light within it is singly or doubly refracted.

The phenomenon of double refraction can be easily observed by the aid of special instruments. The detection of its presence or absence is a valuable aid in identifying and discriminating precious stones in the cut condition. Thus by the aid of an appropriate instrument we can decide whether a certain red stone is a doubly refracting ruby or a singly refracting spinel, two stones which, though very similar in appearance, are very dissimilar in

rarity and costliness.   It is also possible by this means to distinguish glass imitations, which
are always singly refracting, from genuine precious stones, which are for the most part doubly
refracting.

The kind of refraction, single or double, exhibited by a body is a necessary consequence
of the crystalline structure of its substance, and varies in the different crystal systems.   All
amorphous bodies, together with all those which crystallise in the cubic system, are singly
refracting, while all other crystals, without exception, namely, those included in the
hexagonal, tetragonal, rhombic, monoclinic, and triclinic systems are doubly refracting.   It
is thus possible from the behaviour of a stone with respect to the refraction of light to
learn whether, on the one hand, it is amorphous or crystallises in the cubic system, or
whether, on the other hand, it crystallises in one of the five remaining crystal systems ; and
this observation can be made on a very small irregular fragment of the mineral.   Thus in
the example just quoted we know that the singly refracting spinel must crystallise in the
cubic system, while the doubly refracting ruby crystallises in one of the remaining five
systems, namely, the hexagonal.

Since the observation of the kind of refraction, whether single or double, exhibited by
a stone is a step towards determining to which of the crystal systems it belongs, and moreover is frequently a decisive test of its identity, it is important to be acquainted with the
method of making this observation.   In the third part of this book, dealing specially with
the determination of precious stones, considerable use will be made of this method, and it
will also be mentioned under the description of each species of precious stone.

In some substances the phenomenon of double refraction is directly observable, for an
object, when viewed through a plate of the substance, will appear double instead of single,
as is more usually the case, for example with a plate of glass.
Each of the two refracted rays *BO* and *BE* (Fig. 21) gives
an image of the object ; these two images are, as a rule, very
close together, but in some few minerals they may be so widely
separated as to be both distinctly visible.

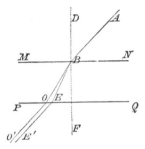

In Fig. 22, let *MNPQ* be a plate of doubly refracting
substance with the surface *MN* parallel to the surface *PQ*.
The incident ray of light *AB*, striking the surface *MN* at *B*,
enters the plate and is split up into the two rays *BO* and
BE ; these emerge from the surface *PQ* in the directions *OO′*
and *EE′* both parallel to *AB*.   Each of these rays *OO′* and

FIG. 22.   Path of light through
a doubly refracting plate.

EE′ gives rise to an image of the source of light, and an eye
placed at *O′E′* will see one image along *O′O* and another
along *E′E*.   Other conditions being equal, these two images will be the more widely
separated the thicker the plate is.

A substance which shows the phenomenon of double refraction to a very marked degree
is calcite or Iceland-spar, which on this account is also called doubly refracting spar.   If a
crystal, or, better still, a transparent cleavage rhombohedron of Iceland-spar is placed over
an object, such, for instance, as the page of a book, the letters, when viewed through the spar,
will appear double, as shown in Fig. 23.

In calcite the two refracted rays are inclined to each other at a comparatively large angle,
much greater than in the majority of other minerals.   The greater the angle of separation
of the two refracted rays (*OBE* in Fig. 21) the greater the double refraction of the mineral,
and different substances differ considerably from each other in this respect.

The double refraction of the majority of precious stones is not very strong ; and as usually only a small thickness of such substances is available for examination, the two images of an object viewed through the stone will be very close together or partly overlap

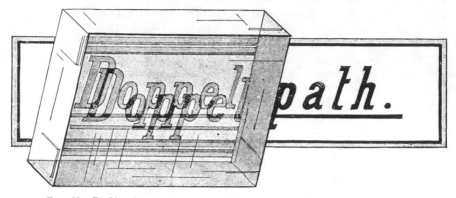

FIG. 23. Double refraction by calcite or doubly refracting spar (German, Doppelspath).

and then tend to appear as a single image. They would thus, by simply viewing an object through a thin plate, appear to be only singly refracting, whereas in reality they are doubly refracting.

It is possible, however, in such cases to bring about a wider separation of the two images by using a prism instead of a parallel-sided plate of the stone. This is illustrated in Fig. 24, where, as in the case of single refraction (Fig. 18), the rays of white light are

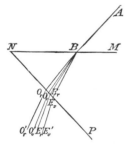

FIG. 24. Path of light through a doubly refracting prism.

decomposed into rays of differently coloured light. The ray of light AB, coming from a small flame at A, on entering the prism is split into two rays travelling in the directions BO and BE. In consequence of dispersion, each of these rays is separated into its coloured components BO_r to BO_v and BE_r to BE_v; and on passing out at the second surface of the prism NP, are again refracted, and thus emerge still more widely separated. To an eye placed at $O'_rE'_v$, two images of the flame, $O'_rO'_v$ and $E'_rE'_v$, will be visible close together or partly overlapping. Each image shows the columns of the spectrum of white light as did the image seen through a prism of singly refracting substance; moreover, if thrown on a screen, the red ends, O'_r and E'_r, of both spectra will be nearer the refracting angle of the prism, and the violet ends, O'_v and E'_v, further away.

Fig. 25 gives a perspective view of the path of light in a doubly refracting prism, similar to the one given by Fig. 19 in the case of a singly refracting prism. The two faces of the prism $MNM'N'$ and $NPN'P'$ are inclined together at the refracting angle MNP and intersect in the refracting edge NN'.

A ray, AB, emitted by the centre of the candle flame, A, strikes the face $MNM'N'$ of the prism at B, and is refracted along BO and BE. These two refracted rays pass out at the second face $NPN'P'$, and take the directions OO' and EE'. To an eye placed at $O'E'$, two images of the candle flame will be visible in the directions OOA^o and $E'EA^e$. Many precious stones show the two images A^o and A^e quite close together, often, indeed, overlapping more or less. As was the case with the single image given by a singly refracting prism (Fig. 19), each of the double images has a red margin r and a violet margin v.

Now every facet at the front of a cut transparent gem forms with any facet at the back (provided they are not parallel) a prism ; and through every such pair of facets can be seen, when viewed in the proper direction, an image of a small flame. As a matter of fact, a large number of such images will be seen, since for any one facet at the front of the stone there will be several at the back, each of which may form with the front facet a prism and give rise to an image. The images given by singly refracting stones are single, as in Fig. 19, while doubly refracting stones give two images very close together, as shown in Fig. 25. This difference enables us to distinguish a singly refracting from a doubly refracting stone.

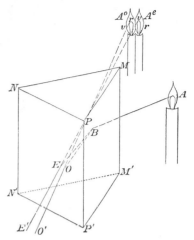

FIG. 25. Path of light through a doubly refracting prism. (Perspective view.)

For this purpose the stone should be held with the largest front facet, namely, the table, close to the eye, and a small flame viewed through it. On turning the stone about, a position will be arrived at when numerous coloured images of the flame become visible, each being single if the stone is singly refracting, or double if it is doubly refracting. Each image, whether double or single, has originated by refraction through a prism formed by one of the facets at the back of the stone and the table at the front. The images seen through a doubly refracting stone are shown in Fig. 26a, while those seen through a singly refracting stone are shown in Fig. 26b.

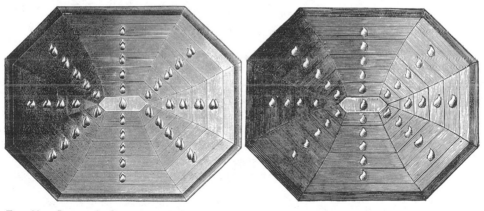

FIG. 26a.   Images of a flame observed through a doubly refracting stone.

FIG. 26b.   Images of a flame observed through a singly refracting stone.

This experiment is best performed in a dark room so that no light other than that from the small flame passes through the stone.

Instead of using a flame, however, any other convenient object may be observed through the stone, and for this purpose a needle may be used. When the needle is placed in the proper position relative to the stone, there will be seen several coloured single images of it in the case of singly refracting stones, while doubly refracting stones will give coloured double images of the needle.   Contrary to the previous case, this experiment must be performed in a lighted room.

When a stone thus examined shows unmistakably double images the fact may be regarded as a decisive proof of the doubly refracting nature of the stone ; when, however, single images only are observed the stone cannot be stated to be singly refracting on these grounds alone, for stones which have only feeble double refraction may give double images so close together, or may be overlapping, that to recognise the double character of such images is a matter of considerable difficulty.

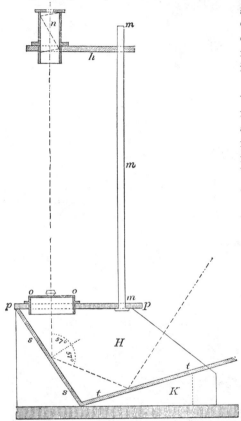

The investigation by the direct method of the kind of refraction possessed by a stone thus requires a certain amount of skill, which is only acquired by practice.   On this account the refraction of stones is often investigated by an indirect method, which has the advantage of being applicable to stones with rounded surfaces, and also to small and irregular fragments of material, neither of which could be used with the method of direct observation.   Further, very small cut stones are easily examined by the indirect method, while their examination by the direct method would present difficulties.

The instrument used for the indirect observation of the singly or doubly refracting character of a stone is known as the **polariscope**.   A simple form of this instrument, sufficient for the present purpose, is shown, one-third the actual size, in Fig. 27.

This consists of a wooden box, H, into the cover, pp, of which fits the circular object-carrier, oo ; the latter consists of a plate of glass in a brass setting, and may be easily rotated.   From the box rises the vertical brass rod, mm, which carries, on the horizontal arm, h, a Nicol's prism,

FIG. 27.   Polariscope for observation in paralle light.   (One third actual size.)

n, constructed of Iceland-spar.   This is placed in the same vertical line with the centre of oo, and is capable of being rotated in the arm, h.   In the box, H, is fixed, at an angle of 33° with the vertical, a sheet of unsilvered glass, ss, or better still, a large number of thin glass plates arranged in a pile.   The box also contains an ordinary mirror, tt, the inclination of which can be varied by means of the wooden wedge, K.

Rays of light from a clear sky enter the open side of the box, as indicated in the figure by the dotted line, and are reflected from the mirror, tt, on to the glass plate, ss, at an angle of 57° with the normal to the plate, whence they are again reflected in a vertical direction through the object-carrier and the Nicol's prism to the eye of the observer.

Ordinary daylight, after reflection from the glass plate, ss, at the particular angle mentioned above, becomes endowed with special properties, and is said to be *polarised*.   n other words, the rays of ordinary light which strike the plate, ss, are reflected from it as rays of polarised light, and as such reach the Nicol's prism, n.   On rotating the Nicol's prism, it will be found that in certain positions it does not allow the light reflected from *ss* to pass through, and in these positions the field of view becomes dark, while in other positions it is light. On turning the Nicol's prism through a complete revolution, that is 360°, it will be observed that there are four gradual changes from maximum lightness to maximum darkness or *vice versâ*; the four positions of maximum lightness and maximum darkness being separated by angles of 90°. If the Nicol's prism be turned into one of the two positions in which the field of view has maximum darkness, the polariscope will afford a means whereby singly and doubly refracting stones can be distinguished from each other with ease and certainty.

The different behaviour of singly and doubly refracting substances when examined with the polariscope is as follows : When a singly refracting substance, such as a piece of glass, is placed on the object-carrier, *oo*, and observed through the Nicol's prism, the whole of the field of view will be dark and will remain dark during the rotation of the object-carrier. It should be mentioned here that in this, as well as in all other observations, it is advisable to shade the side light from the object with the hand, or, better still, by means of a tube of black paper placed on the object-carrier and round the object; otherwise light will reach the eye which has been reflected from the surface of the stone without passing through it.

When a doubly refracting body is examined in the same way, it is found that in certain positions the portion of the field which it occupies becomes light. This is due to the fact that the polarised light, which before was unable to pass through the Nicol's prism, becomes so modified by its passage through the doubly refracting substance that it is now capable of passing through the Nicol's prism, when it will reach the eye of the observer. As the object is turned round through 360°, there will be eight changes from maximum lightness to maximum darkness or *vice versâ*; there being four positions of the stone in which the lightness is a maximum, and four in which there is maximum darkness, an interval of 45° lying between each. Through the complete rotation of the object, however, the portion of the field not occupied by it remains dark so long as the Nicol's prism is undisturbed.

There is thus an essential and important difference in the behaviour of singly and doubly refracting stones when examined in polarised light. A singly refracting stone remains dark in the dark field of the polariscope, while a doubly refracting stone changes from light to dark as it is rotated with the object-carrier.

Even in this method, however, there are certain liabilities to error which must be carefully avoided. In all doubly refracting substances the strength of the double refraction is not the same in all directions. Thus the two images of a flame or needle seen through a doubly refracting stone will be further apart when viewed in some directions than in others, while in certain directions a single image only is to be seen. The substance is therefore, along these particular directions, not doubly refracting but singly refracting.

Those directions in a doubly refracting body along which there is only single refraction are known as **optic axes**. All doubly refracting stones can be grouped into two classes: the one containing stones having one optic axis, described as being **optically uniaxial**; and the other containing stones having two optic axes, and described as being **optically biaxial**. The optic axes of any given substance are closely connected both in number and direction with its crystalline form. Thus all hexagonal and tetragonal crystals are uniaxial, and the optic axis of these crystals coincides in direction with the principal crystallographic axis. All rhombic, monoclinic, and triclinic crystals are biaxial, and in the case of rhombic and monoclinic crystals definite relations exist between certain crystallographic and optical directions. Crystals belonging to the remaining system, namely the cubic, are, as mentioned above, optically isotropic, that is, singly refracting.

The fact that doubly refracting crystals are singly refracting along their optic axes must not be forgotten in making observations with the polariscope. A doubly refracting stone, placed in the instrument so that its optic axis coincides with the line of vision, will behave as if it were singly refracting, and will remain dark during a complete rotation of the carrier. A single observation of this kind is therefore not sufficient to prove the singly refracting character of the stone. The probability of a stone being placed in the instrument in this position will, as a rule, be small ; when a stone gives the indications of single refraction, explained above, it should be placed on the carrier in another position and re-examined. A second indication of the same kind may be regarded as conclusive, though in case this second indication should be due to the second optic axis coinciding with the line of vision— a very improbable chance—the stone may be examined in a third position. As a rule, an examination of the stone in two positions will be sufficient to establish its singly refracting character. If, on the other hand, a stone should at the first trial give the indications of a doubly refracting substance, the observation may be regarded as conclusive and further examination is superfluous.

When a stone in the polariscope gives the appearances peculiar to singly refracting substances after examination in two, or even in three, positions, it has been stated above that it may legitimately be concluded that the stone is really singly refracting. We have now to show that, under certain circumstances, such is not the case, and that the stone may be in reality doubly refracting. When a cut stone is examined in the polariscope, the facets on the side turned towards the observer will not be parallel to the facet upon which the stone lies, but they may be very steeply inclined to it. Light travelling vertically upwards from beneath will always be able to enter the stone by the facet on which it lies ; it may, however, strike the upper, steeply inclined facets so obliquely as to be totally reflected and pass out at the sides of the stone, thus never reaching the eye of the observer. The error which this may lead to may be avoided in several ways.

The majority of cut stones have, as shown in Plates II.–IV., a large facet, the table, on one side, and a small facet, the culet, parallel to the table, on the opposite side. If the stone be examined through these two parallel faces, there will be no possibility of internal total reflection. With this object in view, the stone should be placed on the object-carrier so as to rest upon the culet, the table being uppermost ; should the culet be very small the stone may be supported by pieces of wax. With the stone in this position, the light entering it will strike the table perpendicularly and there will thus be no chance of reflection from this facet. Moreover, the position in which the stone rests upon the culet has the further advantage that the whole area of the table is available for the egress of the light which enters the stone by the culet ; whereas, in the reverse position, much of the light which enters by the table will fail to escape by the culet, but will be totally reflected from the side facets.

When a stone, examined in the polariscope in the position just described, gives the indications of a singly refracting substance, this observation, as explained above, cannot be regarded as conclusive, and the stone must be re-examined in another position. In any other position, however, there is a possibility of a doubly refracting stone appearing to be singly refracting owing to total reflection of the light within it. This possibility may be avoided by the following simple device :

The stone is completely immersed in a strongly refracting liquid contained in a small glass vessel placed on the object-carrier. The difference between the index of refraction of the stone and of the surrounding medium will be much less than when the stone was in air, and the result will be, as has already been explained, that a larger proportion of the light will escape total reflection and will pass out of the stone. The amount of light totally reflected will in any case be considerably diminished, but total reflection will not be entirely absent, unless the refractive index of the liquid is exactly the same as that of the stone. In the case of diamond, however, this state of affairs will not exist, since there is no liquid with so high a refractive index.

Liquids used for this purpose should be transparent, not deeply coloured, and of high refractive index. One which fulfils these conditions, and has been already mentioned, is methylene iodide. It is one of the most strongly refracting liquids known, having at the ordinary temperature of 15° to 20° C. an index of refraction of 1·75 for the middle rays of the spectrum ; this value for the refractive index is exceeded by only few precious stones, notably the diamond, the index of refraction of which is 2·43. From the upper faces of a feebly refracting stone immersed in methylene iodide there will be no total reflection of light, but this will still take place if a diamond is substituted for the feebly refracting stone. All rays of light, forming with the normal to the surface from which they emerge an angle greater than 46° 19′ (Fig. 13) ; in other words, all rays travelling vertically upwards will be totally reflected from facets inclined to the horizontal object carrier of the polariscope at an angle greater than 46° 19′. Total reflection in diamond is thus not eliminated but considerably diminished in amount, the corresponding critical angle for diamond in air being 24° 24′.

A drawback to the use of methylene iodide for this purpose is its high price ; monobromonaphthalene is a much cheaper liquid, with a refractive index almost as high as that of methylene iodide ; it can therefore be used as a substitute for the latter in optical determinations, but not in determinations of specific gravity, being too light a liquid for this purpose.

It will be well at this point to review the method of using the polariscope for determining the singly or doubly refracting character of a precious stone. The stone is placed on the object-carrier in the dark field of the polariscope and the carrier rotated. If now, all side light being carefully screened off, the field shows alternations of lightness and darkness the stone may be considered to be without doubt doubly refracting. Should the field remain dark during the rotation, the stone must be placed in another position on the object-carrier and again rotated. If this rotation results in alternations of lightness and darkness in the field, the stone is certainly doubly refracting ; but should the whole of the field still remain dark we cannot conclude that the stone is singly refracting until it has been proved that the absence of light has not been due to total reflection within the stone. With this object the stone must be examined for the third time, either in the position described above, resting upon the culet and with the table horizontal and uppermost, or immersed in a strongly refracting liquid to diminish or eliminate total reflection. If on rotation the whole of the field still remains dark the stone must be singly refracting. For the examination of stones cut in a spherical form, or those with a rough and irregular surface, it will often be necessary to immerse them in liquid at first. Observations made with the polariscope require no special skill, and with practice and attention to the necessary precautions are very reliable.

Before leaving the subject of refraction certain anomalous cases must be considered. Many singly refracting substances, such, for example, as diamond, occasionally show the appearances peculiar to doubly refracting substances. When this is the case, such substances are said to possess **anomalous double refraction**. The phenomenon is frequently due to internal strains set up on the solidification of the substance or brought about by subsequent causes. These internal strains may be so great in certain crystals, for example those diamonds known as " smoky stones," as to cause such stones to fly to pieces without any apparent reason.

Anomalous double refraction is usually, however, only feeble, and the alternations of lightness and darkness exhibited in the polariscope by minerals exhibiting this character are much less marked than is the case with truly doubly refracting stones. Moreover, the illumination of a doubly refracting body is uniform over the whole of its area ; but in the case of a body showing anomalous double refraction, portions of its area will remain dark during the complete rotation of the carrier, while the portions which become alternately light and dark during the rotation appear as stripes and bands, or in variously shaped sectors. It is thus a comparatively easy matter to distinguish between anomalous and true double refraction.

The phenomenon now under consideration sometimes affords a means whereby a glass imitation may be distinguished from a genuine precious stone. While glass under ordinary conditions is singly refracting, being an isotropic substance, unannealed glass⋅ possesses anomalous double refraction. Thus if a fairly thick plate of glass be first strongly heated and then suddenly cooled, internal strains will be set up, and when examined in the polariscope it will show a more or less regular black cross, with the two arms at right angles, sometimes surrounded by coloured circles. Should a supposed precious stone show this or a similar appearance, the observer may regard it as conclusive evidence against the genuineness of the stone.

The refraction of a precious stone is expressed, as explained above, by a number known as its refractive index. In the case of singly refracting substances there is, for monochromatic light, only one refractive index, which is constant for every direction in the stone. The index of refraction of a doubly refracting substance, however, varies according to the direction. It is greatest in one particular direction and least in a direction at right angles to this. These maximum and minimum values vary slightly for differently coloured light, but are constant for monochromatic light. The greater the difference between the greatest and the least values of the refractive indices of a precious stone, the greater will be its double refraction, which is measured by this difference.

The number which thus expresses the strength of the double refraction of a substance is constant for, and characteristic of, that substance, and could be made use of for purposes of identification. Its exact determination is, however, a matter of considerable difficulty and requires special and costly instruments, as well as suitable preparation of the stone to be examined. This method is therefore of no practical value to jewellers.

The refractive indices of the more important precious stones are given in the following table, the values for singly refracting stones being indicated by n, and the greatest and least values for doubly refracting stones by n_g and n_l respectively. In both cases the values apply to the middle rays of the spectrum. The strength of the double refraction of each stone is indicated by $d = n_g - n_l$, that is, by the difference between the greatest and the least refractive indices of the stone.

(a) Singly Refracting Precious Stones.

	n			n
Diamond . . .	2·43		Spinel	1·72
Pyrope . . .	1·79		Opal	1·48
Almandine . . .	1·77		Fluor-spar . . .	1·44
Hessonite . . .	1·74			

(b) Doubly Refracting Precious Stones.

	n_μ	n_l	d
Zircon	1·97	1·92	0·05
Ruby			
Sapphire	} 1·77	1·76	0·01
Chrysoberyl	1·76	1·75	0·01
Chrysolite	1·70	1·66	0·04
Tourmaline	1·64	1·62	0·02
Topaz	1·63	1·62	0·01
Beryl	1·58	1·57	0·01
Quartz	1·55	1·54	0·01

5. COLOUR.

An important character of precious stones not yet touched upon is colour. The beauty of opaque and lustreless stones, such as the turquoise, depends wholly upon this character. Every shade of colour is represented among minerals used as precious stones and for ornamental purposes. As has already been mentioned, stones which are perfectly colourless and also perfectly transparent are described as being water-clear, or of the first water.

In the majority of cases colour is a very variable character; there are, however, examples amongst minerals in which the colour is a fixed and essential character, appearing the same whether the mineral is in the largest masses or reduced to the finest powder. Such a stone would be described as being **idiochromatic**, it having a colour of its own, which is essential and characteristic of the mineral. As an example of an idiochromatic stone, malachite, which is always green whether in mass or in powder, may be quoted.

The majority of precious stones are, if perfectly pure, completely colourless. But this purity of composition is, as a rule, not attained, and thus, owing to the admixture of foreign colouring-matter, such stones occur much more frequently coloured than colourless. The colour, being thus due to accidental impurities, may vary in different specimens of the same stone, or even in different portions of the same specimen, and must therefore be regarded as a non-essential character. Stones in which colour is a variable character are distinguished as **allochromatic**, and the foreign matter, to which their colouring is due, may be regarded as pigment. Different pigments give rise to differently coloured specimens of the same mineral species. Such specimens show their colour best in fragments of some thickness; in very thin splinters, or in fine powder, they appear only faintly coloured or even completely colourless.

The variety of colour exhibited by quartz well illustrates the fortuitous nature of this character when due to impurities. Thus, rock-crystal is transparent and water-clear quartz, smoky-quartz is brown, amethyst is violet quartz, citrine is yellow quartz, green quartz is known as plasma, blue quartz as sapphire-quartz, and there are still other coloured varieties, with special names. Again, the mineral corundum, which sometimes occurs colourless, is known as ruby when red and as sapphire when blue; it is also found of many other colours, which will be mentioned in the special description of corundum. Though diamond, in its most valuable condition, is water-clear, yet specimens of every shade of colour are found.

The range of colour shown by any one allochromatic mineral is known as its **suite of colours.** Thus the suite of colours shown by quartz includes brown, violet, yellow, green, blue, &c.; that by corundum includes red and blue and many others. The suite of colours shown by any one mineral will not usually be shown by any other; in nearly every case certain colours will be unrepresented in the suite.

In minerals in which the lustre is other than metallic, a group which comprises nearly all precious stones, eight principal colours may be recognised for the purposes of descriptive mineralogy; these colours are white, grey, black, blue, green, yellow, red, and brown. Intermediate colours may be described by terms compounded of the names of the eight principal colours; as, for instance, reddish-white, greenish-blue, bluish-black, &c. The different shades or tints shown by each of the principal colours are indicated by a descriptive prefix; as, for example, sulphur-yellow, grass-green, indigo-blue, smoke-grey, carmine-red, &c. The colour of a mineral can be judged more correctly by observing it close to the eye, when small differences in colour will be more apparent.

The character of the colour shown by a mineral depends partly on the lustre and transparency of the specimen; it may be described by terms in use in ordinary language, such as lively, warm, fresh, dull, delicate, soft, dirty, dusky, &c. The intensity of the colour shown by a mineral also varies in different specimens; it may be described as deep or dark, when approaching to black; high or full, when pure and intense; light, when approaching to white; finally, as pale, when more nearly approaching to white. In speaking of some precious stones, for example the ruby, it was formerly the custom to describe specimens with a deep or full colour as "masculine," and those with a lighter colour as "feminine." These terms have now, however, fallen into disuse.

Intensity of colour depends on the amount of colouring-matter present; the greater this is, the deeper will be the colour of the stone. When the pigment of a stone is distributed equally throughout its mass, the stone will be uniformly coloured. If, on the contrary, the pigment is present in some parts and absent in others, or present in varying amounts in different parts of the stone, the latter will show corresponding differences in colour.

One and the same stone may be differently coloured in different parts owing to the presence of different pigments; thus sapphire often shows blue spots or patches on a colourless background, and amethyst may show violet areas also on a colourless background. The irregular distribution of colour in such stones detracts considerably from their beauty; specimens of precious stones in which the colour is intense and distributed with perfect uniformity are therefore specially valuable.

The distribution of colour in any one kind of stone is sometimes remarkably constant, appearing repeatedly in a large number of specimens. Thus in the four-sided columns of diopside from the Zillerthal in the Tyrol, which are sometimes used as gems, one end is colourless and the other of a fine, dark, bottle-green colour. In the same way the hexagonal prisms of red, green, or almost colourless tourmaline from Elba frequently have a black termination (so-called negro-heads). A regular arrangement of different colours in the same crystals is sometimes seen in tourmaline, as illustrated in Plate XV., Figs. 8 and 9, where the central portion is red and the external portion green, the two colours being sharply separated from each other. The beauty of agate is due to the arrangement of its various colours in bands. The following terms are used in describing colour distribution: spotted, mottled, clouded, veined, marbled, striated, banded, &c.

Brown or black arborescent, or tree-like, markings are frequently seen in certain specimens of chalcedony, and are described as *dendritic markings*. Stones showing such markings are known as dendrites. They are cut and polished with the object of bringing out the markings as prominently as possible (Fig. 89). Dendrites, among which is moss-agate with its peculiar and moss-like distribution of green colouring-matter, will be further considered in dealing with opal, chalcedony, &c.

The various **pigments** to which the colouring of precious stones is due may be organic or inorganic, and differ much in character. They may exist in considerable amount, but more frequently they are present in such small quantities that very exact chemical analysis is necessary for their detection. In the latter case, the colouring-power of the pigment must be comparable to that of carmine and some other pigments, an extremely small quantity of which is capable of giving a decided colour to an enormous quantity of a colourless substance.

The precise nature of the colouring-matter of many precious stones it has been impossible as yet to determine. Large quantities of the precious stones would be needed to yield an amount of colouring-matter sufficient for a reliable analysis, and here lies the chief obstacle to the investigation. In spite of this difficulty it has been possible in some cases to determine definitely to what substance the colouration is due. Thus, for example, the emerald owes its green colour to the presence, in small quantity, of a compound of the metal chromium, while the apple-green colour of chrysoprase is due to a compound of the metal nickel. Other stones are coloured by compounds of iron or copper; while the brown colour of smoky-quartz is due to an organic substance, which can be distilled off as a dark brown oil, possessing an empyreumatic odour.

The colouring-matter of precious stones is frequently distributed so intimately and uniformly through their substance that it is impossible, with the strongest magnification, to distinguish single particles of the pigment. The relation between the substance of the precious stone and the pigment seems analogous to that which exists between a solvent and a substance dissolved in it. In such cases, it is inferred that the colouring-matter is not an essential constituent of the substance of the stone from the fact that specimens of other colours, or devoid of colour, are known. This intimate association of the pigment with the ground substance of the stone exists, for example, in the green emerald, in the blue opaque turquoise, the colour in the latter case being due to compounds of copper and iron; as also in diopside, the green colour of which is given by a compound of ferrous oxide.

In most of these cases we are dealing with something more intimate than a mere mechanical mixture. The colouring-matter is isomorphous with the ground substance of the precious stone, that is, it has the same type of chemical formula, and when this is the case, the intermixing of the two substances involves not microscopically small particles of each but the ultimate particles or molecules of each substance. Thus to take diopside as an example, we must picture the molecules of the compound of ferrous oxide distributed uniformly between the molecules of the ground substance of the stone and imparting to it its characteristic green colour. The same may perhaps be said of emerald which belongs to the mineral species beryl, specimens of which sometimes occur colourless; also of turquoise and many other precious stones.

In other stones, on the contrary, the colour is due to vast numbers of minute coloured particles, with definite boundaries, mechanically intermixed with the colourless ground-mass of the stone. These particles may be large enough to be just perceptible to the naked eye, or so small as to require a lens or microscope for their detection; they may have the form of grains, scales, fibres, or needles. Small blue grains distributed in large numbers through the colourless ground-mass of lapis-lazuli give to this precious stone its fine blue colour. Green needles and fibres of the mineral actinolite give rise to the green colour of prase, a variety of quartz, which of itself is colourless. Felspar is sometimes coloured red by minute scales of iron-glance (hæmatite), and is then used as an ornamental stone under the name of sun-stone: chalcedony, coloured by a similar red pigment, is the much used carnelian.

Stones coloured in this manner, by the mechanical intermixture of particles of pigment, are more or less cloudy or even opaque; those, on the contrary, in which a more intimate or chemical relation between the colouring-matter and the ground-substance is possible, are clear and transparent.

The **apparent change of colour** shown by many precious stones when exposed to different kinds of illumination is worthy of remark. In most cases the colour seen in clear day-light is the most beautiful, the appearance by artificial light being less pleasing. Thus amethyst by day-light is of a beautiful purple colour, but in candle-light it appears dull grey. Purple corundum or " oriental amethyst," on the contrary, shows its fine colour as well by candle-light as by day-light. Specially peculiar in this respect is the variety of chrysoberyl, known as alexandrite, which, as we shall see later on, is green in day-light and red in candle-light. Yellow diamonds retain their colour in the electric light, but appear colourless in candle-light. Many other stones afford similar examples of a change in colour accompanied by a loss of beauty in artificial light, a property which naturally diminishes the value they might otherwise possess.

The possibility of temporarily masking the colour of yellowish diamonds has, in recent years, frequently led to fraud. Since the discovery of the South African mines, yellowish diamonds are fairly abundant, and therefore comparatively cheap, while perfectly colourless stones command a high price. By giving these yellowish stones a very thin coating of some blue colouring-matter, they can be made to appear colourless, the mixture of blue and yellow light rays producing on the eye the effect of white light. As soon, however, as the blue coating is worn off the fraud becomes apparent.

Not only an apparent, but an **actual change of colour**, may be experienced by some precious stones. As a rule, the colours of precious stones are extremely lasting, only disappearing with the destruction of the stone itself, this being, for example, the case with the yellow diamond, the ruby, emerald, and others. The colour of other stones, however, is less constant, and may be completely destroyed, the substance of the stone undergoing no change in the process. The colouring of such a stone will frequently disappear when the stone is raised to a red heat or even less; this will invariably happen if the colouring-matter is organic in nature, since it will be decomposed at such a temperature. Brown smoky-quartz and reddish-yellow hyacinth behave in this way, becoming completely colourless when heated to redness. Other stones on being heated experience not a loss but a change of colour; thus the violet amethyst becomes yellow, and the dark yellow topaz becomes rose-red in colour. These particular changes in colour are sometimes brought about intentionally in order to obtain yellow quartz (" burnt amethyst ") and rose-red topaz (rose topaz), both of which are used as cut stones, but occur in nature to only a small extent.

Many stones show characteristic changes in colour during the progress of a rise and fall in temperature. Thus the red ruby, at a high temperature, is colourless; on cooling it first becomes green, after which it gradually assumes its original fine red colour. The red spinel behaves somewhat differently under similar conditions; at a high temperature it becomes colourless, and on cooling it regains its original colour, so far resembling the behaviour of the ruby, but at the intermediate temperature it assumes not a green but a violet tint. A high temperature is not invariably necessary to effect a change of colour in precious stones; some stones are so sensitive that their colour fades or disappears merely on exposure to light and air. Certain topazes behave in this way, and after a few months exposure will be recognisably paler in colour; the same phenomenon may be observed in green chrysoprase and in rose-quartz, as well as in some blue turquoises, the colour of the latter of which may gradually change to green. Obviously the value of such stones will be considerably diminished, since it is difficult, if not impossible, to avoid a rapid loss of colour, and therefore of beauty, when they are used under ordinary conditions. Colour lost in this way may sometimes be restored by keeping the stone in darkness, by burying it in moist earth, or by treating it with certain chemicals, all of which devices are made use of by unscrupulous dealers. As a contrast to the behaviour of such stones on exposure to light, it may be mentioned here that amber, instead of being bleached by exposure, is darkened, gradually becoming of a dark, reddish-brown colour.

The **artificial colouring** or recolouring of precious stones, which was known and practised to some extent among the ancients, is of some importance. At the present day agate and similar stones are most frequently subjected to this treatment, the exact methods of which will be dealt with under the special description of these stones. The capacity for absorbing the liquid which imparts its colour to the stone, often even to the central portions, depends on the porous nature of its substance.

Streak.—In speaking above of idiochromatic and allochromatic minerals, we have seen that in the former the fine powder of the mineral is also coloured, the colour being characteristic of the mineral. For the purpose of quickly and easily obtaining a mineral in the state of fine powder, it is rubbed on a plate of rough, unglazed, " biscuit " porcelain. The line of powder left upon the plate by the mineral is known as its streak, the colour of which can be easily observed on the white background. The streak is often characteristic of a mineral, and thus the observation of the streak is a step towards the determination of the mineral. The character is not of much practical value in the determination of precious stones, since on account of their hardness they are much more likely to scratch the porcelain than to leave a streak upon it ; moreover, the streak of most precious stones, as in other allochromatic minerals, is white, and therefore not a distinguishing feature.

6. Dichroism.

An important optical property of many precious stones is that known as dichroism or pleochroism. A stone possessing this property, when observed in different directions will show different colours or shades of colour which may resemble each other more or less closely, or may, on the other hand, differ considerably. A mineral sometimes used as a cut stone, and known as " water-sapphire," exhibits this phenomenon to such a marked degree that it has received the name dichroite, although at the present time it is usually known to mineralogists as cordierite. A crystal of this mineral, when viewed in three particular directions, perpendicular to each other, appears of three distinct colours, namely, a fine dark blue, light blue, and greyish-yellow. In intermediate directions are seen intermediate tints, which approach one or other of the three principal colours according to the direction of the line of view. The three particular directions along which these maximum differences in colour are observable are definitely related to the crystalline form of the mineral ; they are in fact the three crystallographic axes of the rhombic crystal.

The difference in colour shown by cordierite when viewed in different directions is very great, but it is perhaps even greater in some kinds of tourmaline. In this mineral the colour, as seen in different directions through a crystal, varies from yellowish-brown to asparagus-green, or in other crystals (of the same mineral) between dark violet-brown and greenish-blue, or again in others between purple-red and blue, &c. Some dichroic precious stones show only very small differences in colour when viewed in different directions ; the yellowish-green chrysolite is an example. Indeed the majority of pale coloured stones are only feebly dichroic, stronger dichroism being exhibited by minerals having a deeper tone of colour.

Finally there are other minerals, such as garnet and spinel, which show no differences of colour in different directions ; these precious stones behave in this respect like their glass imitations.

Taking refraction as the basis of classification, we have seen that all minerals can be divided into two groups, namely, those which possess single refraction and those which possess double refraction. The former group will include those minerals which are not possessed of dichroism, while all dichroic minerals fall into the latter. Thus amorphous substances and those which crystallise in the cubic system are characterised by single refraction and absence of dichroism, while all coloured minerals included in the remaining five crystal systems are dichroic, and all without exception are doubly refracting.

The phenomenon of dichroism then, furnishes us with additional aid in distinguishing singly from doubly refracting stones. A body showing this character to even the feeblest degree cannot be amorphous nor can it be a cubic mineral. The apparent absence of dichroism however must be considered only as negative evidence in favour of single refraction, since dichroism may be present but so feeble as to be detected only with difficulty. It has been shown above that the phenomena of double and of single refraction enables us to distinguish a ruby from a red spinel, and this is made still more easy from the fact that the hexagonal ruby is distinctly dichroic, while the cubic spinel does not possess this property. In the same way an imitation ruby of red glass could not be confused with the genuine stone, since the former, being amorphous, is not dichroic and shows the same colour in all directions.

The detection of dichroism usually requires the use of a special instrument. The most convenient instrument for the purpose is that devised by the Viennese mineralogist Haidinger, and known as a **dichroscope.** This instrument is inexpensive and easily used, and should

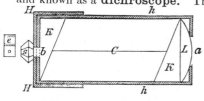

FIG. 28. The dichroscope. (Actual size.)

be in the hand of every one who buys or sells precious stones, since a single glance through it is sufficient to establish the presence or absence of dichroism in a stone.

This instrument is shown in section, and of its actual size, in Fig. 28. It consists essentially of a cleavage rhombohedron, C, of Iceland-spar (calcite), which is longer in one direction than in others. At each of its oblique ends is cemented a glass prism or wedge, K, the outer surfaces of which are perpendicular to the long edge of the calcite rhombohedron. A brass tube, h, encloses these essential portions of the instrument, and has at one end a small square aperture, b, and at the other a circular aperture, a. Between the circular aperture and the glass prism, K, is placed a lens, L, of such a focal length that on looking through the instrument in the direction ab a sharp image of the square aperture, b, will be seen. This image will, however, owing to the intervention of the doubly refracting calcite, not be single, but double. The instrument is so proportioned that these two images, o and e, will appear side by side, in contact but not overlapping. The image o will be only slightly displaced from the axis of the instrument and will be quite colourless ; the image e is rather more displaced, and shows a narrow red border on its inner edge and a narrow blue border on its outer edge, as indicated by striations in the small figure at the side, otherwise the image is colourless. The instrument is so constructed that the distance of the square aperture from the lens can be varied, which enables the images to be sharply focused and adjusted so that their edges are in contact.

In using this instrument, the precious stone to be tested for dichroism is placed over the square aperture, b, and, the instrument being directed towards a clear sky, the observer places his eye close to the round aperture, *a*. An object-carrier, *H*, is sometimes provided for the purpose of more conveniently holding the stone. This has the form of a brass tube fitting loosely over the tube, *h*, and having the closed end perforated by an aperture somewhat larger than the square aperture, *b*, over which the stone can be fixed with wax, as shown in the figure. This arrangement allows the carrier *H*, with the stone attached, to be rotated while the calcite rhombohedron remains unmoved. Should this carrier not be provided, the stone may be fixed by wax to a glass-plate, or simply held in the fingers in front of the square aperture, and the instrument rotated in the hand.

If the stone under examination is not dichroic, the two images *o* and *e* will be of the same colour and will show no variation while the instrument or the stone is rotated through 360°. If, for example, a red garnet, which crystallises in the cubic system, is examined, the two images *o* and *e* will both be of the same red colour as is the garnet itself when viewed without the aid of the instrument.

The images *o* and *e* given by a dichroic stone, on the contrary, will be in general differently coloured. In four particular positions, however, at 90° apart, the colours of the two images are identical. On rotating the stone or the instrument a difference between the colours appears, which gradually increases and reaches a maximum at 45° from the original position. Further rotation will again result in a gradual decrease in the colour difference, and at 90° from the original position the colours of the two images once more become identical. The same changes occur during the rotation through the remaining quadrants, and thus a complete rotation of 360° is accompanied by eight changes from identity of colour in the two images to maximum difference in colour between them and *vice versâ*. The juxtaposition of the two images makes it possible to detect the smallest differences in colour, and consequently the slightest degree of dichroism.

We have previously seen that doubly refracting crystals are singly refracting along the direction of an optic axis; similarly dichroic crystals exhibit no dichroism in these directions. To prove the absence of dichroism in a crystal it is therefore necessary to examine it in two directions, or, as an additional precaution, in a third direction also. After each observation the stone must be fixed on the holder, *H*, in a new position and again rotated. The absence of dichroism can be conclusively proved only after an examination of the stone in at least three different positions. The dichroism of a stone may be so feeble that it is not possible to detect it even with the aid of a dichroscope; moreover, it must be borne in mind that a coloured doubly refracting stone is not necessarily dichroic, and this feature is naturally absent in colourless doubly refracting stones. The real or apparent absence of dichroism in a stone is therefore no proof of its singly refracting character, but the presence of dichroism is, on the contrary, a conclusive proof of the doubly refracting nature of the stone.

The degree of dichroism in a crystal varies according to the direction through which the crystal is observed. The colours of the two images seen in the dichroscope in the examination of all dichroic stones become the more nearly identical as the optic axis of the stone becomes more nearly coincident with the axis of the instrument. Conversely, the greater the angle between the axis of the dichroscope and the optic axis of the stone the more marked will become the difference in colour between the two images. The two colours between which there is the maximum difference are known as the principal or axial colours; these colours, as seen through the dichroscope, differ in tint from the colours the stone shows when observed with the naked eye in the same direction. Uniaxial dichroic crystals, such as tourmaline, show two principal colours, while biaxial crystals, such as cordierite, show three. The pairs of colours, other than the axial or principal colours, shown by a precious stone in the dichroscope are due to various combinations of the principal colours. The axial colours of each precious stone will be given below along with their special descriptions.

The detection of dichroism in a coloured stone is a simpler matter than the observation of double refraction, and the dichroscope is a less expensive instrument than the polariscope, hence the former is the more often used. The polariscope may also be used to detect the presence or absence of dichroism by removing the Nicol's prism, n (Fig. 27), and placing the stone to be tested on the object-carrier. If the stone is not dichroic, as for instance spinel, there will be no change of colour as it is rotated with the object-carrier. If, on the contrary, a dichroic stone, such as ruby, is examined in this way, the colour will be seen to change as the stone is rotated, varying between two extremes ; the change from one extreme to the other will occur four times during a complete rotation of 360°. These two colours are identical with those seen when the stone is examined in the same position in a dichroscope ; the advantage of the latter instrument lies in the fact that the two colours can be seen side by side, and thus small differences between them more easily detected. Just as in the use of the dichroscope, several observations must be made before the absence of dichroism can be considered to have been conclusively proved. As explained before, in using the polariscope the portion of the field occupied by the stone may remain dark owing to the total reflection of light within the stone. This can be avoided as before, by placing the stone in a certain position or immersing it in a strongly refracting liquid. Care must also be taken that all side light, which might be reflected from the surface of the stone, is screened off with the hand, or by means of a paper tube placed around the stone.

As we have already seen, dichroism is a character of which important use can be made in identifying precious stones, and in distinguishing them from each other and from glass imitations. Moreover, its observation does not necessitate a mounted stone being removed from its setting, which would often be necessary in the observation of other optical characters.

Dichroism is a character of precious stones which is important also from other points of view, such as that of the lapidary. A stone in which dichroism is strong must be so cut that the rays of light received by the observer have passed through the stone in a direction such that they will appear of the finest colour possible. Such a stone as cordierite, for example, must be so cut that the dark blue colour is prominently brought out, which will give a far more pleasing effect than if the light blue or yellowish-grey were predominant. The beauty and, consequently, the value of two pleochroic stones of the same size and quality will accordingly depend upon the manner in which they are cut, and hence a knowledge of the dichroic properties of stones is of importance to the gem-cutter.

Dichroic stones are sometimes cut and mounted in a manner which will bring out this character as prominently as possible. With this object in view a cube is fashioned out of the stone, the faces of which are perpendicular to the directions in which the greatest differences in colour are exhibited. Such cubes are pivoted at one corner, so that on being turned round the different colours will successively come into view. Cordierite, andalusite, and other stones are cut and mounted in this way, as will be explained in more detail later.

7. Special Optical Appearances and Colour Effects.

In this section we shall consider certain optical peculiarities and colour effects of a special and more or less abnormal kind ; these features are not shown by every specimen of a particular mineral species, but only by isolated examples. These appearances are governed by the ordinary laws of reflection and refraction of light, and are due to the peculiar and unusual conditions present in each case. The exact nature of these conditions, and the manner in which they cause the abnormal appearances, is not completely known in every case ; as full an explanation as is possible will be given in the description of each particular case.

The **play of prismatic colours** exhibited by the diamond has already been dealt with in detail ; we have here the simplest case of the refraction and dispersion of light.

The prismatic colours produced by cracks in the interior of a transparent stone, and best shown in colourless examples, gives rise to the appearance usually known as **iridescence.** The irregular fissures, or more frequently the plane cleavage cracks, inside a crystal represent narrow crevices, which may be vacuous or filled with air ; these films give rise to the brilliant prismatic colours known as Newton's rings, or as the colours of thin films or plates. These colours, which are shown to perfection by soap-bubbles, are independent of the colour of the substance itself, or of any colouring-matter contained in it, but are due to purely physical causes connected with the passage of white light through the film. The phenomenon which thus gives rise to the appearance of prismatic colours is known to physicists as the interference of light.

Some iridescent stones, such as rock-crystal, are occasionally cut so as to bring the crack, to which the display of prismatic colours is due, near the surface, and thus render them the more striking. Especially beautiful prismatic colours are shown by some kinds of colourless opal, namely, the so-called noble or precious opal (Plate XVI., Figs. 6–9). These colours are not shown over the whole surface of the stone, but in small discontinuous patches, closely aggregated. This appearance in opal is certainly a kind of iridescence, but as to its exact cause there is still a difference of opinion.

The translucent or semi-transparent variety of potash felspar or orthoclase, known as adularia, sometimes shows a bluish or milky reflection of light, not from the whole surface but only from certain crystallographic planes. This opalescent appearance is specially prominent when the stone is cut and polished with a rounded convex surface, over which, when the gem is moved, a streak or wave of such reflected light passes. This appearance, being specially pronounced in adularia, is sometimes known as **adularescence.** It has been compared to the soft light of the moon, and specimens showing it to perfection are called moon-stones (Plate XVI., Figs. 4 and 5), and are often used for ornamental purposes. The pearly opalescence of adularia is due to reflection of light from internal platy fractures or planes of separation, and from microscopically small crystal plates embedded in the adularia along these planes.

A similar appearance is shown by some specimens of chrysoberyl, which are also valued as precious stones under the name of cymophane or cat's-eye (Plate XII., Fig. 11), since the sheen of this green, yellowish-green, or brown stone recalls the appearance of the eye of the cat. We shall see later on that there is a variety of quartz having this same appearance ; the chrysoberyl variety is therefore distinguished as true or " oriental cat's-eye."

The brilliant colours shown on certain faces of labradorite, a felspar from Labrador (Plate XVI., Fig. 2), as well as by a potash-felspar from Fredriksvärn in southern Norway (Plate XVI., Fig. 3), is also, like adularescence, due to the presence of numerous minute crystal plates enclosed in the felspar and arranged parallel to these planes. The appearance resulting from the peculiar structure of these minerals is known as **change of colours** or **labradorescence.** In most positions these minerals are dull grey and unattractive looking, but certain faces in reflected light, and at a certain inclination to the light, show the most brilliant shades of green, blue, violet, red, yellow, &c. The small plates, which give rise to the reflection of coloured light, may be seen under the microscope embedded in the material ; they consist of an unknown substance which is very feebly refracting, but it is possible, however, that some are mere vacuities. The whole of the polished surface of such a mineral may reflect the same colour, or different areas of the surface may show different colours. Except with a particular inclination of the light, however, no colour of any kind is seen. On account of this beautiful exhibition of colour, Labrador felspar is more often used for ornamental purposes than is the felspar of Fredriksvärn, the colour of which is less brilliant and variable.

On certain faces of the minerals hypersthene, bronzite, and diallage, when viewed in a particular direction in reflected light, is to be seen a **metallic sheen**, uniform in character over the whole surface. These minerals consist of non-metallic substances, and the metallic lustre seen on certain faces is due to the presence of numerous minute plates embedded in the substance of the mineral, parallel to certain directions. When cut and polished with a plane or curved surface parallel to these directions, these minerals are sometimes used as ornamental stones. Hypersthene shows a fine, dark, copper-red reflection ; while in the other minerals grey, yellow, green, and brown colours are predominant.

A red metallic glittering sheen is also exhibited by avanturine-quartz ; here, however, it is not distributed uniformly over the whole surface, but occurs in numerous small isolated points. These can be seen with the naked eye to be due to small scales of mica enclosed in the quartz ; in the same way, avanturine-felspar, or sun-stone, encloses small plates or scales of hæmatite.

Beautiful effects due to the modification of light are sometimes seen in minerals which possess a more or less pronounced fibrous structure. Such stones when cut with rounded surfaces in the direction of the fibres exhibit a wave of milky light travelling over the surface of the stone as it is moved about. Ordinary cat's-eye, also known as quartz-cat's-eye (Plate XVIII., Figs. 4a and 4b), consists of quartz enclosing numerous fibres of asbestos all arranged in the same direction ; the asbestos may sometimes have been weathered out, in which case the quartz will be penetrated by numerous fine hollow canals. These fibres or canals cause much the same appearance as that seen in adularia and cymophane (oriental cat's-eye), and in this case it also is known as **opalescence** or **chatoyancy**. Quartz-cat's-eye may be green, brown, or yellow, and is similar in appearance to the true or oriental cat's-eye. This similarity does not extend to the structure which is the cause of this appearance, for the sheen of quartz-cat's-eye is in reality of the nature of a fine silky lustre, such as is often shown by minerals possessing a fibrous structure, the character of which is modified, however, in the present case, by the nature of the quartz itself. Another variety of fibrous quartz is tiger-eye which is often used for cheap jewellery ; it shows a fine golden reflection, and has a marked tendency to metallic lustre (Plate XVIII., Fig. 5).

The appearance known as **asterism** belongs to the same class of phenomena ; it is most frequently seen in ruby and sapphire among precious stones, but is not confined to these. When one of the hexagonal crystals of ruby or sapphire (Fig. 53, e–i) has a plane or curved surface cut at the ends, a six-rayed star may be seen by viewing a flame through the stone, or by observing the milky reflection from the surface of the stone. Such stones are known as star-, or asteriated-sapphires or rubies as the case may be, or simply as star-stones or asterias. The effect is produced by reflection of light from a multitude of extremely fine, hollow and long canals. These canals lie in one plane, and are arranged in three directions inclined to one another at 120°. The planes which contain the canals are perpendicular to the principal crystallographic (vertical) axis of the crystal, that is, they are parallel to the plane in which the stone must be cut. According to another view, the star is due to the reflection from numerous twin-lamellæ arranged in three sets.

Finally, the phenomena of fluorescence and phosphorescence must be briefly described, though they are of little importance in the case of minerals used as precious stones.

Fluorescence is shown to a marked degree by fluor-spar from the lead mines of Cumberland ; the phenomenon, indeed, takes its name from this mineral, which is known to mineralogists as fluor or fluorite. A fluorescent substance appears of one colour in transmitted light and of quite another in reflected light ; thus fluor-spar from Cumberland is green in transmitted light and purple in reflected light. This mineral is, however, very little used as a precious stone ; one which is used more frequently for this purpose is amber, and specimens from certain localities, namely from Sicily and Burma, show a remarkable fluorescence. Amber from these localities varies in colour from yellow to brown in transmitted light, and from green to blue in reflected light. The rounded polished surface of such specimens shows a peculiar sheen, which, according to present tastes, diminishes the beauty of the stone, and consequently its value.

Substances in which the phenomenon of **phosphorescence** is seen emit, when submitted to certain external influences, a soft, white or coloured light which is often only distinctly visible in a dark room. In some cases the emission of light persists for some time, while in others it lasts for a much shorter time, perhaps for only a few moments. Phosphorescence is exhibited by several precious stones, and may therefore aid in their recognition or discrimination. Two pieces of rock-crystal (quartz) phosphoresce when rubbed one against another ; diamond shows a marked phosphorescence when rubbed on cloth ; even when lightly rubbed on a coat-sleeve it will be seen in the dark to phosphoresce brilliantly. Many diamonds also phosphoresce after being exposed to the direct rays of the sun ; they store up the sunlight, as it were, in order to give it out again when placed in darkness. Lapis-lazuli from Chili phosphoresces when warmed to a temperature considerably less than that of red-heat ; white topaz, some diamonds, and other minerals behave in the same way. In many minerals the phosphorescence induced by warming lasts only for a short time, but may be produced again and again on reheating.

E. THERMAL, ELECTRICAL, AND MAGNETIC CHARACTERS.

There is nothing of special importance in the behaviour of precious stones when exposed to the influence of heat, or in their electrical and magnetic characters.

1. THERMAL CHARACTERS.

Different minerals differ very considerably in their **conductivity for heat,** and this character may serve in some cases to distinguish minerals similar in appearance from each other. The majority of precious stones are good conductors of heat, and on this account they are cold to the touch, since the heat of the hand is quickly conducted away. Glass is a somewhat poorer conductor and hence a glass imitation is not so cold to the touch as a genuine stone, since the warmth of the hand is not so quickly conducted away. The difference in the power of conducting heat of genuine stones and their imitations may thus, under certain conditions, afford a means of distinguishing between them ; the specimens tested must not, however, have remained long in the hand nor have been otherwise warmed, neither must they be too small. It is said to be possible for an expert to select, by the sense of touch alone, a diamond out of a bag containing a large number of pieces of glass of similar size and shape.

Amber is one of the feeblest conductors of heat ; its conductivity is much less than that of glass, hence a piece of amber can be easily distinguished from its imitation in yellow glass, since it feels so much warmer to the touch.   Another substance having a feeble conductivity for heat is jet, a variety of coal, which is frequently made use of for mourning ornaments. An opaque, black glass is often used for the same purpose, but a single touch of the finger tips is all that is needed to enable an expert to distinguish between the two.

A device for distinguishing a genuine from an imitation stone, depending upon the power of conducting heat, is to breathe upon the stone.   The moisture of the breath will condense upon the genuine stone with more difficulty than upon glass, and when condensed will disappear again much more rapidly, since the precious stone is both more rapidly warmed and more rapidly cooled than is glass.

For the purpose of distinguishing rough stones, their **fusibility** before the blowpipe may sometimes be made use of.   All glass imitations are easily fusible before the blowpipe, while few of the minerals used as precious stones can be so fused.   Red garnet is one such mineral, and can be easily distinguished from other red stones which are infusible before the blowpipe, such as ruby and spinel.   The application of this test is naturally limited to rough stones, splinters of which can usually be detached for examination.

2.   ELECTRICAL CHARACTERS.

Many precious stones when exposed to certain external influences acquire a greater or less charge of electricity.   They differ from each other in the length of time this charge can be retained, some retaining it for a considerable time, others for a less time, perhaps only a few minutes.

The French abbé, Haüy, the founder of modern scientific mineralogy, attempted to make extensive use of these characters as a means of identifying stones and distinguishing them one from another.   In his book, published in 1817, *Traité des caractères physiques des pierres précieuses*, he devoted seventy-two out of a total of two hundred and fifty-three pages to the consideration of electrical characters, while the optical characters are dismissed in thirty-two pages.   A comparison with the number of pages devoted to the treatment of these two branches in the present volume, shows how much more important to-day is the consideration of the optical characters of minerals.

The examination of the electrical, as of the optical, characters of a stone, has the advantage that no injury to the stone results therefrom.   The observation of electrical characters, however, requires a certain amount of skill and practice; for the detection of the very small electrical charges acquired by most precious stones is difficult; and, further, these observations must be conducted in a perfectly dry atmosphere, a condition not always easy to obtain.   Any charge located on the surface of a stone is rapidly lost in a damp atmosphere, and a stone which retains its charge in dry surroundings will rapidly lose it in the presence of moisture.   The length of time a stone retains its charge, a test to which Haüy attached great importance, depends largely therefore upon external conditions.

At the time Haüy was engaged on his researches the methods of electrical investigation were, at least for his purposes, fairly well developed, while methods for the optical investigation of minerals had received little or no attention.   Observers had indeed noticed that some minerals were singly refracting and others doubly refracting, but there was no polariscope to give precision to their observations and the phenomenon of dichroism had yet to be discovered.   We can thus readily understand why Haüy attached so much more importance to the electrical than to the optical characters of minerals and precious stones. With the discovery of the dichroscope and a convenient polariscope the optical characters of minerals assumed their true importance, while their electrical characters became a minor consideration, as may be gathered from their brief mention in this place.

For the purpose of demonstrating the existence of a charge of electricity upon the surface of a stone an instrument known as an electroscope may be used; for very feeble charges an electrometer of complicated construction will be necessary. Haüy employed for this purpose an " electrical needle"; it consisted simply of a brass rod, terminated at either end by a small brass ball, and balanced on a vertical fine steel point, on which it could turn freely like a magnetic needle. An electrically charged body, when presented to either of the balls, would attract it. By giving an electric charge to the balls, they would be attracted or repelled on the approach of a body according as its charge was unlike or like that of the balls. The electric pendulum, consisting of a pith ball suspended by a silk thread, served the same purpose. With the help of such instruments it is easy to demonstrate that minerals, including precious stones, become, under various conditions, charged with electricity; the fact of itself is, however, of little note.

After rubbing on cloth all precious stones, like glass, become positively electrified. Topaz and tourmaline become strongly electrified after such treatment, diamond less strongly, and the majority of precious stones only feebly. Smooth faces are more susceptible of electrification than are rough ones, and hence cut stones furnish the most favourable material for this purpose. In perfectly dry air, some precious stones retain an electrical charge for a comparatively long period; this is especially so in the case of topaz, the electrification of which can be detected after an interval of thirty-two hours; sapphire will retain its charge for from five to six hours, and diamond for half an hour. Colourless topaz, colourless sapphire, and diamond may be distinguished by this difference in their behaviour; after imparting a charge by rubbing with a cloth, the stones should be laid on a metal plate and their electrical state tested from time to time. The majority of precious stones lose their charges with great rapidity, some, indeed, after only a few moments.

Amber, like other resinous substances, becomes negatively electrified on rubbing, and so strongly that it attracts to itself any light bodies, such as pieces of paper. These, after contact with the amber, themselves become charged and are then repelled by it. This particular character of amber is of value as a means whereby it may be distinguished from its imitations, which will be mentioned later under the special descriptions.

The electricity developed on some precious stones when under the influence of changes of temperature is known as **pyroelectricity.** The charge produced in this way on the surface of a stone varies in sign at different areas of the surface, the charge at one point being positive while that at another is negative. Those parts of the surface which become positively electrified on heating become negatively electrified on cooling, and *vice versâ*. Tourmaline and topaz are remarkable for the strength of the pyroelectrical charge they acquire; and this distinguishes them from other precious stones, which when exposed to the same influences acquire but feeble charges, or none at all. Thus, with the help of one of the electrical instruments mentioned above, a red tourmaline can be distinguished from a ruby, and a greenish-blue topaz from an aquamarine of the same colour; for the former in each case will show a strong pyroelectrical charge, and the latter none. During the gradual cooling of tourmaline after being heated, it assumes the power of attracting light bodies to itself, as does amber after being rubbed.

3. MAGNETISM.

Some minerals, such as magnetite, are magnetic and respond to the influence of a magnet, being attracted by it. Magnetite has a black metallic lustre, and a certain titaniferous variety, namely iserine, takes when polished a very brilliant lustre, and is sometimes used for ornamental purposes. The magnetic character of magnetite distinguishes it from other black stones, all of which are either non-magnetic or very feebly influenced by a magnet.

D. OCCURRENCE OF PRECIOUS STONES.

A complete account of precious stones must include a consideration of the localities at which they are found and the conditions under which they occur in nature. These subjects will be dealt with in a general way here, and again more in detail with the special description of each precious stone.

Precious stones, like other minerals, have two distinct modes of occurrence. They may, on the one hand, be found at that spot in the earth's crust where they had their genesis, or, on the other hand, owing to the weathering and breaking down of the rocks and the action of transporting agencies, we may find them in secondary deposits far from their original home.

Precious stones, in their primary situation, frequently form a constituent of the rocks which make up the earth's crust at that place. They are embedded in the so-called mother-rock, and were formed at the same time as the other constituents of the rock. Under such conditions, stones sometimes show regularly developed crystal-faces, but more frequently their boundaries are irregular and distorted. A perfectly developed crystal of red garnet (almandine) embedded in its mother-rock of gneiss is shown in Plate XIV., Fig. 3, while Fig. 69 shows the completely developed crystal after being isolated from the mother-rock.

Many precious stones and minerals, however, are found not completely embedded in the rock-mass but attached to the walls of cavities in the rock and projecting freely into the interior space. These cavities may be either completely enclosed by the rock, in which case they are of various shapes and sizes, or they may partake more of the nature of cracks and fissures penetrating the rock and varying in width and length between wide limits. The formation of minerals found inside such cavities is always of later date than that of the rock-mass itself. Such later-formed minerals may completely fill a cavity or fissure, or they may form a more or less thick incrustation on its walls.

Such cavities lined with crystals are known as *drusy cavities* or *druses*. Crystals detached from a drusy cavity will show a broken surface at the end by which they were attached to the wall of the cavity, but in other directions they will be perfectly developed in accordance with the type of symmetry peculiar to them. These *attached crystals* differ in this respect from the *embedded crystals*, mentioned above, which latter are equally developed on all sides.

The quartz crystals shown in Figs. 85 *b—d*, are examples of attached crystals broken away from their underlying matrix, while Fig. 85 *a* is a representation of an embedded quartz crystal, equally developed on all sides. In Figs. 85 *b—d*, the irregularly broken point of attachment of each crystal is directed downwards and is fairly large; it is sometimes, however, quite small and may be hardly observable. A group of crystals, of the variety of quartz known as rock-crystal, such as frequently occurs in crevices and fissures in the gneiss of the Alps, is shown in Plate XVII.

More important than the occurrence of precious stones in primary rocks is their presence in loose, secondary deposits, which have been derived from the weathering and breaking down of primary rocks, and are known as **gem-sands** or **gem-gravels.**

The mother-rock, in which the precious stones were originally formed, has been exposed to the action of atmospheric agencies, rain, frost, &c., and has become weathered at the surface. Some of the constituents of the rock are dissolved in water and carried away and thus the cohesion of the mass is destroyed. The more or less loose, clayey, or sandy residue is the weathered product, and this will contain the precious stones which were present in the original rock, since, as a rule, they are unattacked by weathering agencies. The precious stones will be present in the weathered product in relatively greater numbers than in the original mother-rock.

It will be readily understood that it is more profitable to work weathered material than the unaltered primary rocks for precious stones, for not only is the former relatively richer in gem-stones than the latter, but it allows of the stones being easily separated or washed out. The extraction of a gem-stone from solid rock involves much labour and patience, and, even when every care is taken, may result in serious damage to the stone.

The loose, incoherent material which results from the weathering of a rock, when it contains a mineral worth extracting for technical purposes, is known generally as a *sand*, and as a *gem-sand* when it contains precious stones. It is in such sands which, wherever they occur, cover the solid rocks and form the outer portion of the earth's crust, that the most valuable precious stones are found, such, for instance, as diamond, ruby, and sapphire. They are separated from these masses of detritus by the process of *gem-washing*, in which the heavier stones and larger fragments remain behind, while the lighter clayey and sandy constituents are washed away.

When the weathered material has not been carried away by the various transporting agencies, but remains near the parent rock, the precious stones and other minerals it contains will preserve intact the sharp edges and the crystalline form they possessed when embedded in the solid rock. Such cases, however, are rare ; much more frequently the whole of the loose material is transported in streams and rivers, and is finally deposited in a lower part of the valley, far away from its original resting-place. In such river sands and gravels, which are known as alluvial deposits, the mineral fragments, and even the precious stones, in spite of their hardness, become so rubbed by mutual friction during their travels that all angularities are lost, and they present the appearance of smooth, rounded pebbles or grains.

The presence or absence of this smooth water-worn appearance in the mineral fragments of rock detritus is conclusive proof in the one case that water has been the transporting agency, and in the other that it has not. The greater the hardness of the precious stone transported in gravels by water, the less will be the rounding it undergoes ; even diamond, the hardest of materials, may show traces of rounding if the action of other softer stones is only continued long enough.

The precious stones, found in such water-worn materials, are frequently superior to specimens which have not been subjected to the action of running water, and are still to be found in their parent-rock. Such stones are frequently traversed by fissures, often scarcely visible, but enough to make them unfit for use as gems, since, as has been mentioned before, they have a tendency to break along these fissures. Precious stones which have been rolled about and ground together in the bed of a river during long ages have undergone a fairly severe trial ; any which have a tendency to fragment will be reduced to splinters at an early stage of their journey ; those, on the other hand, which survive may be considered to have proved their durability.

As regards the **geographical distribution** of precious stones, it may be mentioned that in former times the most valuable came from India and other parts of the " Orient." It was therefore believed in the Middle Ages that the glowing sun of tropical countries was essential to the development of those qualities in precious stones which are so highly prized, and that specimens from colder countries were deficient in these qualities. Every good stone, of which the locality was not certainly known, was for this reason assumed to have come from the " Orient." A remnant of this belief still lingers in the application of the terms " oriental " to the more valuable, and " occidental " to less valuable stones. It has now long been known that the habitat of the finest of precious stones is by no means confined to the " Orient " and hot countries, such as India, Ceylon, Burma, Siam, Brazil, Colombia, &c., but that equally fine stones may also be found in North America, the Urals, and other Northern Countries. The terms " oriental " and " occidental," as now applied, have no longer a geographical signification, but refer simply to the quality of the stones to which they may be applied. Thus to distinguish cymophane from the more common quartz-cat's-eye it is termed " oriental cat's-eye "; in the same way yellow sapphire is known as " oriental topaz," and yellow quartz as " occidental topaz." The various localities in which precious stones are and have been found will be considered in detail, along with the special description of each precious stone.

II. APPLICATIONS OF PRECIOUS STONES.

The use to which a precious stone is put depends in the first place on its appearance, and in the second on the hardness it possesses. Should it possess beauty of appearance combined with a fair degree of hardness, it may be used as a personal ornament, while if it possesses hardness alone, there are various technical purposes it may serve.

A. TECHNICAL APPLICATIONS.

The technical applications of precious stones are not numerous, and need be but briefly mentioned here.

Since the year 1700 the pivot-bearings of watches and delicate chronometers have been made of some hard precious stone, since this material will best withstand the continual wear of the steel axis. The stones commonly used for this purpose are known as " rubies," but are in reality chrysoberyl, topaz, spinel, or indeed any stone the hardness of which is greater than that of steel. The true ruby would of course answer this purpose, but a stone so valuable for ornamental purposes would naturally not be used when cheaper substitutes are available. Any precious stone which has the required degree of hardness, and which from cloudiness, opacity, or any such blemish, is unsuitable for use as a gem, may be utilised for the purpose.

The pivot-supports of other delicate instruments, such as balances, &c., are made of agate or some other hard stone; by this means the wear is reduced to a minimum, and the delicacy of the instrument preserved unimpaired for long periods. In the manufacture of very fine gold and silver wires, the hole through which the wire is drawn is usually made in some hard precious stone; this will withstand the continued friction, and thus avoid the possibility of gradual increase in the diameter of the hole, and consequently in that of the wire. Tools used in polishing metals and for similar purposes are also made of hard stones, preferably of agate.

Those precious stones which have the greatest technical importance as abrasive agents are naturally those which are at the same time the hardest of stones, namely, diamond and corundum. For such purposes the latter is used in its impurest state, when it is known as emery. The use of these materials in the cutting, grinding, and polishing of precious stones will be dealt with below in the special description of these processes.

The diamond, on account of its extreme hardness, has many other technical applications, which will be noticed in detail further on under the special description of this stone. It is used for engraving and boring precious stones and other hard materials, while its use in rock-drills for mining and other operations is scarcely less extensive than its use as the glazier's diamond.

B. APPLICATION AS JEWELS.

The use of precious stones as gems is much more extensive and varied than for any other purpose. In their rough state they have not, as a rule, a pleasing appearance, and therefore are unsuitable for this purpose ; it is only after cutting and polishing that their beauty appears in all its fulness.

The process of cutting aims at giving each stone such a form as will best display its natural lustre and beauty. Thus the form in one case may be rounded, in another bounded by small faces or facets, the latter being very frequently used. The various modes of cutting in vogue at the present day, each of which is best suited to the idiosyncrasy of the particular stone to which it is applied, are the results of centuries of trial and observation on the part of gem-cutters. Thus the form in which transparent stones are cut differs from that best suited to opaque stones; and in the same way, the form in which dark-coloured stones are cut differs from that given to lighter or colourless specimens. The appearance of each would suffer if it were given any other than its own appropriate form.

The amount of refraction and dispersion exercised upon light by a transparent stone greatly affects its appearance, as has been shown in the case of diamond. It has also been shown that to obtain a maximum effect, the greater part of the light which enters by the front facets of a cut stone must be reflected from the back facets, and must again pass out by the front facets. Since the path of a ray of light in a stone varies with the refractive index of the stone, and this character is different in different stones, it follows that the form of cutting must be adapted to the requirements of each particular case. It is thus the task of the gem-cutter to give to each stone that form which is calculated to bring out and display its beauties to the greatest possible advantage, and which, at the same time, involves the least possible waste of valuable material.

Gem-cutters, by prolonged experience, have arrived at certain empirical rules which are always applied, and which are modified to suit particular cases. In colourless stones, for example, there must be a fixed proportion between their breadth and their thickness, and there should be also certain relations between the shape of the back and that of the front of the stone.

Too great depth in a cut stone is as inimical to the full effect of its beauty as is too great shallowness. In the one case a stone is said to be *thick* or " lumpy," and in the other *thin* or " spread." Of two similar stones, one of which errs on the side of too great depth, and the other on that of too great shallowness, the latter is to be preferred. The facets at the back of the stone must occupy a certain position relative to the front facets, otherwise the light entering by these will not be totally reflected from the back.

The same rules apply also to coloured stones. In this case the depth the stones are cut is important, and this must vary with their intensity of colour. A deeply coloured stone if too thick will appear dark or almost black, while a pale coloured stone will not exhibit a sufficient depth of colour unless it is cut of some thickness.

So long as the mutual relations of the facets of a cut stone are correct, the direction these take relative to the faces of the natural crystal is in most cases immaterial. In a few special cases, however, the directions of the cut facets must bear a definite relation to certain crystallographic directions in the stone. Thus the special colour effects of labradorite, moon-stone, &c., are only manifest in certain directions; if cut in other directions, the beautiful effects for which these stones are prized would be lost. This is also true in the case of dichroic stones, which, as we have already seen, vary in colour in different directions. Other cases of the same kind will be mentioned with the special description of each precious stone.

In the cutting of any given rough stone, not only must it receive the form best calculated to display its special beauties, but the facets must be cut in such positions as to involve the least possible waste of material, thus obtaining the largest possible size for the cut stone. In considering the positions in which the facets are to be cut relative to the boundaries of the rough stone, there are still other points which may require attention. Thus the rough stone may have a flaw, and in this case the facets should be so placed that the faulty material will be cut away altogether, or, at least, so located in the cut stone that the beauty of the latter is impaired as little as possible.

With rough material containing flaws, a question will often arise as to whether, in the cut stone, size should be sacrificed to beauty, or *vice versâ*. European gem-cutters are generally unanimous in the opinion that such a specimen should be cut so as to attain the highest possible degree of perfection and beauty even if this should involve considerable loss of material. A small stone, all the beautiful features of which are displayed to their full advantage, is more highly prized than a larger stone, the beauty of which is less perfectly developed. In every rough stone, the aim of the gem-cutter is to obtain a cut stone of the largest possible weight combined with the greatest possible beauty, since, the latter condition being fulfilled, the price obtained for the cut stone varies with its weight. The earnings of a cutter of precious stones depend largely upon his skill in treating each stone so as to obtain the greatest effect with the least waste of material.

These principles have not always been followed, for in earlier times the aim in gem-cutting was to reduce the size and weight of the stone as little as possible. This is the case even at the present day in India and the East generally, as well as in various remote parts of the world where precious stones are found. Stones so cut have their facets very irregularly grouped, and consequently much of their beauty is undeveloped. Such stones are unfit for use in European jewellery, and are frequently re-cut according to modern principles; the increased beauty of their appearance so obtained more than compensates for the loss of the material cut away.

We now pass to the consideration of the various shapes and forms in which precious stones are cut with a view to their use in jewellery.

A. FORMS OF CUTTING.

The various forms of cutting which have been found by experience to be most effective for gems and which are at present exclusively used, at least for valuable stones, fall naturally into two groups. The one includes all forms having *facets*, the other embraces forms of a rounded or *cabochon* shape. All faceted forms may be referred to one or other of four types according to the number and arrangement of the facets; forms intermediate between these types may also be met with.

The facets of a cut stone may be more or less uniformly distributed on all sides, or, again, they may be all located on one side, the other side being occupied by a single large facet. In the latter case we have the form of cutting known as the rosette or rose. A cut stone provided with facets on all sides is represented in Fig. 29, *a* and *c* being views from above and below respectively and *b* from the side. When such a stone (Fig. 29) is set as a jewel the side turned towards the observer is known as the upper portion or *crown*, while the opposite side or lower portion is referred to as the *culasse* or pavilion. The facets of the crown and of the culasse meet in the edge *RR* (Fig. 29 *b*), which is known as the *girdle* or edge, and is the portion of the stone which is fixed in the setting. The whole forms, as it were, a double pyramid with truncated summits, each pyramid having a common base in the girdle.

Of the four types of faceted stones the rose or rosette type has been already mentioned; the remaining three are known as the brilliant-cut, step-cut, and table-cut. The number, arrangement, and grouping of the facets differ in these three types, but each has a crown, a culasse, and a girdle.

These different forms of cutting, which are illustrated in Plates II.–IV., must now be considered more in detail. In these plates, the same figure-number is given to different aspects of the same stone, the addition of the letter *a*, *b*, or *c* to the figure-number indicating that the stone is represented as seen from the side, from above, or from below respectively; the same letters are also used when only one or two of the three aspects are represented. Plate II. gives a series of forms of the brilliant, and Fig. 1 of Plate III. belongs to the same series. The other figures of Plate III. represents variations of the step-cut, while Plate IV. shows various kinds of rosettes, table-stones, and stones cut *en cabochon*.

The expense involved in a complicated form of cutting with regular facets, grouped in the way experience has shown to be most effective, is very considerable; such perfection in cutting is never bestowed upon cheap material, but only upon more valuable stones which will repay the outlay. In the cutting of less valuable stones, they receive the correct form, but the facets are reduced in number and less attention is paid to their regular and precise distribution; by these means the expense of cutting is considerably lessened though the appearance of the stone suffers.

1. **The Brilliant.**—This form of cutting is said to have been originated by Cardinal Mazarin, and was first employed at the time this minister was endeavouring to revive the diamond-cutting industry in Paris. Mazarin caused twelve of the largest diamonds of the French crown to be cut in this form, and these stones have since been known as the twelve " Mazarins." The existence of only one of these stones, however, is now known, and the genuineness even of this is doubted. The superiority of the brilliant over all other forms of cutting for diamond and other colourless, transparent stones, and also for some coloured stones, is now so firmly established that it is at present by far the most generally used. Only in quite exceptional cases is a good diamond cut in a form other than that of the brilliant; indeed, so generally is this form given to diamonds, that they are often referred to colloquially as " brilliants." Coloured, transparent stones are very frequently brilliant-cut, but not so invariably as is the case with diamonds.

The upper portion or crown, *OO*, of a brilliant (Fig. 29) bears a broad facet, *b*, known as the *table*, while the lower portion or culasse, *UU*, bears a much smaller facet, *B*, known as the *culet* (or collet), both being parallel to the girdle, *RR*. Of other facets, those meeting the table in an edge and lying wholly in the crown of the stone, are known as *star facets* and are lettered *d* in the figure. The *cross facets*, lettered *f, g, E,* and *D* in the figure, meet the girdle in an edge; some lie in the crown of the stone and some in the culasse. Between the star and cross facets, which are triangular in shape, lie other larger facets having four or five edges; those which lie above the girdle are lettered *a* and *c* in the figure, while those which are below are lettered *A* and *C;* these facets are not, however, invariably present in the same

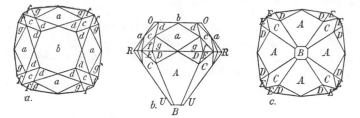

FIG. 29.   Brilliant (triple-cut).   *a,* view from above ; *b,* from the side ; *c,* from below.

number. The girdle, *RR,* always lies in a plane, and forms the boundary of the stone as seen in Figs. 29 *a* and 29 *c.*

Several varieties of the brilliant-cut are distinguished according to the number of facets present. The *double-cut brilliant,* shown in Plate II., Fig. 1 *a,b,c,* has four triangular star facets arranged so that their four upper edges form the boundaries of the square table, while the four opposite angles of each lie in the girdle. The space between each pair of adjacent star facets is occupied by three cross facets, the central one of each group having the form of an isosceles triangle, and the cross facet on either side having the form of an oblique triangle. On the crown or upper portion of such a stone, therefore, there are sixteen facets besides the table ; these facets are arranged in two series, hence the term "double-cut brilliant." The under portion consists of twelve triangular cross facets, which are the same in number and arrangement as the cross facets in the upper portion ; between these lie four five-sided facets, intersecting the small culet in short edges.

The *English double-cut brilliant,* differing somewhat from the double-cut brilliant just described, is shown in Plate II., Fig. 2 *a, b, c.* Here the table is the centre of an eight-rayed star, formed of eight triangular star facets, which alternate with eight triangular cross facets. The facets of the lower portion are similar to those of the ordinary double-cut brilliant (Fig. 1 *c*) ; the corner cross facets having the shape of isosceles triangles are, however, occasionally absent (Fig. 2 *c*).

The number of facets present in the forms of double-cut brilliants does not allow of the perfect development of the brilliancy and the play of prismatic colours of the stone. Such forms are therefore given usually to small and less valuable stones ; for large stones the *triple-cut brilliant* is more appropriate. Here three series of facets lie one above the other on the upper part of the stone ; the total of thirty-two facets, exclusive of the table, is made up of eight triangular star facets, sixteen triangular cross facets, and eight four-sided facets. The arrangement of these different facets is shown in Fig. 29, and in Plate II. Figs. 3 and 4. The under portion of the stone has also sixteen cross facets, while the small culet is surrounded by eight large, five-sided facets. The form, shown in Fig. 29, and Plate II., Fig. 3 *a, b, c,* in which the girdle has a roughly square outline, is now somewhat out of date ; since the eighteenth century the form shown in Plate II., Fig. 4 *a, b, c,* has received more favour. The facets of this form are the same in number and arrangement, but are more nearly equal in size, and the outline of the girdle approximates very close to a circle. The outline of the girdle is not, however, by any means constant, it depends largely upon the natural form of the stone before it is cut. In Fig. 5 *b, c,* it is oval, in Fig. 6 *b, c,* it is pear-shaped, and in Fig. 7 *a, b, c,* it is roughly triangular in outline. The last case is

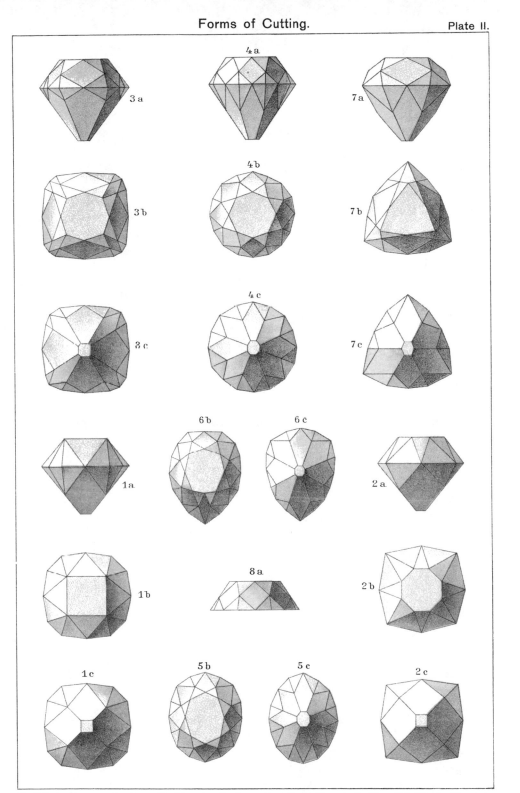

Brilliant Forms. 1a, b, c. Double-cut. 2a, b, c. English double-cut (double-cut brilliant with star).
3a, b, c. Triple-cut, old form. 4a, b, c. Triple-cut, new form, round. 5b, c. The same, oval.
6b, c. The same, pear-shaped. 7a, b, c. The same, triangular. 8a. Half-brilliant.

also noticeable from the fact that the facets, instead of being in multiples of four, are in multiples of three.

The forms just described may be regarded as typical brilliant forms, and are used far more frequently than any other. They are nevertheless subject to certain modifications, not, however, deviating far from the normal types. These modifications usually take the form of variations in the arrangement and number of the facets; in the latter case further small facets are introduced in groups, which are placed symmetrically relative to the other facets. The majority of the large historical diamonds are cut in the brilliant form, as an examination of Plates X. and XI. will show. On comparing these forms with the normal forms of Plate II. a strong general resemblance, accompanied by differences in minor details, will be noticed.

To bring out all the beauties of a stone, and to display them to the greatest possible advantage, involves infinite care and precision in the cutting. The facets must be regularly and symmetrically grouped, and corresponding facets must be of precisely the same size; moreover, it is of the greatest importance that all the different parts of the stone should be correctly proportioned. In this connection may be mentioned the following rules which are generally observed, and which are only departed from where there are special reasons for so doing:

The height of the upper portion of the brilliant above the girdle must be one-third, and that of the lower portion must be two-thirds, of the total thickness of the stone from table to culet. The diameters of the table and culet must be respectively five-ninths and one-ninth of the diameter of the girdle; hence the diameter of the table is five times that of the culet. Few of the best cut and most beautiful brilliants show any essential deviation from these dimensions; the exceptional cases mentioned above occur when the rough stone is of such a shape that to give it these proportions would involve too great a waste of material; or, again, in the case of a coloured stone, where the thickness is varied in order to obtain the particular depth of colour desired.

The " Koh-i-noor," the famous diamond now in the English crown jewels (Plate X., Fig. 5), on account of the former reason departs considerably from the typical form. The " Regent," a large brilliant in the French crown jewels, is perhaps the most perfectly beautiful stone of its kind existing at the present day (Plate XI., Fig. 8). It conforms with the greatest precision to the proportions laid down above, and consequently far surpasses the " Koh-i-noor " in brilliancy and play of colours, although the two stones are equal in quality.

It remains to be mentioned that the girdle of a brilliant is sometimes left with sharp edges (Plate X., Fig. 5), as is the custom of English gem-cutters; or the edges may be ground down (Plate XI., Figs. 8 and 9), as is done in Holland. The former plan improves the appearance and effect of the stone, but the sharp edges are liable to get chipped, which is not the case when they have been rounded off.

Mention may be made here of the *half-brilliant* (Plate II., Fig. 8a), or brillonette, which is very occasionally made use of. It is essentially an ordinary brilliant, the under portion of which is replaced by a single large face, which forms a base to the upper portion as in rosettes. This device is occasionally resorted to when the rough stone is very flat, but the appearance of a stone so cut is far inferior to that of a complete brilliant.

The *star-cut*, which is closely related to the brilliant form, was devised by the Parisian jeweller, Caire, at the beginning of the nineteenth century, and is illustrated in Plate III., Fig. 1 *a, b, c*. In this form Caire aimed at combining the advantages of the brilliant with those of the rosette. As may be seen from the diagrams, the facets are arranged in multiples of six, and are distributed with great regularity, which serves to enhance the appearance of the stone. This form of cutting was devised principally for diamonds, to which it gives a very effective star-like or rayed appearance, very little inferior to that of the ordinary brilliant. The cutting of this form from the majority of rough stones is attended with but little loss of material; the form is, however, not in general use.

2. **Step-cut** or **trap-cut.**—The different types of step-cut stones, together with the various modifications of this form, are illustrated in Plate III., Figs. 2 to 8, Figs. 2 to 4 being typical forms. In one of these (Fig. 2 *b*) the girdle is square; in another hexagonal (Fig. 3 *b*); in a third it is eight-sided (Fig. 4 *b*, *c*); while it may be occasionally twelve-sided. The outline of the girdle may be such that all its diameters are approximately equal, or it may be more elongated in one direction. Above the girdle rises the upper portion of the stone, bearing a large table of the same outline as the girdle (Figs. 2 *b*, 3 *b*, 4 *b*); the lower portion terminates in a small culet (Figs. 2 *a*, 4 *c*), or in a point (Figs. 7 *a*, *c*). On both portions lie a series of facets arranged in such a way that their edges of intersection are parallel to the corresponding edges of the girdle. In passing from the girdle to the table or to the culet, the facets become successively less and less steeply inclined (Fig. 2 *a*, &c.). The upper portion has two, or sometimes three, series of facets, each series differing but slightly in their inclination to the table. The facets of different series may be of the same width (Figs. 2 *b*, 3 *b*), or the facets of the lower series are wider than those of the uppermost bordering on the table (Fig. 4 *b*). On the lower portion of step-cut stones, there are usually from four (Figs. 8 *a*, *c*) to five series (Figs. 2 *a*, 4 *c*, &c.) of facets. None of the facets of these lower series differ in width.

The step-cut is the form employed for less deeply-coloured stones when they are not cut as brilliants. It brings out the colour and lustre of the stone to great advantage; it must, however, be specially proportioned, particularly in the lower portion, to suit the stone to which it is applied. The brilliancy and colour of the stone do not attain their full value with an insufficient number of facets; there are scarcely ever less than four or five series of facets on the lower portion of the stone, and in faintly-coloured stones this number may be increased. In such faintly-coloured stones the lower portion is rather deep, as is shown in the figures, while in stones of a deep colour it is flat, sometimes very flat.

While certain insignificant modifications of the lower portion of step-cut forms are effective in varying the depth of colour of the stone, the upper portion may undergo more marked modifications, a few of which are illustrated in Plate III., Figs. 5 to 8. Here we find the step-like facets of the upper portion replaced by an arrangement of facets similar to that of a brilliant. These forms are therefore, to a certain extent, combinations of the step-cut and the brilliant-cut, and are in general specially suited to stones of a pale colour. The *mixed-cut* (Fig. 5 *a*, *b*) is a form in frequent use; it bears on the upper portion a series of triangular star facets and of similarly shaped cross facets, separated by a series of four-sided facets. The mixed-cut brings out in light-coloured stones a stronger brilliancy and lustre than does the typical step-cut. The outline of the girdle in this form need not necessarily be circular, as in Fig. 5, but may be square, hexagonal, &c.

Fig. 6 *a*, *b*, shows the *cut with double facets*, a form which differs from the mixed-cut in that several single facets of the latter are replaced by two facets; the arrangement of these facets in two series can be easily made out from the diagrams without further explanation. The cut with double facets is no more effective than is the mixed-cut; it is used simply for the purpose of removing, or rendering inconspicuous, any faults which may exist in the rough stone. In the *cut with elongated brilliant facets* the arrangement of the facets on the upper portion is much the same as in the previous form; the facets, however, are much elongated

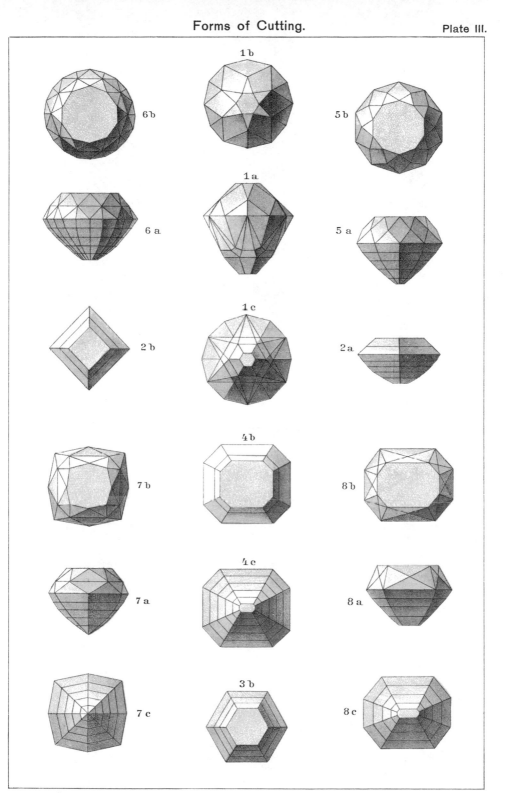

1 a, b, c. Star-cut (of M. Caire). 2 a, b. Step-cut, four-sided. 3 b. Step-cut, six-sided. 4 b, c. Step-cut, eight-sided. 5 a, b. Mixed-cut. 6 a, b. Cut with double facets. 7 a, b, c. Cut with elongated brilliant facets. 8 a, b, c. Maltese cross.

or shortened (Fig. 7 *a*, *b*, *c*). The outline of the girdle may approach that of a square, as in the figure, or it may be oblong. This form of cutting is peculiarly adapted to stones of an elongated shape, and it brings out their lustre to a marked degree; the elongated brilliant facets seem to compensate for any lack of depth in the lower portion of the stone. Another similar form is that known as the *Maltese cross* (Fig. 8 *a*, *b*, *c*), so called from the cruciform arrangement of its facets. Other similar forms exist, differing but slightly from those already described; a detailed account of these is therefore unnecessary.

3. **Table-cut.**—This term includes a number of forms, all of which are more or less related to, and may be derived from, a four-sided double pyramid or regular octahedron. This octahedral form is the natural crystalline form of many diamonds, and it may sometimes be seen in the stones of jewellery which dates back to the time when no cutting of the rough stones was attempted, but the preparation of the stones for ornamental purposes was confined to the polishing of the natural faces of the crystals. Such stones date back to very ancient times, and are known as *point-stones*. The table-cut, and other forms related to it, are derived from the octahedron by the greater or less truncation of two opposite corners (Plate IV., Figs 11 to 16); a few additional facets may be given to the upper portion of the stone (Figs. 11, 13, 14, 16).

The typical table-stone is derived from an octahedron by cutting two opposite corners to an equal amount. The upper and lower portions of the stone are then exact replicas the one of the other, and the table is of the same size and shape as the culet, the outline of which may be either square or oblong. Fig. 15 *b* shows a view from above of a square table-stone, and Plate XIV., Fig. 2, shows an epidote cut as an elongated table-stone. This form of cutting is not, as a rule, specially effective; it is, however, advantageously used for several coloured stones, including the emerald. The effect of additional facets on the upper portion is to increase the brilliancy and lustre of the stone. With this object in view, the four edges of the table may be replaced by narrow facets (Fig. 11 *a*, *b*), or the four edges between the pyramidal facets may be more or less truncated so that the table becomes eight-sided (Fig. 16 *b*). Again, the upper portion may be of the brilliant form (Fig. 14 *a*, *b*), though the arrangement of the facets in a typical brilliant need not be exactly reproduced.

The two opposite corners of the octahedron may be truncated to a greater or less degree. In the former case the result will be quite a thin table, which is known as a *thin-stone*. This can be modified by the addition of further facets in the manner described for table-stones (Figs. 12 *a*, 13 *b*). A table-stone in which the culet is larger than the table is described as *half-grounded*, while one in which the reverse relation holds is known as a *thick-stone*. Such stones, in which the table is usually double the size of the culet, are described as *Indian-cut*, and many precious stones from the Orient, and especially from India, are of this form; they are usually re-cut in Europe into a more effective form. The thick-stone is, in a way, as already explained, the ground form of the brilliant; all the modifications described for table-stones may be applied equally well to thick-stones.

FIG. 30. Rosette (viewed from above).

4. **Rosette** or **Rose-cut.**—In this form of cutting the stone is bounded on its underside by a single large and broad face, which forms a base to the whole. This form, which consists of an upper portion only, the lower portion being entirely absent, is pyramidal in shape, the uppermost facets meeting above to form a more or less sharp solid angle. A rose of the ordinary type, as seen from above, is shown in Fig. 30, and Plate IV., Fig. 1 *b*. The facets

are in multiples of six, and are arranged in two groups: the upper group, of which the facets are lettered *a*, constitute the *crown* or *star*, while the series lettered *b* and *c* are known as the *teeth* (*dentelle*). The star facets, *a*, and the cross facets, *b* and *c*, are both, as a rule, triangular in shape, as in Fig. 30, but in special cases the cross facets may be four-sided (Plate IV., Fig. 5 *a*).

The appearance due to this arrangement of facets has been compared to that of an opening rose-bud, hence the name applied to this form of cutting. It has been in vogue since about the year 1520, principally for diamonds of small thickness, from which comparatively small brilliants only could be obtained, and these with considerable loss of material. This form of cutting for the diamond is second in importance only to the brilliant, and a diamond cut in this manner is frequently referred to as a rose or rosette. The rose-cut well displays the brilliancy of the stone, but is inferior to the brilliant form in bringing out the play of prismatic colours. It is also applied to coloured stones, for example pyrope or Bohemian garnet, but less frequently than to diamonds.

The number and arrangement of the facets of rosettes may be considerably varied. Some of the modifications which result are distinguished by special names, and are represented in Plate IV., Figs. 1 to 7. The description of Fig. 30 applies to the typical *Dutch rose*, or crowned rose (Plate IV., Figs. 1 *b*, 3 *a*), with six star facets and eighteen cross facets. The character which distinguishes this from other rose-cuts is the height of the pyramid above the base. This height is, as a rule, half the diameter of the whole stone. Further, the distance from the base to the crown should be three-fifths of the total height, while the diameter of the base of the crown should be three-quarters of that of the whole stone. This is the form of rose-cut ordinarily employed; its base is usually round, but it is occasionally oval or pear-shaped (Fig. 2 *b*).

Among the other forms of roses which are much less used is the *Brabant* or *Antwerp rose*. This differs from the Dutch rose in that the star facets form a much lower pyramid, while the cross facets are somewhat more steeply inclined to the base (Fig. 4 *a*); the number and arrangement of the facets is otherwise the same as in the Dutch variety. Two modifications of the Brabant rose with its low crown are shown in Figs. 5 *a* and 6 *a*. Of these the former has six triangular star facets and six four-sided cross facets, while the latter has twelve cross facets in addition to the six star facets. A form with a larger number of facets, the *rose recoupée*, is shown in Fig. 7 *a*, *b*; it has twelve star facets and twenty-four triangular cross facets, the apices of the latter being directed alternately upwards and downwards.

Closely related to the typical roses are a few forms, illustrated in Plate IV., Figs. 8 to 10. Fig. 8 *a*, *b*, shows the very rare form known as the *cross-rosette*, in which the facets are arranged in multiples of eight. A cinnamon-stone, cut in this form more than a hundred years ago, has been recently brought to light, and described by Professor Schrauf. Fig 9 *a* shows a form which may be regarded as two roses joined base to base. This is the *double rosette*; also sometimes known as the briolette or the pendeloque; the latter names are, however, more frequently applied to stones with a pear-shaped outline, to be mentioned presently. This form, which was formerly much used for ear-rings and watch-chain pendants, was given by L. van Berquen, the originator of the modern process of diamond cutting, to the first diamonds cut by him. These included, among others, the " Florentine " and " Sancy " diamonds, both of which are figured in Plate XI. (Figs. 10 and 11).

Next to be mentioned are the *briolettes*, *brillolettes*, or *pendeloques* (Fig. 10), which are bounded by small facets on all sides, and are somewhat elongated in one direction, so that they have a pear-shaped outline. They are often pierced in the direction of their greatest

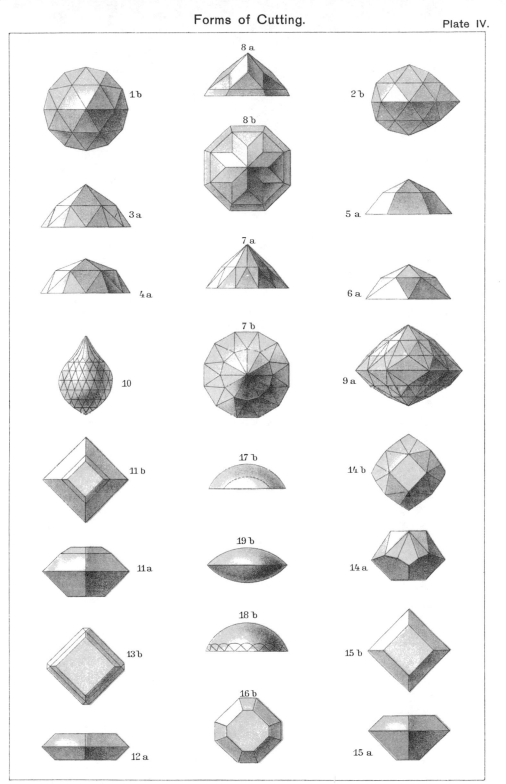

1—8. Rosettes (Rose-cut). 1b. Rose, round. 2b. Rose, pear-shaped. 3a. Dutch Rose. 4a. Brabant Rose.
5a, 6a. Roses of other forms. 7a, b. Rose recoupée. 8a, b. Cross-rose. 9a. Double Rosette
(Pendeloque). 10. Briolette. 11a, b. Table-stone. 12a, 13b. Thin-stone. 14a, b. Table-stone,
with brilliant form above. 15a, b, 16b. Thick-stone. 17b. Cabochon, simple (hollowed).
18b. The same with facets. 19b. Double Cabochon.

length so as to enable them to be used as ear-rings or to be strung on a thread with others. The application of the terms briolette and pendeloque is extremely variable ; as mentioned above, they are sometimes applied to the double rosette. Small stones bounded on all sides by more or less regularly distributed facets, not elongated in any one direction more than another, and bored so that they may be strung on a thread, are known as *beads*.

5. **Rounded forms.**—As a general rule, only transparent stones are faceted ; chalcedony and other translucent stones are occasionally cut in this way ; opaque stones, like turquoise, are never faceted, but always cut *en cabochon*. The rounded forms with convex surfaces, characteristic of this style of cutting, are shown in Plate IV., Figs. 17 to 19. Many deeply coloured transparent stones, such as garnet, are often cut *en cabochon*, as are also those stones whose beauty is due to their peculiar optical effects; such, for instance, as cat's-eye and precious opal. The flat base of these rounded forms is circular or elliptical in outline, and from this rises a more or less convex dome (Fig. 17 *b*). Transparent stones of a deep colour, for example garnet, are sometimes hollowed out at the back, the inner surface having the same curvature as the outer, as is shown by the dotted line in Fig. 17 *b*. Not only is the transparency of the stone increased by this means, but facility is given for removing any faulty portions of the interior of the stone. A stone cut in this manner is known as a *shell*, and, in the case quoted, as a garnet-shell. Frequently the flat base just described is replaced by another convex surface of the same or different curvature as the upper portion ; the stone has then the form of a double convex lens, as shown in side view in Fig. 19 *b*. The form with one curved surface is known as the *single cabochon*, while that with two is referred to as a *double cabochon ;* a much flattened form having only a slight curvature is described as *tallow-topped* (*goutte de suif*).

The convex dome of transparent or translucent stones cut *en cabochon* sometimes has a border of small facets arranged in one or several series (Fig. 18 *b*). Not infrequently, in the cheaper coloured stones and in glass imitations, the table of the brilliant-, the step-, and the table-cuts is given a slight curvature.

Bastard forms are those which are not pure examples of any of the typical forms described above, but combine in themselves portions of any of these types. Rare and costly gems are never so cut, but bastard forms are common enough in inferior stones and glass imitations. Such stones are often cut also in quite irregular and unorthodox forms, which are not subject to any definite rule, but depend solely upon the caprice of the cutter. Of such forms it is obviously neither possible nor desirable to give detailed descriptions. The facets of these capricious forms are sometimes regularly and symmetrically arranged, but at other times this is not the case, and the stone is then described as *cap-cut*.

The form of cutting specially suited to each individual variety of precious stone will be mentioned with the special description of that stone.

B. PROCESS OF CUTTING.

The method by which a rough stone is transformed into a faceted gem is in principle very simple. That part of the surface of the rough stone at which it is desired to place a facet is rubbed with a harder stone or with some other convenient substance. The harder stone abrades small fragments from the softer, and the surface of the latter is gradually worn away and replaced, if the operation has been suitably performed, by a plane face, the so-called facet. This operation is repeated until all the facets required for the particular form the stone is desired to take are produced. A rounded surface is obtained by a method essentially the same as the one described.

A notice of all the technical details connected with the cutting of precious stones would be entirely out of place here.   Only the main principles of the methods which are applicable to all precious stones will here be considered.   The special modifications of general methods which are necessary for certain stones, particularly diamond, will be considered when we come to the special description of such stones.

In the process of **grinding**, the harder stone or cutting material, by means of which the rough stone is fashioned into a gem, is almost invariably used in the form of a fine powder.   This grinding powder is mixed to a paste with olive-oil (in the case of diamond powder) or with water (in the case of emery, &c.), and placed near the edge of a circular disc or lap about a foot in diameter and an inch in thickness.   This disc, which is usually of metal, revolves with great velocity in a horizontal plane about a vertical axis.   The precious stone to be ground down is pressed against that part of the disc on which the grinding powder lies.   The pressure of the stone against the revolving disc causes the powder to become embedded in the soft metal of the disc.   This then acts as a file, the hardness of which is equal to the hardness of the grinding powder.   By the gradual abrasion of the material of the stone over the area which is being ground the facet develops, the length of time occupied depending on the hardness of the precious stone and of the grinding material.

At short intervals during the progress of the work the gem-cutter must ascertain the size to which the facet has grown, so as to avoid making it too large.   A facet which exceeds the other corresponding facets in size is said to be over-ground.   Such an irregularity in a stone greatly diminishes its value.   Great care is also necessary to avoid the over-heating of the stone during the operation of grinding.   Neglect of this precaution results in the development of small cracks, known as " icy flakes," in the interior of the stone, which cause it to be dull and so diminish its beauty.   Before beginning the actual work of grinding, the particular form in which the stone is to be cut, and also its most favourable orientation with respect to the boundaries of the rough stone, must be decided so that the precise position and direction required for each facet may be given to it by the cutter.

One of the most obvious essentials in the grinding of a stone is a means whereby it may be retained constantly in one position.   For this purpose a special kind of holder, the so-called *dop* (or dopp) is used.   This consists of a small hemispherical cup of copper attached by the convex side to a stout copper rod.   The cup is filled with an easily fusible alloy of tin and lead, which is fused and allowed to cool ; immediately before solidification sets in, the stone is placed in the cooling alloy in the position desired, and so that about one-half is embedded in the alloy and the other half projects out of it.   By this means the stone is fixed in the holder in an unalterable position.   In the case of stones of comparatively little value the holder just described may be replaced by a stick of cement, or by a rod of metal or wood, to which the stone may be fixed by a cement consisting of pitch, resin or shellac, and the finest brick-dust.

The rod of the dop or the stick of cement is fixed in a clamp at one end of a small bar, perpendicular to which at the other end are two short legs.   This apparatus is placed so that the two legs rest on a fixed table, while the precious stone in its holder rests upon the grinding disc, which is parallel to and a little higher than the table.   The stone is pressed against the upper surface of the disc by loading the bar with leaden weights, which are larger or smaller according to the hardness of the stone to be ground.   To avoid irregularly loading the grinding disc, two stones are ground at the same time, these being placed at opposite ends of a diameter of the disc.   In the case of a stone of little value the rod to which it is fixed is simply held in the hand until the facet is ground down to the required size, but the result is naturally less perfect than when a mechanical clamp is used.

When a facet is completed the dop is fixed in a new position in the clamp in such a way as to ensure the second facet being correctly placed. By repeating this operation the half of the stone not embedded in the alloy is faceted all over, the different positions of the facets being attained by varying the inclination of the dop or holder in the clamp. Formerly the position of the holder in the clamp necessary to produce any given inclination between the facets was judged of by the eye; this rough method was naturally not susceptible of any great degree of accuracy, and, indeed, frequently gave results very far from what was desired. More recently various appliances furnished with graduated arcs have made it possible to turn the holder through any required angle, which thus ensures that the facets are accurately located in the desired position.

When the projecting portion of the stone has been faceted as far as possible, the alloy must be re-fused and the stone embedded in a new position, so that a fresh portion will be exposed for grinding. This will usually require to be done several times before the stone has received its full number of facets, when it is finally freed from the alloy or cement and cleaned.

The completely faceted stone thus obtained is by no means ready for use as a gem. Its facets are dull and rough, and when examined through a lens they will be seen to be beset with scratches and pits marking the places from which the surface material has been abraded by the cutting powder. To render the stone bright and shining, the roughnesses and irregularities in the surface of the facets must be removed by polishing. This process is not commenced until the whole of the grinding is completed.

The process of **polishing** is precisely the same in principle as that of grinding; the same machine and apparatus are employed, but the cutting material, now known as the polishing material, is softer than before. It may be of about the same hardness as that of the stone to be polished or it may be softer. The stone is placed in a holder, and its rough facets brought to bear one by one upon the polishing disc supplied with the polishing material, which have replaced the grinding disc and grinding powder used in the previous operation. The abrasive effect of the polishing disc is much less marked; it gradually obliterates the small striations and pits, which are the cause of the dull appearance of the stone, and the facets become brighter and brighter. Finally, a point is reached when their brilliancy is not increased by continued polishing; this marks the maximum polish of which the particular stone is susceptible and any further efforts are superfluous. It is important that the maximum degree of polish should be imparted to every stone, otherwise its full beauty will not be apparent.

If the softer polishing powder is used for cutting instead of the grinding powder, the facets will not require any final polishing, but will leave the disc in the most perfectly polished condition. This is the case, as we shall see later, in the cutting of diamonds, since no harder cutting material can be found than their own powder; the process of cutting is, under these circumstances, extremely lengthy and proportionately expensive. It is for this reason that stones other than diamond are cut with a grinding material of greater hardness than that of their own substance. The process of cutting is much shortened by this means and the roughness of the facets can be easily removed by a final polishing with softer material.

As a preliminary to the grinding and polishing of a stone an operation known as *rounding* is performed, with the object of obtaining a first rough approximation to the form the stone is finally to assume. The worker holds the stick of cement or other holder in his hand and presses the stone on the grinding disc in the approximate position for each facet. The rough form so obtained is then ground down more exactly as explained above. With diamonds the preliminary shaping of the stone is performed by a process known as *bruting*. This consists in rubbing together two diamonds, each being cemented at the end of a stick or holder, until the desired form is obtained. The operation is performed over a trough, so that the particles detached shall not be lost.

In order to obtain the cabochon-cut or other rounded forms the dop or the stick of cement is held in the hand and constantly turned so that the stone performs a rolling motion on the surface of the rotating disc. The production of such rounded forms requires a special skill. The harder and more quickly abrasive grinding powder will naturally be used in these cases also and the stone afterwards finished with a suitable polishing material.

The material of which the **grinding disc** or lap is made is varied according to the hardness of the stone to be cut, harder metals being used for the harder stones and *vice versâ*. The disc may be made of iron, steel, copper, brass, tin, lead or pewter; wooden discs also are sometimes used. The upper surface must be perfectly plane, but roughened somewhat, at least near the margin where the grinding takes place. The disc is usually driven by water- or steam-power; it rotates at the rate of two or three thousand revolutions per minute, and sometimes at an even greater rate. The harder the stone to be cut the greater is the rate at which the disc is rotated, as the rapid motion greatly intensifies the action of the grinding powder; so much is this the case that powder of the same substance as the stone to be cut can be employed for grinding, provided the velocity of the disc is sufficiently high.

For the grinding of softer stones, especially varieties of quartz, a grindstone or disc of sandstone is employed without grinding material other than the substance of the sandstone itself. Other details of this process will be given later when we deal with the cutting of agates.

The material of the **polishing disc** or lap may be the same as that of the grinding disc used for the same stone, but as a rule a softer material is employed. Polishing discs are frequently made of wood covered with leather, cloth, felt or paper.

The most important **grinding** or **abrading material** is corundum, the hardest of all minerals except diamond. The pure and transparent varieties of this constitute the ruby, sapphire, and other costly precious stones, but it occurs also in nature in an opaque form in large masses; and the finely granular, black variety known as *emery* is specially abundant. This mineral owes its black colour to the inter-mixture of softer minerals which, however, do not seriously affect its hardness. Emery occurs in large masses, especially in Asia Minor and in the Island of Naxos in the Grecian Archipelago; also at Chester in the State of Massachusetts, U.S.A., and at other places. Commercial emery is largely obtained from Naxos and from Asia Minor, and, in accordance with the various uses to which it is put, is ground to various degrees of fineness. Crystallised common corundum is used for similar purposes; it occurs in a few localities in large masses of comparative purity, and not being mixed with softer minerals possesses a hardness greater than that of emery. Other hard minerals, such as topaz and garnet and sometimes even quartz, are occasionally employed as grinding materials.

In recent years an artificial grinding material, called *carborundum*, has been largely used as an abrasive agent, especially in the United States. It was first prepared in large quantities by the firm of Acheson of Pittsburg in Pennsylvania; the method employed being simply the fusing together of quartz-sand and coal at the enormously high temperature of the electric furnace. The substance obtained is a compound of carbon and silicon having the chemical formula SiC, and consisting of 30 per cent. of carbon and 70 per cent. of silicon. It frequently crystallises in very distinct hexagonal plates, which are translucent, greenish-yellow in colour and very brilliant. It is considerably softer than diamond, but scratches corundum with ease; its great brittleness renders it easily reduced to powder in spite of its great hardness. Carborundum can be obtained cheaply and with little difficulty in crystalline blocks weighing a hundredweight, and it will doubtless gradually replace emery for abrasive purposes.

Finally, the hardest and most important grinding material is *diamond*. Diamond often occurs so impure as to be useless as a gem; this impure material is the so-called bort, which when finely powdered becomes an important abrasive agent. The black, opaque, finely granular variety of diamond known as carbonado, when finely powdered also furnishes a useful grinding material. In spite of the fact that even these varieties of diamond are of very high price, their use in grinding precious stones greatly cheapens the process, for, on account of their enormous hardness, a considerable saving in time and labour is effected. Since the discovery of the South African diamond fields, the price of diamond dust has been considerably lowered, and it is frequently advantageous at the present time to use this material, instead of emery-powder, in the cutting of many precious stones; the economy effected in time and labour more than compensating for the higher price of the grinding material. In cutting the diamond no choice of grinding material is at present possible, since diamond-powder is the only substance hard enough for the purpose.

The **polishing material** is varied according to the nature of the precious stone to be operated upon. In the form of the very finest powder, such diverse substances as tripolite, rotten-stone, jeweller's rouge, pumice-stone, putty-powder (tin oxide), and sometimes a variety of clay known as bole, are all employed for this purpose. The polishing material, like the grinding material, is made into a paste before being applied to the lap or polishing disc; water is generally used for this purpose, but with tripolite sulphuric acid is sometimes employed.

It has already been mentioned that different precious stones are worked with different grinding powders and on laps of different material, these two conditions being varied to suit the particular characteristics of the stone, that of hardness receiving special attention. The choice of material for the lap and of grinding and polishing powder lies between certain somewhat arbitrary limits, thus the same method for one and the same kind of precious stone is not always exactly followed. All precious stones may be conveniently grouped in a few classes according to their hardness; those of each group may be worked in the same manner and with the same materials. The following grouping shows such an arrangement.

a. **Very hard stones.** Ruby, sapphire, and other varieties of corundum.

Ground on an iron, brass or copper lap with diamond-powder. Emery works only very slowly.

Polished on a copper disc with tripolite.

b. **Hard stones.** Spinel, chrysoberyl, topaz.

Ground on a brass or copper disc with emery. (With topaz a tin or lead disc even may be used.)

Polished on copper with putty-powder or tripolite.

c. **Stones of medium hardness.** Emerald, beryl, aquamarine, zircon, tourmaline, garnet, rock-crystal, amethyst, agate, jasper, chalcedony, carnelian, chrysoprase, &c.

Ground on copper or tin or lead with emery.

Polished usually on tin with tripolite or on zinc with putty-powder, sometimes also on wood.

Garnets of fairly considerable size for use as larger ornaments are ground with emery or garnet-powder on a leaden disc and polished on a tin disc with tripolite and sulphuric acid. Smaller garnets, on the contrary, which are strung as beads after being pierced with a fine diamond point, are ground on a disc of fine sandstone with emery and olive-oil, and polished on a wooden disc with tripolite and water or on a tin disc with tripolite and sulphuric acid. Rock-crystal and amethyst are ground on a copper or lead disc with emery, and polished on a tin disc or on a wooden disc covered with felt, with putty-powder, tripolite or bole. For grinding agate, jasper, chalcedony, carnelian, and chrysoprase a disc of copper, tin or lead is often used with garnet or topaz-powder instead of emery; while for polishing, pumice-stone, jeweller's rouge or putty-powder on a tin disc, or pumice-stone on a wooden disc is employed. Agate and other varieties of quartz are, however, often worked in another manner, which will be considered under the special description of agate.

d. **Soft stones.** Obsidian, chrysolite, opal, adularia, turquoise, lapis-lazuli.

Ground with emery on a disc of lead or tin.

Polished with tripolite on tin or hard wood, or sometimes with pumice-stone on wood.

e. **Glass imitations**. These are usually ground and polished on wooden discs; emery being used as the grinding material, and tripolite as the polishing material.

Not infrequently, before the process of grinding a precious stone can proceed, a preliminary preparation of the stone is required. Many precious stones, especially the most valuable, such as diamond and ruby, occur in nature in relatively small but perfectly pure fragments. In such cases no preliminary preparation is needed, and the stone can be at once cut into any desired form. It is otherwise, however, with many precious stones, such, for example, as aquamarine; the naturally occurring crystals and fragments of such stones may be too large for a single cut stone, or may include material which is cloudy or faulty and unfit for cutting. In such cases the specimen must be divided into several pieces of convenient size, and the faulty and undesirable portions removed by a process less lengthy and costly than that of simply grinding them away.

This operation is performed with the aid of a thin metal disc, usually of soft iron, the edge of which is charged with some hard cutting powder, preferably diamond. This *cutting* or *slitting disc* usually rotates in a vertical plane on a horizontal axis; by pressing the stone against the cutting edge of the rapidly rotating disc it will be slowly cut through by the small splinters of diamond which are embedded in the metal. The division of the stone can also be effected by means of a wire smeared with cutting powder, stretched in a bow and used as a saw. This method is, however, very slow, and is now rarely practised.

The superfluous or undesirable material of cheap stones, which occur in large masses, is not cut away, but simply broken off by the blow of a hammer; a mode of procedure which cannot be recommended for valuable stones, unless, as in diamond and topaz, they possess certain known directions of easy cleavage. Stones which can be cleaved with such facility may, as we have already seen, be readily reduced in certain directions with a chisel and hammer with no fear of losing valuable material, so that by this means the work of cutting and grinding is considerably lessened. This subject will be again considered, especially with reference to the diamond, under the special description of each stone. The fragments of valuable material thus broken or cut away are carefully collected and preserved; they may be utilised for the fashioning of smaller gems, or, if of sufficient hardness, pulverised and used as a cutting and grinding material.

As the treatment of precious stones must be varied to suit the different nature of each stone, each requiring different contrivances and methods, so different branches of the gem-cutting industry can be identified, each establishment dealing, as a rule, with but one kind of stone, to the exclusion of others. In diamond-cutting works diamonds only are cut, while the work of other establishments is limited to the treatment of other precious and semi-precious stones to the exclusion of diamonds; such establishments are known as precious-stone or fine-stone-cutting works. Another branch of the industry is concerned exclusively with the large ornamental stones with plane surfaces, or only a small number of facets, such as are suitable for use as stones for signet-rings, crosses, and so forth: here the less valuable stones only are used, such as agate, chalcedony, jasper, &c. In addition to the objects just mentioned, articles such as letter-weights, cups, vases, boxes, etuis, inkstands, stick and umbrella knobs and handles, knife-handles, &c., are fashioned at these large-stone-cutting works, and such materials as granite, marble, serpentine, &c., which do not fall under the head of precious stones, are made use of in the industry beside the minerals mentioned above. The work carried on respectively at the large- and the small-stone-cutting establishments is now to a large extent regarded as different branches of the same trade, and is often directed by the same firm, but the diamond-cutting industry is a distinct and specialised calling, which is never combined with other branches of the trade.

C. BORING.

Not infrequently precious stones, such as garnets, are strung together and worn as beads, which necessitates that they shall be pierced. In the cases in which precious stones have a technical application, such, for instance, as the making of pivot-supports for watches, and the orifices through which very fine gold and silver wires are drawn, it is often necessary to bore a hole through the centre of the stone. The boring is usually effected by the rapid rotation of a fine diamond point, fixed in a metal holder, which acts like a drill. The diamond point of the drill is often replaced by a steel point charged with diamond powder moistened with oil. For convenience in conducting this operation, special machines are constructed by means of which hard stones are bored with ease and rapidity.

D. WORKING ON THE LATHE.

Many stones, especially the softer ones, occurring in large masses, furnish material for the manufacture of balls and other rounded objects. Such articles may be turned on the lathe; this machine is more appropriate for the class of work carried on in the large-stone-cutting works, and is seldom used in the working of precious stones. Nevertheless the hardest stones used for gems may be worked in this way, the steel turning-tool being replaced by a diamond point; as, however, the turning of precious stones is but rarely undertaken, it need not be further considered here.

E. ENGRAVING.

Precious stones are not only cut in various forms, but they are also engraved with devices such as figures, crests, monograms, or with inscriptions. In the cutting of a stone, the lapidary strives to give it such a form as will best display its natural beauty; this beauty depends essentially on the characters of the stone, not on the form given it by the lapidary; this form being simply a means to an end, namely to develop to the uttermost the natural beauties of the stone. The engraver, on the other hand, aims at producing a work of art of value in itself; the material upon which the artist works is a secondary consideration, for he will probably be able to produce just as fine an engraving on some other stone of an entirely different character.

The art of engraving and cutting precious stones is very ancient, much older indeed than that of faceting.   It is mentioned by writers of the earliest historical times, and specimens of this art dating back to very early times are preserved in our museums.   The art of engraving precious stones is still practised, especially in Italy, but the popularity, at the present day, of engraved gems is not comparable with that of faceted stones.   The cutting of precious stones with a view to the production of an engraving is known as the glyptic art, and is thus distinguished from the grinding of facets upon the stone, which can scarcely rank as an art.

Engraved precious stones are generally known as *gems*.   The device engraved upon a gem is either sunk in the stone so as to lie below its surface, in which case the gem is known as an *intaglio ;* or the device is in relief so as to lie above the surface of the stone, a gem so engraved being known as a *cameo*.   Intaglios are frequently used for seals or signets to produce a raised cameo-like impression ; for this purpose they are commonly engraved with a crest or monogram.   Cameos, on the other hand, have no application of this kind, but are used merely as ornaments.   The art of engraving intaglios is known as *sculpture*, while that of producing cameos is known as *tornature*.   Of the two arts, the former is the more ancient, but the antiquity of the latter is well established by the number of cameos, in the shape of a beetle—the so-called scarabs—found in Egyptian tombs.

All kinds of stones, whether hard or soft, opaque or transparent, have been and are employed in the production of **intaglios.**   The harder the stone the sharper and clearer will be the engraved figure, and the more irksome and lengthy the process of engraving. The engraving of the diamond, in spite of its hardness and the consequent difficulty in working, is occasionally effected ; ruby and sapphire have also been sometimes worked in this manner ; the harder precious stones are, however, much less frequently engraved than those which are softer and offer less resistance to the process.   Specimens of this art, dating from the earliest times, are still in existence ; these comprise engravings executed on emerald, aquamarine, topaz, chrysolite, turquoise, rock-crystal, amethyst, plasma, chalcedony, carnelian, agate, heliotrope, opal, lapis-lazuli, nephrite, obsidian, magnetite, and many other stones.   At the present day the materials most frequently used are quartz and chalcedony, together with such varieties of these as agate, onyx, &c., also hæmatite and a few others. The material of the intaglio shown in Plate XX., Fig. 6, is carnelian , another example of intaglio is shown in Fig. 92.

Transparent stones cut as **cameos** are very rarely seen ; opaque stones of fine colour are usually chosen for the purpose, those constituted of differently coloured bands, as in onyx and sardonyx, being specially suitable.   Different portions of the device of a cameo cut in such stones may lie at different levels ; thus, supposing for example that the device is a human figure, a white layer may be worked to form the face and hands of the figure, and a black layer for the hair and garments.   In the cameo shown in Plate XX., Fig. 7, the white figure and the red background are at different levels in the red and white-banded stone ; other specimens of this kind of work are represented in Figs. 93 and 94.   Opaque stones of one colour throughout, such as turquoise and malachite, are also used in the cutting of cameos ; the material of the Egyptian scarabs is very frequently not a precious stone at all, but serpentine or some similar stone.   In Italy, where the industry of cameo-cutting especially flourishes, the shells of certain marine molluscs are employed instead of the stones mentioned above ; in these shells, as in many agates, red and white layers occur in regular alternation.   The majority of cameos sold, for example, in Naples, are made of such material ; they may be readily distinguished, however, from a genuine cameo, cut from onyx, &c., by the fact that a shell-cameo is soft enough to be scratched with a knife, and will effervesce when touched with a drop of acid.

The tool used in cutting or engraving is a very small iron wheel fixed at the end of a rotating axis in a lathe. The small wheel, which may be conical, hemispherical, or disc-shaped, and often not more than a twelfth of an inch in diameter, is charged with moistened diamond-powder. The stone to be engraved, after being cut to the required form and polished, is placed in a holder and its surface pressed against the rotating wheel. The out-line of the device to be engraved is obtained by moving the stone about while the cutting is taking place, more or less prolonged working gives a deeper or shallower engraving. The final polish must, as a rule, be given after the cutting process is completed; this and the last touches to the engraving are given by hand with a graving tool furnished with a diamond point.

Etching.—An alternative to the difficult, lengthy, and costly process of engraving is furnished by the simpler, quicker, and cheaper process of etching. The process is, however, only applicable to certain stones, and the figures produced are less clear and beautiful than those created by the engraver. The method requires that the stone used shall be susceptible to the action of some acid; hence none of the more valuable precious stones are available as material, since they are unacted upon by acids. Those stones, however, which consist wholly of silica, namely rock-crystal, chalcedony, agate, &c., are acted upon by hydrofluoric acid as easily as is glass, and by its means devices may be etched upon them. The polished surface of the stone to be etched is covered with a thin coating of wax, upon which the outline of the device is drawn with some sharp instrument; the surface of the stone along this outline is therefore laid bare while other parts are protected by the wax. The stone is then placed in liquid or gaseous hydrofluoric acid, which eats away the surface of the stone where it is not protected by a layer of wax. There soon appears a hollowing out of the material along the outlines of the device, and this will be deeper the longer the acid is allowed to act. After removing the wax and cleaning the stone the device (monogram, crest, &c.) sketched upon the wax will appear as if cut in the stone.

F. COLOURING AND BURNING.

The methods adopted for altering or improving the natural colours of stones may con-veniently be considered here. Such a change of colour, which has already been stated to be possible under the general discussion of colour, is effected in various ways. We shall not now discuss the change of colour produced in a stone already set and mounted by a surface coating of colouring-matter, but only those methods by which a change in the colour of the whole mass of the stone is effected.

Those precious stones which have a porous structure can be artificially coloured through their whole mass with great ease. The stone to be coloured is allowed to lie in the liquid in which the colouring-matter is dissolved; its porous structure causes it to become after a time saturated with the liquid, which penetrates to the innermost parts of its mass. The stone is then taken out and allowed to dry, during which process the solvent evaporates and the colouring-matter in solution is left behind, lodged in the interstices of the stone, and imparts its colour to the stone. This method is not infrequently practised, and many agates are remarkable for the ease with which they can be coloured by its use. In other cases the use of two liquids is necessary to produce an artificial colour, the *rationale* of the method being that by the interaction of the two liquids a chemical precipitate is formed, which is deposited in the pores of the stone, and imparts to it the desired colour. In this method the stone is first placed in one liquid, in which it is allowed to remain until completely saturated: it is then taken out and dried and placed in the second liquid, when precipitation takes place. The coloured precipitate so produced is regularly distributed in the interior of the stone, so far as the stone is uniformly porous and capable of absorbing liquids. This process of colouring will be described in greater detail under the description of agate. The exact methods of producing the artificial colours, which are black, yellow, blue, green, and brown, are in many cases preserved as a trade secret.

The writings of Pliny show that these arts were known even in ancient times, and the methods then employed for colouring agates are apparently identical with those now practised. At that time, however, a process was known by which certain colours, such for instance as the fine green of the emerald, could be imparted to rock-crystal, a process which is unknown at the present day. Rock-crystal, not being porous, cannot be coloured artificially by the methods described above; the only process known at the present day is to plunge the stone when strongly heated into a cold, coloured liquid; the abrupt change in temperature causes the stone to become penetrated by numerous cracks, into which the coloured liquid enters, and on evaporation deposits its colouring-matter. Rock-crystal so coloured is not, however, very suitable for cutting on account of the cracks developed in it by its sudden cooling; the method therefore is of no practical significance as compared with the colouring of agates, which is of great commercial importance.

Some precious stones when subjected to the action of heat become either completely decolourised or changed in colour, the original colouring-matter being in the one case destroyed and in the other altered. This process, which is known as **burning,** is often employed for increasing the natural colour of many stones, for rendering it more permanent, for completely changing the colour, or for removing unsightly patches. The heating and cooling must be very slow and regular and all sudden changes of temperature avoided; otherwise the stone may develop cracks, or the change in the original colour may not take place uniformly throughout the whole mass of the stone. For the purpose of attaining a uniform rate of heating and cooling the stone is embedded in some powdered material in a crucible; the material made use of being coal-dust, fine sand, iron-filings, clay, quick-lime or wood-ashes, &c. In some cases only a comparatively slight heating is necessary, but often the temperature must be raised to a red-heat in order to destroy or change the colour.

The pigment to which the colour of a stone is due is in many cases unstable at high temperatures, becoming changed into some other body, which is either colourless or coloured differently from the original substance. Thus the change or destruction of the colour of a precious stone can only be effected by burning when the pigment to which its colour is due is altered on exposure to heat.

The yellow Brazilian topaz becomes rose-red when heated, and is then known as " burnt topaz." Amethyst when exposed for a short time to a gentle red-heat in a mixture of sand and iron-filings loses any darkly coloured patches it may have had; after strong and pro-longed heating to redness the violet colour is changed to brownish-yellow, the stone being then known as " burnt amethyst." Many of the naturally occurring brown carnelians become, when heated, a bright red; in this case the original brown colouring-matter is a hydrated oxide of iron, which is changed into the bright red anhydrous oxide by the action of heat. Burning causes yellowish-red hyacinth to become colourless, and at the same time appreciably increases the lustre of the stone. The blue sapphire also completely loses its colour on heating. Other similar cases might be cited; they will be mentioned, since the change of colour effected by burning is of some technical importance, each in its proper place with the special description of each kind of precious stone. The majority of precious stones, however, undergo no alteration in colour even at the highest temperatures.

G. MOUNTING AND SETTING.

The majority of precious stones are destined, after being cut, to be devoted to purposes of jewellery. The less valuable stones are occasionally bored with holes and strung together as beads for personal ornaments, such as necklaces and bracelets. It rarely happens that the whole of a personal ornament, such for instance as a finger-ring, is cut wholly in stone ; this may be done, for example, with nephrite, a stone which occurs in sufficiently large masses and is possessed of the necessary toughness and firmness for the purpose. Much more frequently the stone is firmly and permanently fixed in a piece of metal of suitable shape and of more or less artistic workmanship ; this is known to jewellers as the *setting* of the stone.

The setting of precious stones and the manner in which different kinds of stones are associated affords scope for the exercise of much taste and judgment ; these matters are naturally, however, regulated to a large extent by the fashion current at the moment. A single stone is sometimes set by itself in an article of jewellery ; when this is the case the stone should be of large size and as perfect a specimen of its kind as possible. Usually, however, the effect of such a stone would be enhanced by a border of small stones of another kind ; thus a large and fine opal is often surrounded by a border of small diamonds, the opacity and opalescent lustre of the one forming a pleasing contrast with the transparency and adamantine lustre of the other. This kind of setting is known as carmoizing. Both for decorative work and for personal ornament different kinds of stones are associated in elegant groups, representing butterflies, flowers, and other objects ; here again practice, discrimination and taste are required in order to produce effects of contrast worthy of the beauty of the individual stones.

Gold and silver are the only metals used as the material for the setting of valuable stones. In cheap jewellery some substitute for these metals, such, for example, as gilded brass, is employed. The beauty of some stones is best displayed in a setting of silver, as is the case with diamond ; while gold is a more effective background for rubies and other stones. The girdle, when present, is the portion of a cut stone which is held by the metal setting ; when there is no girdle, as in rosettes, the stone is fixed by the lower margin ; other forms of cut stones are fixed in a variety of ways.

In the kind of setting described as an *open (à jour) setting*, the metal is in contact with the stone at a few points only along the margin, so that the stone is exposed to view on all sides, it being possible to view an object through it. In another method of setting, the stone is fitted into a metal receptacle of the same size and shape as itself, so that it can be seen only from the front ; the back and margin of the stone being concealed by the metal. With this kind of setting, which is described as a *closed setting*, it is impossible to look through the stone, but sometimes the bottom of the metal receptacle is hinged so that it may be opened and closed, and the back of the stone exposed to view, if desired.

In the open setting (à jour) the stone is surrounded by a ring of metal from which several small metal pins or claws project. These are slightly cleft at their extremities, so that each somewhat resembles a pair of small pincers, which grasp the stone at the girdle or margin and hold it, as it were, suspended. This kind of setting is specially adapted to transparent, colourless stones and to coloured stones of flawless quality whose natural perfections require no improvement.

The mounter of gems must be guided in the choice of a setting for a given stone by the form in which it is cut. For stones such as brilliants with an upper and a lower portion, the open setting is employed, the broad table of the stone being, of course, placed towards the observer. Faulty stones are sometimes set with the table towards the back, as in the so-called Indian setting. Stones, on the contrary, which have no under portion are rarely, and rosettes never, mounted in an open setting ; they are much more effective in a closed setting. A stone is held very firmly and permanently in a closed setting, and this is especially the case when cement is used to fix it to the metal, as is very frequently done. In an open setting, on the contrary, there is much more likelihood of the stone becoming, loosened from the claws which hold it, and therefore of its ultimate loss.

Flawless stones of a good colour, which are destined to be mounted in a closed setting, need only to be simply inserted in the receptacle described above. But those in which the colour, lustre, or other quality leave something to be desired can be improved by certain artifices in mounting and their faults more or less concealed.

The artifice which has been longest practised is designed to conceal any dark patches the stone may have, and is known as " mounting on moor." A black pigment made by mixing burnt ivory and mastic is applied to those parts of the closed setting over which the lighter portions of the stone will rest. The stone, when set, will then appear to be uniformly coloured, the patchiness being effectively concealed.

An artifice more frequently used is designed to improve the lustre and colour of the stone to which it is applied. Thin plates of gold, silver, copper, tin, &c., which are known as *foils*, are laid under the stone ; they may be of their own natural colour or they may be artificially coloured in some suitable manner ; in any case they show their own strong lustre. Instead of metallic foils, pieces of silk with a coloured sheen or cuttings of peacock's feathers and similar substances are sometimes used. An uncoloured, natural foil shows its lustre and body-colour through the stone beneath which it is placed, and causes it to appear more brilliant and of a finer colour than would otherwise be the case. A golden foil will give a deeper yellow tint to a pale coloured stone, while a dull stone may be made to appear brighter by placing beneath it a bright and shining foil. A peculiar variation of this use of the foil exists in the Orient in the mounting of rubies, the back of the stone being hollowed out and filled in with gold, a device which considerably heightens the effect of the stone as regards brilliancy and colour.

In the employment of foils, it is desirable that the stone and the foil should be of corresponding colours. As the natural colours of the metals employed are often unsuitable the foils are sometimes artificially coloured blue, red, yellow or green, these colours being always placed on a white metal. Carmine, saffron, litmus, &c., are some of the pigments used for this purpose, intermediate tints being obtained by mixtures ; the pigment is dissolved with isinglass in water and so applied to the metal.

Instead of using a foil, the pigment may be applied to the inside of the case in which the stone is set, or even to the back of the stone itself. It is actually possible by the judicious application of colour at the back of a colourless stone to give the effect of a coloured stone, or even by applying a coating of various tints, to produce the effect of a play of colours. This latter device is now often applied to rock-crystal or colourless glass with the purpose of imitating the play of prismatic colours characteristic of the diamond. Stones so treated, which have sometimes quite a pretty effect, are often used under the name of " iris " for cheap jewellery, pin-heads, &c. It is especially in the Orient, however, that the artifices under consideration have been developed to great perfection, Eastern jewellers being possessed of marvellous dexterity, which they often make use of for the deception of buyers of gems.

By associating several stones of the same kind and of an exactly identical shade of colour in one piece of jewellery a very fine effect is produced; there is however often considerable difficulty in obtaining naturally coloured stones of exactly the same shade of colour. This difficulty may be overcome by coating the back of a dark stone with a light pigment or a pale stone with a darker pigment, and thus securing a uniform depth of colour.

A very effective artifice is that of placing beneath a stone a second of the same form of cutting. Large rosettes are frequently treated in this way, a smaller stone and a foil being placed under the rosette in the closed metal case in which it is set; by this device the brilliancy and lustre of the rosette are wonderfully increased. Similar manipulations will be further considered when we come to treat of the imitation of precious stones.

The use of these and similar artifices for increasing the beauty of a stone is naturally attended with less difficulty when the stone is mounted in a closed setting and one side of it entirely hidden from view, than when an open setting is used. Even in the latter case, however, such artifices may be made use of to a certain extent; thus, a thin strip of foil, or a coating of pigment is applied to the inner side of the setting just below the girdle of the stone; this often has the effect of increasing the brilliancy and colour of the stone. A ruby of too pale a tint mounted in an open setting may be treated in this way, the inner rim of the setting being coated with carmine-red enamel, which gives the stone a very beautiful depth of colour. Other precious stones are treated in corresponding ways.

H. FAULTS IN PRECIOUS STONES.

The beauty and value of a precious stone naturally depend on the absence of all disfiguring faults within its substance or upon its surface. Thus a perfect specimen of a gem must be free from cracks and fissures in its interior and its lustre and polish must be uniform and uninterrupted over the whole surface. A transparent stone should be perfectly clear, with no cloudy patches, and free from all enclosures, especially of small, opaque, foreign substances. Colourless stones ought to be quite free from any faintly coloured patches; while the colour of coloured stones must be uniform and regularly distributed, so that the stone shows no light or dark patches, or differently coloured portions of its substance. Exception to the latter rule is, of course, made in such a stone as agate, the effect of which depends on a difference of colour in different portions of its substance. Each imperfection of the kinds just noted, each crack, each dark or cloudy or differently coloured patch, is a fault in the stone, and as such detracts from its beauty. It is in transparent stones that faults are especially noticeable, and in a cut form they are reflected again and again from the facets and thus rendered still more obvious.

Small insignificant faults, when they are few in number, do not render a stone entirely unsuitable for decorative purposes, but they do diminish its value, and that sometimes very considerably. When, however, the appearance of a stone is so disfigured by the presence of numerous glaring faults that its use for ornamental purposes is out of the question, it becomes absolutely worthless, unless indeed its hardness enables it to serve some useful technical purpose.

The more obvious faults of precious stones, such for example as the presence of light, or dark, or differently coloured patches, are easily detected; they are indeed often apparent at the first glance. Frequently, however, the detection of faults requires a practised eye, since a clever gem-cutter will so arrange the facets of a stone that any faults it may have become quite inconspicuous and almost unnoticeable to an unskilled observer. When considering the subject of the refraction of light it was mentioned that the faults of transparent stones may be made more conspicuous by immersing the stone in a strongly refracting liquid, such as methylene iodide or monobromonaphthalene. This device is due to Sir

David Brewster, who for the same purpose made use of Canada-balsam, oil of anise or sassafras-oil.

It is not difficult, by one or other of the artifices already described, to conceal or make inconspicuous the various faults to which precious stones are liable. On this account, therefore, it is a rule to be observed that valuable stones of high price should never be purchased in a setting, but in a loose and unmounted condition, which admits of a thorough and complete examination of the stone being made. Such an examination should be made before the purchase not only of cut-stones, but also of uncut, rough stones. Small faults in the latter are often very difficult to detect, since the roughness of the surface interferes more or less considerably with the transparency of the stone. In such cases it is advisable to place the stone in a highly refractive liquid, which will have the apparent effect of lessening the roughness of the surface and increasing the transparency of the stone. Even when placed in water such stones will appear more transparent than in air. In examining a rough stone it is important to determine whether any fault it may have lies quite in the interior of the stone or near the surface. In the latter case it may often be removed in the process of cutting, and a somewhat imperfect rough stone may be transformed into a gem of flawless beauty.

The nature of the faults which are of the most frequent occurrence in precious stones may be gathered from what has already been said. Those of the nature of coloured patches, enclosures of large foreign bodies and such like, need no further description ; other faults, however, which occur again and again with the same characteristic appearance are distinguished by jewellers with special names and receive special mention below.

1. **Sand.**—Small grains of any foreign substance, of a white, brown or reddish colour scattered singly through the material of the stone are known as sand.

2. **Dust.**—This is the name by which extremely small particles of foreign matter scattered in great numbers through the substance of a stone are known.

3. **Clouds.**—By this term is meant muddy or cloudy patches of various colours— white, grey, brown, reddish, greenish, &c.—which may occur in the substance of a stone and which when brought to the surface in the process of cutting, give it a dull appearance which no amount of polishing will remove. They are most frequent in diamonds and pale rubies.

The three kinds of faults just described are all due to the inclusion of small mineral grains as impurities in the substance of the precious stone. If not too small, they may sometimes be seen with the naked eye or with a good lens, but more frequently the powerful magnification of a microscope is required. Their presence is best demonstrated, however, when the stone is examined in polarised light ; in the dark field of the polariscope such inclusions will sometimes appear bright and vividly coloured.

4. **Silk.**—This term signifies the whitish, shimmering streaks, disposed in certain directions, which sometimes mar the appearance of a stone. Such streaks are in reality strings of microscopically small cavities in the substance of the stone, which may be quite empty or may, on the other hand, contain a liquid. Such cavities are not at all uncommon in precious stones ; they are sometimes of quite appreciable size, as in topaz, sapphire, &c., when they may be seen with the naked eye or with a lens. The cavities, to which the fault known as "silk" is due, are however definitely arranged in bands and strings, and only become visible as single objects under strong magnification. By scattering at their surfaces the light which should pass directly through the stone, these minute cavities produce the dull, whitish shimmer. They have the same effect as "clouds" in that, when they occur on the surface of a stone, a perfect polish over that area is impossible.

5. **Feathers.**—Under this name are included cracks and fissures which may be present in all kinds of precious stones, and which exert a disturbing action on the path of the rays of light passing through the stone. They may be of large or of almost microscopic size, and may occur singly or aggregated together in large numbers. They are especially common in stones which possess a very perfect cleavage, such as diamond and topaz; they then have the direction, regularity and flatness peculiar to cleavage cracks. Faults of this nature are, however, also present in stones which possess no marked cleavage, such as quartz and garnet, but here they are irregularly curved and bent. In such cracks, brilliant, iridescent colours are often to be seen, and when this is the case the cracks are more noticeable. The existence of feathers, which do not exhibit such interference colours, is very difficult to detect, even with a lens; they may be best demonstrated by placing the stone in methylene iodide.

Feathers are more to be feared than any other kind of fault, since there is always a tendency for these small cracks to extend, thus adding to the disfigurement of the stone, and perhaps in the end causing its complete fracture. This is especially likely to happen during the grinding, owing to the vibration consequent on the operation, or it may be brought about by subsequent careless handling, and often even with no apparent cause. It is therefore very desirable before going to the labour and expense of cutting a stone, to make certain that it is free from such faults. This may be done as mentioned above, by immersing the stone in a strongly refracting liquid and inspecting it with a lens. Another method is to heat the stone and then quickly cool it by immersing it in cold water; any incipient cracks will be made to develop by this treatment and thus become more distinctly visible, or the stone may even fracture along the cracks; this operation is of course risky and should not be attempted with valuable stones.

6. **Icy flakes.**—These are small cracks which are not of natural occurrence in precious stones, but are developed during the process of grinding if a stone has been allowed to become too hot. Their presence is manifested as dull cloudy areas on the surface of the stone which are incapable of receiving a good polish. Such faults can be avoided by keeping the temperature of the stone down during the process of grinding.

I. ARTIFICIAL PRODUCTION.

As in the case of many minerals, it has been possible to produce the majority of precious stones by artificial means. These artificial products are in every respect identical with naturally occurring precious stones—namely, in their chemical composition, crystalline form, and in all their physical characters. When it is possible to obtain such artificially formed precious stones of sufficient size and of the clearness, transparency, and fine colour of the naturally occuring precious stones, they will be equal in value to the latter, and equally applicable to decorative purposes. To consider artificially formed stones inferior to natural stones is nothing but baseless prejudice. They differ from the latter in no respect save origin, having been produced under artificial instead of natural conditions; they are therefore truly genuine stones, and in no sense must they be regarded as imitations.

Many have been the experiments made with the object of producing minerals by artificial means, and in numerous instances such efforts have been successful. In the case of precious stones the results, although of the greatest scientific interest, have had, up to the present, little or no practical importance, since the stones obtained have been but of very small size, often microscopically minute, and thus useless as gems. Only in the case of two of the more valuable precious stones, the ruby and turquoise, have results of importance been yet obtained. The French chemist, Fremy, has prepared crystals of ruby, which, though not large, are yet of sufficient size to be mounted as gems. It appears also that a method has been discovered for the artificial production of turquoise. The details of this method are, however, kept secret. These subjects will be reconsidered when we come to treat specially of ruby and turquoise.

It may probably be safely asserted that the artificial production of every precious stone in a form suitable for decorative purposes is only a question of time. But the possessor of natural gems need not fear on this account a depreciation in the value of his jewels, since the artificial products hitherto obtained are but just within measurable distance of the required standard. Moreover, research in the direction of the preparation of artificial rubies has shown that the cost of artificial production, owing to the expense of raw material of the necessary purity and the costly nature of the apparatus required, is quite equal to the price commanded by natural stones.

It is probably possible, even at the present day, to so improve on the apparatus and methods of manipulation in the artificial manufacture of rubies that fine crystals may be produced at less cost. Naturally such researches have been frequently undertaken, although possibly often in secret, owing to the value a happy discovery might have. The possession of the secret by which costly precious stones could be prepared at comparatively small cost, in a condition equal to that of natural stones, would indeed be a source of wealth! The artificial production of precious stones in large quantities would, of course, very soon bring their price down to a minimum, and would also depreciate the value of natural stones. Thus it would result that precious stones which had previously been rare and costly objects, and their acquisition possible only to the rich, would come to be within the reach of all classes. This being so, their possession would cease to distinguish the upper and wealthy classes from less-favoured individuals, and hence precious stones would lose the attribute to which at the present time a large part of their value is due.

J. COUNTERFEITING.

In dealing in objects of such value as precious stones, it is not surprising to find that efforts are often made to substitute for a genuine and costly stone one of similar outward appearance, but of less value, in order to deceive an inexperienced buyer. In place of fine stones of high price, attempts are made to pass off stones of less value or glass imitations of the same colour; or, in place of faultless specimens, genuine stones disfigured by the presence of faults, which are hidden by one device or another as completely as possible. Often two small stones are cemented together so that they appear as a single large one; or, again, the upper portion of a cut stone may be genuine while the lower is false.

The inventive genius of dishonest dealers in precious stones is responsible for many other methods by which the unwary purchaser may be deceived. Any person desirous of obtaining a genuine precious stone of any considerable value must be prepared to exercise the greatest caution, unless he is dealing with a well-known and reliable man whose integrity is above suspicion. The more costly the stone the greater is the caution necessary, for the possibility of greater gain is more inducement to fraud. In such cases fraudulent devices are concealed with greater dexterity, and for their detection a sharp eye is necessary as well as expert knowledge, and such experience can only be acquired by familiarity with the trade.

The deceptions practised by Eastern dealers in precious stones are notorious. Many a traveller in India, Burma, Ceylon, &c., with no thought of suspicion in his mind, has bought some apparently beautiful stone, only to learn when too late that he has acquired some utterly worthless object—perhaps a piece of cleverly prepared bottle-glass! There is still more scope for fraud in mounted stones, the setting of which may be used to conceal all kinds of deceptions. In this connection there will be no harm in repeating the rule, that costly and valuable precious stones should never be purchased in a mounted condition.

It is not to be denied that certain artifices, which have been already described, amount to an illusion, inasmuch as they make the stone appear better than it actually is. Such artifices, however, cannot be considered fraudulent, since they are openly practised and are known to all persons concerned. Moreover, a lower price will be asked, at least by a respectable dealer, for a stone that has been so treated than for one which stands in no need of artificial improvement. There are indeed many devices, similar to those which have been described, which are adopted quite openly and in all good faith, and are made use of by every fair-dealing jeweller. Such devices, which fall under the head of recognised and allowable manipulation, are not hidden from an intending purchaser, nor is a higher price set on the stone than its natural qualities justify.

We may contrast with such a transaction one in which a yellowish diamond has received a thin coating of blue colouring-matter, and has then been sold as a colourless, water clear diamond at a correspondingly high price. There can be no two opinions as to the nature of this latter transaction. There are many cases, however, where it is difficult to draw the line between an artistic device and a fraudulent artifice; the decision in such cases will turn on the behaviour of the dealer, whether he gives a genuine description to the stone and asks a price corresponding to its natural qualities, or whether, on the other hand, he conceals its deficiencies and demands a correspondingly higher price.

The buying and selling of precious stones is then, as we have seen, a trade in which fraudulent practices are particularly easy. It would not be practicable to detail every single possibility of fraud, especially as the oldest tricks appear again and again in new dresses. Such frauds as are most frequently practised will be described below:

1. **Substitution of less valuable stones.**—This can only be effected when the cheaper stone resembles to a certain extent the more valuable one in colour, lustre, and general appearance, and if possible approaches it also in some essential character, such as specific gravity or hardness. In such cases a little special knowledge of the subject is necessary in order to distinguish between the two stones. In this way a colourless topaz may be substituted for a diamond, since both stones are colourless, and both the lustre and the specific gravity of the topaz approach those of the diamond. To give another example, either colourless zircon or colourless sapphire may be substituted for diamond, and according to Mawe, a London jeweller, these stones at the beginning of the nineteenth century commanded a high price since they were especially suitable for selling as diamonds at a still higher price. The variety of the yellow quartz, known as citrine, is substituted for yellow topaz, and red spinel is often offered for ruby; other similar cases might be mentioned.

Some of the stones mentioned above cannot in their natural condition be sold as substitutes for more valuable stones of another kind. Thus hyacinth (zircon) is naturally of a yellowish-red colour, and only becomes colourless and acquires a stronger lustre after it has been heated; similarly, blue sapphire is rendered colourless by heating. For the same purpose, stones are not only decolourised but also artificially coloured. A fine blue colour is imparted to chalcedony, so that, to a certain extent, it resembles lapis-lazuli and may be used instead of that mineral in cheap jewellery and ornaments of various kinds.

An experienced eye will usually be able to detect such fraudulent attempts at the first glance, but there are cases in which this is not possible without a more thorough examination of the stone according to the methods of scientific mineralogy. These methods will be detailed in the third section of this book, and under the description of each kind of precious stone it will be stated how that stone may be distinguished from others which resemble it in general appearance.

2. **Doublets.**—A cut stone, which consists of an upper and a lower portion cemented together so as to present the appearance of a single stone, is known as a doublet. There is less deception here when the two portions consist of genuine material, for example diamond, the two small stones forming together a large and apparently single stone, which, if really single, would be of greater value than the two stones mounted separately. Such a combination of two small, genuine stones may be referred to as a *genuine doublet*.

Very frequently, however, the upper portion only of the doublet is genuine, the lower portion being cut from comparatively worthless material, such as quartz or glass. When skilfully contrived, such a doublet has the appearance of a single stone of the same material as that of the upper portion, and at a first glance shows all the beauty such a stone would have. Such combinations are known, from the material of their upper portion, as diamond-, ruby-, sapphire-doublets, &c. It is said that in Antwerp at the present time diamonds and colourless sapphires are often combined in this way. The advantage to be derived from such a proceeding is very evident; the dealer is able to sell what appears to be a large stone but which in reality consists of comparatively little genuine material. When the lower half of a *semi-genuine doublet* of this kind consists of glass, this may be fused to the upper portion and a more intimate and permanent union effected than if the two portions were fixed together by cement or mastic.

A fraud of this kind is difficult to detect when the stone is mounted in a closed setting. Detection is easier when the stone is unmounted, for with the help of a good lens the line of junction of the two portions may be seen, or sometimes the prismatic colours of thin films may be visible where air has penetrated along the plane of junction. The two portions of a doublet cemented together by mastic will fall apart when the stone is placed in hot water, but this of course will not happen when the two parts are fused together. The compound nature of a doublet made, of glass, and of a doubly refracting stone such as ruby, is easily recognised by the different behaviour of the two parts in polarised light. The difference in the refractive index of the two parts will also serve the same purpose, especially with doublets of colourless stones such as diamond and rock-crystal. For the purpose of demonstrating this difference the stone is immersed in a strongly refracting liquid, such as methylene iodide, and this diluted until one or other part of the stone becomes invisible. As explained in the section on optics, this will happen when the index of refraction of the liquid is the same as that of the stone. In the case of a doublet of quartz and diamond immersed in methylene iodide diluted with benzene the quartz will become invisible, while the more strongly refracting diamond will still preserve its sharp outlines.

Such elaborate devices for the detection of a doublet are, however, only necessary when it is very cleverly made, and when the substances of the two portions are well matched. Indian jewellers are specially expert in the production of good doublets. When less carefully contrived and put together, there will be sufficient contrast between the two portions to make the doublet easily recognisable as such.

Doublets, of which the upper portion consists of rock-crystal or other colourless stone, and the lower part of a coloured glass, are known as *false doublets*. Here the colour of the lower portion is imparted to the upper harder portion. The same effect is obtained when between the upper and lower uncoloured portions is placed a thin layer of coloured material, a plate of metal, or even a piece of coloured gelatine-paper. When the two portions are differently coloured, the stone may be instantly recognised as a doublet by holding it up to the light and viewing it from the side, or observation with a lens will disclose the coloured layer between the two colourless portions. Here, as in all cases, when examining a doublet, the stone must of course be unmounted.

The construction of *hollow doublets* is somewhat peculiar; these consist of an upper portion of rock-crystal or colourless glass, which is hollowed out below, the walls of the hollow being finely polished. The cavity so hollowed out is filled with a coloured liquid, and closed with a plate of rock-crystal or glass, or by a complete lower portion of the same material. The whole doublet if viewed from above, that is from the table, appears of the same colour as the liquid; when viewed from the side, however, the boundary of the cavity containing the coloured liquid is plainly visible.

3. Glass imitations.—The manufacture of glass imitations of precious stones has reached a high degree of perfection. The varieties of glass suitable for this purpose are known as paste, and this name is also applied to the imitations themselves, which are often substituted for genuine precious stones. This species of fraud, which is common enough at the present day, was known and practised by the ancients, and attention was drawn to it by Pliny, who gave eloquent warnings on the subject.

The manufacturer of such kinds of glass aims at producing a substance which will, as far as possible, exhibit the more beautiful and valuable characters of genuine precious stones, and which, at the same time, will be in price as far removed as possible from the latter. The method at present followed is to produce a mass of glass which shall be as clear, transparent, and colourless as possible. When a material for coloured imitations is required, the colourless glass must be again fused with some metallic oxide capable of producing the desired colour.

The majority of precious stones can be so successfully imitated in glass that only a very practised eye can distinguish without more detailed examination a genuine stone from its paste imitation. Some of these artificial glasses possess not only the same clearness and transparency, but also in a large measure the lustre and high index of refraction and dispersion of a diamond of the purest water. Others again may exhibit a colour comparable with that of the finest ruby, sapphire, emerald, or topaz, &c.

The one point, however, in which all artificial imitations fail is hardness; they have the hardness of glass ($H = 5$), and are, as a rule, softer than ordinary window-glass. In spite of this fact they take a high polish, which, however, after use is soon lost; neither is the sharpness of their corners and edges retained for any length of time. Although when new these glass imitations are very similar in general appearance to genuine stones, and may be substituted for them with, in most cases, but little fear of detection, yet after they have been in use a short time they become dull and anything but beautiful objects. If it were possible to give these artificial glasses the hardness of true precious stones, they would in many cases be almost as suitable for personal ornament as are the latter, since the objection to their use, which has been just mentioned, would not exist. This drawback to the utilisation of the soft artificial glasses as precious stones is frequently overcome by the use of the semi-genuine doublet, the lower and larger portion of the doublet being of glass, and the upper smaller portion of some hard stone, such as quartz.

An artificial glass may, in almost all cases, be distinguished from a genuine stone by its lack of hardness. Glass is, as we have already seen, easily scratched by a hard steel point, which will not touch the great majority of precious stones. In addition to other methods, an aluminium pencil has recently been used for the purpose of distinguishing between genuine stones and glass imitations; the point of the pencil when drawn over glass leaves a shining, silvery line, which is not the case when the material is a true precious stone.   Glass, being an amorphous substance, is singly refracting, while many precious stones are doubly refracting ; this difference may, with the help of the polariscope, prove useful in distinguishing the two. Again, singly refracting glass is never dichroic, hence a stone which, when examined with the dichroscope shows different colours, cannot be glass.   Some precious stones, for example diamond, are, however, singly refracting, and not dichroic : these could not be distinguished from ordinary glass by the aid of the polariscope alone, unless, indeed, the black cross shown by unannealed glass, as already mentioned, should happen to be observed.   There is usually a difference in the specific gravity of a stone and of its glass imitation ; with very heavy glasses, however, the specific gravity of which may be as high as 3·6 to 3·8, this character may in certain cases approach more or less closely that of the precious stone it imitates.   Finally, it is to be noted that the manufacture of a glass absolutely free from small air bubbles and other irregularities, such as streakiness, banding, &c., which never occur in precious stones, is very difficult.   Observation with a lens, or when necessary with a microscope, will often result in the detection of such bubbles and streaks, the presence of which will prove the false character of the stone.   Moreover, there may often be seen at the edge or girdle of a faceted specimen the marked, conchoidal fracture of glass, which is often present in a characteristic manner, different from anything seen in a genuine precious stone.

The material used for the production of imitations of precious stones is, in most cases, a readily fusible, colourless glass, rich in lead, and known by the names *strass*, paste, or Mainz flux.   The qualities which this substance must show before everything else are the most perfect transparency and clearness and freedom from colour ; it is therefore of importance that the raw materials used in its manufacture should be of the greatest possible purity.   The constituents of strass are, as a rule, the same as those of ordinary glass with the addition of one or two other substances, especially of red-lead.   The most important constituent is quartz, which must be quite free from iron, and is best suited for this particular use in the purest form of rock-crystal.   Potash is used in the form of potassium carbonate (potashes), which must also be as chemically pure as possible ; potassium nitrate often replaces the carbonate, since this salt can be more easily obtained in a pure condition ; for the same reason another salt, potassium tartrate, is sometimes used.   Potash, as a constituent of strass, is sometimes replaced altogether by thallium, which can be used in the form of any of its salts, the product thus obtained being known as thallium-glass.   Lead is employed in the form of red oxide (red-lead), which is prepared from chemically pure metallic lead.   A little white arsenic is sometimes added, but this is not an essential constituent, and because of its poisonous nature is often omitted.   For the purpose of increasing the fusibility of the mixture, a little borax or pure boracic acid is added as a flux ; this, however, does not enter into the composition of the glass, but is volatilised by the heat of the glass-furnace.

These materials, after being powdered finely and intimately mixed, are fused together in a Hessian crucible, and kept at as constant a temperature as possible, which should not be higher than that just sufficient to produce complete fusion.   The fused mass, which should then be homogeneous, and as free from bubbles as possible, is allowed to remain for about twenty-four hours in the furnace, during which time it cools gradually and slowly.   Any disturbance of the fused mass must be avoided in order to guard against the introduction of air bubbles, which cannot be again expelled, and which render the product unsuitable for the purpose for which it is intended.

The constituents mentioned above are not used in the same proportions in all cases.

The amount of lead especially varies very considerably, and is sometimes entirely absent, though a glass free from lead cannot be correctly termed a strass. Many recipes are given for the preparation of glasses suitable for the imitation of precious stones. A few of the best containing varying amounts of lead are given below :

3 parts of fine quartz-sand, 2 of saltpetre, 1 of borax, $\frac{1}{2}$ of white-arsenic.

9 parts of quartz, 3 of potassium carbonate, 3 of fused borax, 2 of red-lead, $\frac{1}{2}$ of white-arsenic.

8 parts of white glass free from lead, 3 of rock-crystal, 3 of red-lead, 3 of fused borax. $\frac{2}{3}$ of saltpetre, $\frac{1}{6}$ of white-arsenic.

$7\frac{1}{2}$ parts of quartz, 10 of red-lead, $1\frac{1}{2}$ of saltpetre.

A mixture which is frequently used is 32 per cent. of rock-crystal, 50 of red-lead, 17 of potassium carbonate, 1 of borax, and $\frac{1}{3}$ per cent. of white-arsenic.

The greater or less the amount of red-lead in the mixture, the more or less rich in lead will the resulting glass be, and the other constituents will vary accordingly. The amount of silica in lead-glasses varies between 38 and 59 per cent., potash between 8 and 14 per cent., and lead oxide between 28 and 53 per cent. As an example of the chemical composition of a lead-glass (strass) used to make an imitation of diamond, the following analysis may be given: silica (SiO_2) 41·2, potash (K_2O) 8·4, lead oxide (PbO) 50·4 per cent.

The physical characters of these glasses vary very considerably with the chemical composition, the amount of lead present having a specially marked influence in this direction. When this element is present in smaller amount the hardness of the glass is rather greater, but the specific gravity, as well as the index of refraction and the dispersive power, are lower. These latter properties are increased with an increase in the amount of lead present ; a glass very rich in lead, such as the one of which the percentage chemical composition is given above, has an index of refraction and a dispersive power comparable with those of diamond, and will therefore have the brilliancy and play of prismatic colours characteristic of this stone. This, indeed, is the object of the addition of lead to glasses which are to be used as imitations of precious stones ; the increase in the amount of lead also raises the specific gravity of the resulting glass, this being sometimes as high as 3·6 or 3·8, higher, that is to say, than that of diamond.

The play of prismatic colours is even finer in glasses in which thallium replaces potassium than in those in which lead is the only heavy metal present. The presence of this heavy metal as a constituent of the glass very considerably increases its dispersion and index of refraction ; such thallium-lead-glasses are, as regards their optical characters, much superior to the ordinary strass of the composition mentioned above. The specific gravity is also higher and reaches 4·18 to 5·6, increasing with an increase in the amount of thallium. A glass containing a moderate proportion of thallium, and with a specific gravity of 4·18, has a dispersion of 0·049 ; that of ordinary lead-glass (flint-glass of Fraunhofer) is only 0·037, the dispersion of diamond being 0·057.

Different glasses, varying in their physical characters according to their chemical composition, may therefore be employed for different purposes. A stone which is to imitate the diamond must have a high index of refraction and a high dispersion, and for this purpose a glass rich in lead, or, better still, a lead-glass containing thallium, will be used. A precious stone possessing only a low index of refraction may, on the other hand, be imitated by a strass containing but little lead, or even by one from which lead is altogether absent.

The recipes given above should produce, when the materials are quite pure, a perfectly colourless glass. A coloured glass is obtained by the addition to the strass of a colouring substance. The substances usually employed for this purpose are metallic oxides, which must be in as pure a state as the other constituents of the strass.   In the manufacture of a coloured glass, the colourless strass already prepared, and the requisite amount of metallic oxide, are both reduced to a state of fine powder, and are then intimately mixed by being passed through a sieve.   The mixture is then fused at a moderate temperature, and allowed to remain in this condition for about thirty hours, after which it is very slowly cooled. Only a very small amount of the metallic oxide is necessary to produce any required colour, the actual amount differing with different oxides.   The depth of colour imparted by any given oxide will, of course, depend on the amount of it used ; a light colour will be given by a very small amount, while a larger amount may produce a colour so deep that thick pieces of the glass will appear black and opaque.   Between these two extremes every gradation of colour is possible.   As an example of the intense colouring power possessed by some metals, it may be stated that one part of gold will impart a vivid ruby-red colour to 10,000 parts of strass, while this same amount will impart an unmistakable rose colour to 20,000 parts of strass.

The substances employed for the production of differently coloured strass are many and varied.   Cobalt oxide or smalt produces a blue colour, to which a tinge of violet may be given by the addition of a small quantity of manganese oxide.   Yellow is produced by silver oxide or chloride, and by antimony oxide or the so-called red-antimony, which is a mixture of the oxide and sulphide.   The addition of a small amount of coal also produces a yellow colour, the intensity of which varies from a light honey-yellow to yellowish-brown according to the amount used ; a beautiful golden-yellow is obtained by adding in addition to the coal a little manganese oxide.   The use of coal for the production of a yellow colour is, however, only possible in glasses free from lead.   Chromium oxide and copper oxide each produce a green colour, to which a bluish tinge may be imparted by the addition of a little cobalt oxide, while a yellowish-green is obtained by adding red-antimony.   A mixture of cobalt oxide and red-antimony produces a green colour, due to the combination of the blue of the cobalt and the yellow of the antimony.   Red may be obtained by the addition of various substances, namely, cuprous oxide (Cu_2O), gold oxide, gold chloride, or purple of Cassius, the last named being used for the production of ruby-glass, so called from the resemblance of its colour to the red of the ruby.   A red colour inclining to violet is obtained by the use of manganese oxide which should be as free as possible from iron, a pure violet being obtained by the use of a little cobalt oxide in addition to this ; a larger amount of cobalt oxide produces a reddish-brown colour.   Black glass, which remains black even in the thinnest layers, is produced among other methods by adding a large amount of tin oxide and afterwards fusing again with a mixture of manganese oxide and hammer-slag from iron-works.

An opaque, white glass, that is an enamel, is obtained by the addition of a small quantity of either tin oxide, calcium phosphate, or bone-ashes.   This opaque glass is capable of colouration by metallic oxides, and hence imitations of opaque stones such as turquoise are possible, the blue colour of this stone being imitated by adding a little copper oxide and cobalt oxide.   The appearance of opal, chalcedony, and other translucent stones, and even to a certain extent the colour bandings of agate, may be imitated in glass by methods very similar to those above described.

It is not by any means to be supposed that pastes can be produced at a very low cost. The production of a strass suitable for making good imitations of precious stones is a comparatively expensive operation, the price of materials of the necessary purity being high, and the appliances and apparatus used in the manufacture necessitating the outlay of some considerable capital.   For this reason, only the more costly precious stones are imitated in this more perfect manner. Poorer imitations, which can be recognised at a glance even by the most unsophisticated, are made from common materials with no special care; the cost of their production is low and they are used in the most inferior so-called ornaments.

The material for the imitation of precious stones having been produced by the methods described above, is ground, polished, and mounted, these processes differing in no wise from the grinding, polishing, and mounting of genuine stones, which has already been described.

Attempts appear to have been recently made to produce a glass imitation of a precious stone which, besides the usual constituents of this substance, shall contain those characteristic of the stone imitated. Thus a rough chemical analysis of this material would be similar to that of the genuine stone. Imitations of emerald have recently come into the market, which contain seven to eight per cent. of beryllium oxide, an essential constituent of emerald but not ordinarily of glass. All the physical characters of the material are those of green glass and not of emerald. Of the place and method of preparation of this material nothing is known.

K. VALUE AND PRICE.

The esteem in which the different kinds of precious stone are held does not by any means depend solely on their beauty, durability, or similar characters, but is influenced by various external conditions. The price demanded for precious stones is therefore fluctuating, since it is regulated, as in the case of any other objects which are bought and sold, by the laws of supply and demand. A large supply and a small demand results in a low price, and vice versâ; when the supply and demand both vary in the same direction, that is, when they both rise or both fall equally, the price remains stationary.

The supply of each kind of precious stone depends essentially on the frequency with which it occurs in nature and on the extent to which it is mined. For reasons which have already been pointed out, precious stones which are of very common occurrence even when possessed of considerable beauty are never held in very high esteem, and consequently never command a high price, the price of the cut stone often only slightly exceeding the cost of cutting. Stones on the contrary which possess the merit of rarity are much sought after and valued far more highly.

The supply of any one kind of precious stone in the market varies at different times and this causes a corresponding variation in price. The exhaustion of a locality, which has formerly supplied a large number of stones, will result in a rise in the value of that particular stone, while the discovery of a new source of supply has an opposite effect. The history of the various discoveries of diamond is an instructive example of this fact. In the seventeenth century, the price of the diamond rose steadily higher and higher on account of the gradual exhaustion of the Indian mines which were the only ones then known. The discovery of the rich Brazilian mines in 1728 caused a rapid and marked fall in the price of diamond; with the gradual exhaustion of the Brazilian deposits, the price again gradually rose until the discovery of diamonds in South Africa in 1867. So large has been the supply of diamonds from this locality that the price, for average material at least, is lower than it has ever been before.

The amount of production is not, however, the only factor which influences the supply. A large accumulation of stock thrown suddenly upon the market has the effect of lowering the prices. Of interest in this connection is the statement made by Kluge in 1860 to the effect that a few years before, at the Easter fair of Leipzig, the price of diamonds suddenly fell fifty per cent. owing to the fact that the Brazilian Government had paid the interest of the national debt in diamonds instead of in cash.

The causes which operate to produce fluctuations in demand are more intricate and difficult to trace. The earning capacity of a nation, and consequently the general prosperity of the people, the general trend of events, the wide-spread existence of a taste for ornament or display, fashion, and such like considerations, are all factors which influence the demand for precious stones. Since precious stones are purely articles of luxury for which there is no absolute need, and which it would be possible to dispense with completely, they are only extensively used in times of peace and commercial prosperity when the necessities of life are abundant and easily obtained. At such times, and especially when public events create and foster a taste for display, the price of precious stones will rise. When, however, the purchasing-power is diminished in consequence of war or of industrial crises retrenchment will first be made, naturally not in the necessities of life but in its luxuries; there will be little or no demand for precious stones, and, moreover, the heirlooms or recent acquirements of old families will be thrown on the market and will help to accentuate its downward tendency.

Many examples of this rise and fall in the value of precious stones may be found in French history. Thus, during the period, of lavish display immediately preceding the French Revolution, when the European Courts vied with each other in extravagance and luxury, the price of precious stones, and especially of diamond, rose high. During the Revolution and the series of miserable wars which followed, precious stones fell in value, only to rise again steadily during the long years of peace which followed Napoleon's fall. This steady rise culminated in the year 1848, the events of which caused a sudden but temporary fall of seventy-five per cent. in the value of precious stones.

The tremendous effect of a commercial crisis is shown, for example, by the statement of the traveller J. J. von Tschudi, that during the great depression in trade and exchange of 1857 and 1858 in Brazil diamonds sank to one-half their original value. At such crises the larger and more valuable stones suffer the greatest depreciation, since the demand for the smaller and more moderately priced stones does not fall away to the extent of that for the highly priced stones. The rise in price of precious stones during a period of prosperity is exemplified in a remarkable manner by their value in the sixteenth and seventeenth centuries when treasure from the rich silver mines of South America came pouring into Europe; and again after the discovery of the Californian and Australian goldfields in 1848 and the following years.

The demand for precious stones regardless of their kind, or for one kind in preference to another, is affected very powerfully by the arbitrary and capricious fashion of the moment. At times of national disaster or commercial depression the effect of fashion on the general demand for precious stones must necessarily be small, but at other times, when the demand has revived, this factor makes itself felt in a preference now for one kind of stone now for another. The most beautiful and costly stones, the diamond, ruby, sapphire, and emerald are indeed always sought after, but it is otherwise with those of less prominent beauty. The recent history of " oriental cat's-eye," a variety of chrysoberyl found in Ceylon, forms an instructive example of the way in which a stone may be suddenly brought into favour. For many years there was no demand whatever for this stone, it was stocked by no jewellers, and its value was correspondingly low. When, however, the Duke of Connaught gave a betrothal ring containing chrysoberyl to the Princess Louise Margaret of Prussia the stone became fashionable, first in England and then elsewhere. So extensively was it used that Ceylon could scarcely supply the demand, while the price of course rose very considerably. The freaks and caprices of fashion afford much scope for speculation in precious stones. As an example of this may be mentioned the acquirement by a French

Company of the so-called topaz mines in Spain. Topaz, which was formerly much worn and therefore prized, is now, in common with other yellow stones, regarded with but little favour and its price is therefore low. The mines mentioned above had been acquired in the hope that sooner or later the topaz may regain its former popularity; it may be stated here, however, that the mineral derived from these mines is not true topaz but quartz of a beautiful yellow colour, which is frequently sold for topaz.

It will not now be surprising to learn that those precious stones, which always have been and always will be most highly prized, have in past times varied greatly in their relative value; in other words, different stones have at different periods been held in the highest esteem. According to C. W. King, to whom we are indebted for much important historical work in connection with precious stones, the diamond was the most highly esteemed of precious stones among the Romans, and also in earlier times in India; in the estimation of the Persians, however, it occupied the fifth place, following after pearl, ruby, emerald, and chrysolite. Benvenuto Cellini placed on record that in the middle of the sixteenth century ruby and sapphire were esteemed more highly than diamond, which had only one-eighth the value of ruby, this latter stone being prized above all others. The Portuguese author, Garcias ab Horto, writing at the same time (1565), placed diamond in a series of precious stones, arranged according to their value, in the third place, giving the first place to emerald, and to ruby, when clear, the second. We find a parallel to this at the present time, for diamond is to-day far exceeded in price by ruby, and is often equalled in price by emerald. These comparisons of course refer to stones of the same size and quality and, when cut, with the same perfection of form.

The value attached to a precious stone depends very largely on the size of the specimen, which is estimated from its weight. The special unit of weight almost universally used is the **carat**. This is supposed to have the weight of a seed of an African leguminous tree, known to the natives as "kuara," a species of *Erythrina* (*E. abyssinica*); the fruit of this tree when dry is characterised by its very constant weight, and is said to have been used in Africa for weighing gold. It is supposed that it was afterwards adopted in India as a standard of weight for precious stones. According to another view, the carat is the weight of a seed from the pod of the locust-tree, its name being derived from the Greek word *keration*, signifying the fruit of the locust-tree. The origins claimed for this standard of weight being so diverse, it is not surprising to find that its value, like that of the old pound and ounce, varies not inconsiderably in different countries. On an average the carat does not differ in value much from a fifth of a gram (200 milligrams), or about $3\frac{1}{6}$ English grains. The exact values in milligrams of the carat at different places are tabulated below:

	Milligrams.		Milligrams.
Amboina	197·000	Paris	205·500
Florence	197·200	Amsterdam	205·700
Batavia	205·000	Lisbon	205·750
Borneo	205·000	Frankfurt-on-Main . . .	205·770
Leipzig	205·000	Vienna	206·130
Spain	205·393	Madras	207·353
London	205·409	Livorno	215·990
Berlin	205·440		

The fractions of the carat used in weighing precious stones are $\frac{1}{2}$, $\frac{1}{4}$, $\frac{1}{8}$, &c., down to $\frac{1}{64}$, smaller fractions than these being neglected; these fractional parts of the carat are usually expressed with a denominator of sixty-four. One sixty-fourth of a carat of 205 milligrams is equal to 3·203 milligrams. The fourth part of a carat is known as a grain, being, however, not an ordinary grain but a " diamond-grain," " pearl-grain," or " carat-grain "; it is a unit but rarely used. In France 144 carats equal one ounce.

The practical inconveniences which result from the discrepancies between the weight of a carat in different countries can easily be imagined. So firmly established, however, is the use of this unit in almost all civilised countries, that there seems no prospect of replacing it by the more convenient gram of the metric system. The change to grams could be effected with no great confusion, since the weight of half a carat very nearly corresponds in all cases to 100 milligrams or $\frac{1}{10}$ gram. The gram, however, finds very little favour with dealers in precious stones, although in Germany since 1872, in Austria since 1876, and in Holland for some time, it has been the lawful unit of weight for precious stones. In 1871, and again in 1876, a syndicate of Parisian jewellers proposed that the carat should universally have the value of 205·000 milligrams, a value it has always had in Leipzig and in the Dutch East Indies. This proposal has met with considerable favour, and it is probable that before long this value for the carat will be universally accepted, and that other values will fall into disuse, since the jewel-dealers of London and Amsterdam, which are the centres of the trade, are at one with their Parisian colleagues in this matter. At the same time, it is proposed to subdivide the carat according to the decimal system instead of the present cumbrous and inconvenient division into sixty-fourths.

In England dealers in precious stones, especially the less valuable so-called semi-precious stones, sometimes make use of troy weight, as it is also employed for precious metals. An ounce troy (480 grains) = 31·103 grams = 151·707 carats (of 205 milligrams). This carat then is equal to 3·165 grains, and inversely a grain = 0·316 carat. Further, a grain troy, avoirdupois or apothecaries', is equal to 1·264 " carat-grains," " pearl-grains," or " diamond-grains "; and one " diamond-grain " is equal to 0·791 grain troy. The word grain is therefore ambiguous, and a weight given in grains is likely to lead to confusion and error, unless it is definitely known what system of units is referred to. This confusion, however, is confined to British weights, for in no other country is the grain troy used for precious stones.

Some other units of weight of little importance are in local use at certain places where precious stones are found. The more important of such units will be briefly mentioned here since they are sometimes to be found in books of travel and in old descriptions of precious stones, as well as in more recent reports of the occurrence of precious stones in various countries, and there is often difficulty in obtaining information concerning such units.

In Brazil the weight of gold and precious stones is estimated in oitavas, an oitava being $\frac{1}{8}$ ounce, and 128 oitavas going to the pound. The oitava, which corresponds in weight to $17\frac{1}{2}$ carats (sometimes given at 18), is subdivided into thirty-two vintems. Sometimes, however, the carat-grain is used as a subdivision of the oitava. Since four carat-grains are equal to one carat, one oitava is equal to 70 (or 72) grains.

While this Brazilian unit of weight has a direct connection with the carat, the unit used in India, especially in former times, is quite independent and distinct from it. This Indian unit varies in different localities, and has also varied in value at different periods. The Indian unit of weight, used principally in Sambalpur, is the masha; it is subdivided into eight ratis, a rati being the weight of the scarlet and black bead-like seed of the plant *Abrus precatorius*; the rati is itself subdivided into four dhans. The value of a rati varies at different places and times between 1·86 and 2·25 grains troy. On an average, therefore, one rati = 2 grains troy = $2\frac{1}{2}$ carat-grains = about $\frac{2}{3}$ carat. In Nàgpur in 1827 one rati was actually equal to 2·014 grains troy; but at the present day it is usually equal to $1\frac{7}{8}$ or 1·88 grains troy, which again corresponds to 2·370 carat-grains or diamond-grains. Tavernier reckoned the value of the rati at $\frac{7}{8}$ carat. The unit of weight at Golconda (Raolconda, Kollur, and Visapur) is the mangelin; this Tavernier valued at $1\frac{3}{8}$ carat.

The miscal is a Persian weight, equivalent to forty ratis; it is usually taken to correspond to $74\frac{1}{2}$ grains troy. Two miscals make one dirhem.

The value of each kind of precious stone varies with the size of the specimen, but in some cases the increase in value is not directly proportional to the increase in size. Some stones, such as topaz, aquamarine, &c., occur frequently in fairly large masses ; of these there is therefore no more difficulty in obtaining a large cut specimen than there is in acquiring one of small size. The value of the stone will then vary directly as the weight, so that a specimen of double the size will cost twice as much. It is otherwise, however, with stones such as diamond and ruby, large specimens of which occur much less frequently than do smaller stones. The latter are more abundant, larger stones are comparatively few, while very large specimens are of great rarity, and cannot be produced when demanded, but must be waited for until they happen to be found. The ratio of the increase in price of such stones is higher than that of their increase in weight ; thus, if the weight of a stone is doubled its value will be more than doubled.

A rule was formerly given by which the price of large specimens of costly stones, and especially of diamond, could be arrived at. This rule, having originated in India, is known as the Indian rule ; it is also referred to as Tavernier's rule, because of its introduction into Europe by the French traveller Tavernier, who travelled as a dealer in precious stones in India and the East in the seventeenth century, his famous *Six Voyages* being published in 1676. It has, however, been pointed out by Schrauf that the rule had been made known in Europe almost a hundred years previously (1598) by the English traveller Lincotius, and that its mention in one of the oldest and most famous books on precious stones, the *Gemmarum Historia*, by Anselm Boetius de Boot, published in Hanover in 1609, was derived from this source.

According to this rule the price of a diamond which exceeds one carat in weight is obtained by squaring its weight in carats, and multiplying by the price of a stone of one carat. If, for example, the price of a stone weighing one carat is £10, then the price of one weighing five carats would be $5 \times 5 \times 10 = £250$. In general, if the price of the carat-stone is p, and the weight of the stone to be valued is m carats, then its price is given by $m \times m \times p = m^2p$.

This rule has, however, never been generally adopted anywhere ; it merely serves to give a rough approximation to the value of large diamonds. In former times the price of smaller diamonds, as given by this rule, was fairly correct and agreed very closely with their actual market value ; later, however, it could not be applied even to stones of moderate size, since it gave them a price higher than that for which they could actually be sold. This disproportion is even greater in the case of larger stones; the original rule has therefore, following the Brazilian diamond dealers, been modified, so that instead of taking the price of a carat-stone equal in quality to the one to be valued, the price of a carat-stone of inferior quality is taken as the multiplier p. Even with this modification the calculated value does not completely agree with the actual market value. Later, Schrauf in 1869 suggested another rule for estimating the value of large diamonds ; here half the weight in carats of the stone is multiplied by its whole weight plus two, and this by the price of a carat-stone. According to this rule, the value of a stone of 5 carats, the value of the carat-stone being taken at £10 as before, would be $2\frac{1}{2} \times 7 \times 10 = £175$, or, in a general expression, $\frac{m}{2} \times (m + 2) \times p = \left(\frac{m^2}{2} + m\right) p.$ At the time this rule was promulgated, it gave results which closely approximated to market prices; since the discovery of the South African diamond-fields, however, large stones have come into the market in much greater numbers than previously, so that even this rule is no longer applicable for trade purposes. The subject of price will be again referred to when we come to treat specially of diamonds and other precious stones.

We must now consider the relative value of cut and uncut stones. A cut stone will naturally be more expensive than a rough stone of the same quality and size. To the value of the rough stone must be added the cost of cutting; and this, especially in the case of the harder stones, and most of all in the diamond, is very considerable. Furthermore, a considerable portion, often one-half or more, of the material of the rough stone is lost in the process of cutting; a cut stone is, therefore, in its rough condition, often double the weight of the same stone when faceted, and this larger weight is taken into account when the stone is sold.

The particular form in which a stone is cut is also an important factor in determining its price, since the cutting of more complicated forms with numerous facets is more expensive than that of simpler forms with fewer facets. Thus the price of a rose diamond of the best quality is only about four-fifths of that of a brilliant of the same weight and quality.

The value of a precious stone varies to a very great extent according as the features on which its particular beauty depends are strongly marked or insignificant; and in this connection small differences of quality, scarcely noticeable to an unpractised eye, are all taken into account. A diamond of the second water, cut in the form of a brilliant and weighing one carat, is usually considered to be two-thirds the value of a similar stone of the first water. Further information of this kind may be obtained from the table published in 1878 by Vanderheym, which is given below in the section dealing with the value of diamonds. The presence or absence of the various faults to which precious stones are liable, and which have been already considered, of course affects the value of a stone to a very large extent, the presence of a large number of faults sometimes rendering an otherwise costly stone absolutely worthless.

III. CLASSIFICATION OF PRECIOUS STONES.

In the present section we propose to consider the various systems of nomenclature and classification adopted for precious stones.

In scientific mineralogy, precious stones are regarded simply as minerals and are classified accordingly. The classification of precious stones adopted by jewellers, however, is more or less arbitrary in nature and differs somewhat widely from the system used by mineralogists. Both scientific mineralogists and dealers in precious stones nevertheless agree in bringing together as of one kind all those stones which resemble each other in their essential characters, and in distinguishing by a special name the stones of that kind from those of another kind. In arranging precious stones into such kinds, the characters which are considered essential by a specialist in jewels may not be so considered by a mineralogist; and, conversely, what a mineralogist considers an essential feature may not have the same importance in the classification of a precious stone specialist. From the mineralogical point of view, a species is defined by the chemical composition and the crystalline form, together with the several physical characters of the stones it embraces. These characters must be constant for the same species, or at least vary within certain limits in a certain definite manner ; characters which are liable to vary in different specimens, such, for instance, as colour, are regarded as non-essential. It is far otherwise, however, in the case of precious stones ; here the application of a particular specimen depends largely on its colour, hence this character plays an important part in the grouping of precious stones according to the second method of classification, while characters, such as chemical composition and crystalline form, having but slight influence on the application of a stone for ornamental purposes, are much less relied upon.

Owing to these differences in the principles of classification, it is easy to understand that many stones which may be brought together in a mineralogical classification, on account of their similar chemical composition and crystalline form, under the same species, and may be known by the same name, may, in the artificial system of classification, be divided among several groups and be known by different names on account of differences in colour. On the other hand, stones of the same colour, which a jeweller may consider of the same kind, and to which he may apply but one name, or at least one with a qualifying prefix signifying small differences of colour or hardness, may by a scientific mineralogist be grouped under different species according to their chemical and crystallographic differences and be recognised by different names.

A good example of the first case is afforded by the mineral species corundum. Mineralogists include in this species all those stones which are composed of pure alumina and which crystallise in the hexagonal system. The stones of this species are all of the same hardness (H = 9) and specific gravity (sp. gr. = 4), while other physical characters are equally constant. It is therefore in accordance with the principles of scientific classification that such stones should be grouped in the same species and be known by the same name. Different specimens of this species of mineral, however, may differ widely in colour ; red, blue, yellow, green, yellowish-green, greenish-blue, yellowish-red, violet, and colourless specimens having all been found. All the colour-varieties mentioned above are not of equal importance for purposes of ornament, but they are considered by the jeweller, in spite of their mineralogical identity, as distinct and separate stones, and as such are distinguished by special names ; these names are given below in the order in which the colour-varieties were mentioned above : ruby, sapphire, " oriental topaz," " oriental emerald," " oriental chrysolite," " oriental aquamarine," " oriental hyacinth," "oriental amethyst," white sapphire (leuco-sapphire). The mineral beryl is another case in point ; the mineralogist includes the deep green, bluish-green, greenish-blue, and yellow specimens in the same species, to which he gives the name beryl, since they all agree in chemical composition and crystalline form, and differ only in colour ; the jeweller, on the contrary, refers to the deep green variety as emerald, to the greenish-blue and bluish-green varieties as aquamarine and to the yellow varieties as beryl.

Another example of the method of classification adopted by jewellers may be given. All light greenish-yellow to yellowish-green transparent stones, whatever may be their chemical composition and crystalline form, are referred to by jewellers as chrysolite. Thus this name comes to include such essentially dissimilar minerals as olivine, chrysoberyl, idocrase, corundum, and the peculiar moldavite or bottle-stone. To distinguish these one from another, such descriptive terms as olivine-chrysolite, opalescent chrysolite for chrysoberyl, " oriental chrysolite " for yellowish-green corundum, &c., are used. The original signification of the prefix " oriental " has already been explained.

Minerals used as precious stones may be classified into groups according to various systems : thus their position may be decided by the feature on which their beauty depends ; by their essential mineralogical characters ; by the frequency of their occurrence in nature ;

or more often, according to their value. Frequently they are divided into two main groups, the true precious stones or jewels, and the semi-precious stones. K. E. Kluge, in his *Handbuch der Edelsteinkunde*, published in 1860, distinguishes five groups of precious stones, characterised by their value as gems, their hardness, optical characters, and rarity of occurrence. Other methods of grouping are, of course, equally possible. There are no sharp lines of division between such groups, which are to a certain extent arbitrary, and there are many stones which would be placed by one authority in one group, and by another authority in another group. As an example of a possible method of grouping, the following **classification by Kluge**, in which the stones are arranged according to their market value, may be given.

1. TRUE PRECIOUS STONES OR JEWELS.

Distinguishing characters are: great hardness, fine colour, perfect transparency, combined with strong lustre (fire), susceptibility of a fine polish, and rarity of occurrence in specimens suitable for cutting.

A. Gems of the First Rank.

Hardness, between 8 and 10. Consisting of pure carbon, or pure alumina, or with alumina predominating. Fine specimens of very rare occurrence and of the highest value.

1. Diamond.
2. Corundum (ruby, sapphire, &c.).
3. Chrysoberyl.
4. Spinel.

B. Gems of the Second Rank.

Hardness, between 7 and 8 (except precious opal). Specific gravity usually over 3. Silica a prominent constituent. In specimens of large size and of fairly frequent occurrence. Value generally less than stones of group *A*, but perfect specimens are more highly prized than poorer specimens of group *A*.

5. Zircon.
6. Beryl (emerald, &c.).
7. Topaz.
8. Tourmaline.
9. Garnet.
10. Precious Opal.

C. Gems of the Third Rank.

These are intermediate in character, between the true gems and the semi-precious stones. Hardness, between 6 and 7. Specific gravity usually greater than 2·5. With the exception of turquoise, silica is a prominent constituent of all these stones. Value usually not very great; only fine specimens of a few members of the group (cordierite, chrysolite, turquoise) have any considerable value. Specimens worth cutting of comparatively rare occurrence, others fairly frequent.

11. Cordierite.
12. Idocrase.
13. Chrysolite.
14. Axinite.
15. Kyanite.
16. Staurolite.
17. Andalusite.
18. Chiastolite.
19. Epidote.
20. Turquoise.

2. SEMI-PRECIOUS STONES.

These have some or all of the distinguishing characters of precious stones, but to a less marked degree.

D. Gems of the Fourth Rank.

Hardness 4–7. Specific gravity 2–3 (with the exception of amber). Colour and lustre are frequently prominent features. Not as a rule perfectly transparent ; often translucent, or translucent at the edges only. Wide distribution. Value as a rule small.

21. Quartz.
    A. Crystallised quartz.
        a. Rock-crystal.
        b. Amethyst.
        c. Common quartz.
            α. Prase.
            β. Avanturine.
            γ. Cat's-eye.
            δ. Rose-quartz.
    B. Chalcedony.
        a. Chalcedony.
        b. Agate (with onyx).
        c. Carnelian.
        d. Plasma.
        e. Heliotrope.
        f. Jasper.
        g. Chrysoprase.

C. Opal.
    a. Fire-opal.
    b. Semi-opal.
    c. Hydrophane.
    d. Cacholong.
    e. Jasper-opal.
    f. Common-opal.
22. Felspar.
    a. Adularia.
    b. Amazon-stone.
23. Labradorite.
24. Obsidian.
25. Lapis-lazuli.
26. Haüynite.
27. Hypersthene.
28. Diopside.
29. Fluor-spar.
30. Amber.

E. Gems of the Fifth Rank.

Hardness and specific gravity very variable. Colour almost always dull. Never transparent. Low degree of lustre. Value very insignificant, and usually dependent upon the work bestowed on them. These stones, as well as many of the last group, are not faceted, but worked by the ordinary lapidary in the large-stone-cutting works.

31. Jet.
32. Nephrite.
33. Serpentine.
34. Agalmatolite.
35. Steatite.
36. Pot-stone.
37. Diallage.
38. Bronzite.
39. Bastite.
40. Satin-spar (calcite and aragonite).
41. Marble.
42. Satin-spar (gypsum).

43. Alabaster.
44. Malachite.
45. Iron-pyrites.
46. Rhodochrosite.
47. Hæmatite.
48. Prehnite.
49. Elæolite.
50. Natrolite.
51. Lava.
52. Quartz-breccia.
53. Lepidolite.

Among the stones enumerated above are a few such as marble, alabaster, &c., which are never worked for personal ornaments, but only for other decorative objects ; these stones will not be considered in the present book. On the other hand, there are certain stones omitted from Kluge's list which will receive attention here, although they are but rarely applied to the use of personal ornament. In the description which is now to follow, the different precious stones are not arranged in classes, but are dealt with one after another in the order of their relative value, combined to some extent with mineralogical characters. Stones belonging to the larger families of minerals are placed in juxtaposition, although individual members of each group may differ considerably in value. The following is a tabular review of the precious stones here dealt with, and the order in which they are taken.

Arrangement of Precious Stones adopted in the present work.

Diamond.

Corundum.

   Ruby, Sapphire including star-sapphire and white sapphire, "Oriental aquamarine," "Oriental emerald," "Oriental chrysolite," "Oriental topaz," "Oriental hyacinth," "Oriental amethyst," Adamantine-spar.

Spinel.

   "Ruby-spinel," "Balas-ruby," "Almandine-spinel," Rubicelle, Blue spinel, Ceylonite.

Chrysoberyl.

   Cymophane ("Oriental cat's-eye"), Alexandrite.

Beryl.

   Emerald, Aquamarine, "Aquamarine-chrysolite," Golden beryl.

Euclase.

Phenakite.

Topaz.

Zircon.

   Hyacinth.

Garnet Group.

   Hessonite (Cinnamon - stone), Spessartite, Almandine, Pyrope (Bohemian garnet, "Cape ruby," and Rhodolite), Demantoid, Grossularite, Melanite, Topazolite.

Tourmaline.

Opal.

   Precious opal, Fire-opal, Common opal.

Turquoise.

Bone-turquoise.

Lazulite.

Callainite.

Olivine.

   Chrysolite, Peridote.

Cordierite.

Idocrase.

Axinite.

Kyanite.

Staurolite.

Andalusite.

   Chiastolite.

Epidote.

Piedmontite.

Dioptase.

Chrysocolla.

Garnierite.

Sphene.

Prehnite.

Chlorastrolite

Zonochlorite.

Thomsonite.

   Lintonite.

Natrolite.

Hemimorphite.

Calamine.

Felspar Group.

   Amazon-stone, Sun-stone, Moon-stone, Labradorescent felspar, Labradorite.

Elæolite.

Cancrinite.

Lapis-lazuli.

Haüynite.

Sodalite.

Obsidian.

Moldavite.

Pyroxene and Hornblende Group.

   Hypersthene (with Bronzite, Bastite, Diallage), Diopside, Spodumene (Hiddenite), Rhodonite (and Lepidolite), Nephrite, Jadeite (Chloromelanite).

Quartz.

   Crystallised quartz: Rock-crystal, Smoky-quartz, Amethyst, Citrine, Rose-quartz, Prase, Sapphire-quartz, Quartz with enclosures, Cat's-eye, Tiger-eye.

   Compact quartz: Hornstone, Chrysoprase, Wood-stone, Jasper, Avanturine.

   Chalcedony: Common Chalcedony, Carnelian, Plasma, Heliotrope, Agate with Onyx, &c.

Malachite.

Chessylite.

Satin-spar (Fibrous Calcite, Aragonite, and Gypsum).

Fluor-spar.

Apatite.

Iron-pyrites.

Hæmatite.

Ilmenite.

Rutile.

Amber.

Jet.

———

Appendix : Pearls and Coral.

SECOND PART

SYSTEMATIC DESCRIPTION OF PRECIOUS STONES

DIAMOND.

THE diamond, although not the most valuable of precious stones, yet unquestionably exceeds all others in interest, importance and general noteworthiness. It is therefore fitting that this stone should stand at the head of the series now awaiting consideration, and should, moreover, receive at our hands more detailed treatment. In hardness, in the perfection of its clearness and transparency, in its unique constants of optical refraction and dispersion, and finally in the marvellous perfection of its lustre, the diamond surpasses all other minerals. For these reasons, and despite the fact that it is not of very great rarity even in faultless specimens of fair size—nine-tenths of the yearly trade in precious stones being concerned with diamonds alone—it is very greatly valued as a gem; moreover, on account of its extreme hardness, it has several technical applications.

A. CHARACTERS OF DIAMOND.

I. CHEMICAL CHARACTERS.

Diamond is distinguished from all other precious stones no less by its chemical composition than by its unique physical characters, for no other gem consists of a single element. It is pure crystallised carbon, its substance is therefore identical chemically with the material of graphite and charcoal. The extraordinary difference in the appearance of diamond and that of other forms of carbon depends solely on the crystallisation of the material and the physical characters consequent on this.

The fact that the one and only constituent of diamond is pure carbon was already known at the end of the eighteenth century, and was suspected even earlier than this. In the year 1675 Sir Isaac Newton had arrived at the conclusion that diamond must be combustible; this conclusion, though correct in itself, was based on theoretical grounds, now known to be mistaken, connected with the high refractive index of the substance. In 1694–5 researches respecting the **combustibility** of diamond were conducted at the "Accademia del Cimento" of Florence, by the Academicians, Averani and Targioni, at the instigation of the Grand Duke Cosmos III. of Tuscany. Diamonds were exposed to the intense heat of a fierce charcoal fire or were placed in the focus of a large burning-glass. A stone so treated did not fuse but gradually decreased in size and finally disappeared, leaving behind no appreciable amount of residue. These experiments proved that the substance of diamond, as such, is destroyed at a high temperature; whether its disappearance was due simply to volatilisation, as in the case of sal-ammoniac, was of course undecided at that early date. Investigations into the chemical nature of diamond and the meaning of its apparent destruction when exposed to heat were undertaken later by the famous French chemist, Lavoisier, as well as by Tennant, Davy and others.

During the year 1772 and later, Lavoisier, the founder of modern chemistry, demonstrated that the disappearance of diamond only took place when it was heated in air, and that it might be exposed to the highest temperatures without loss of weight, provided that any contact with air was prevented. He further showed that the air occupying the space in which a diamond had been heated, and in which it had finally disappeared, possessed the property of turning lime-water milky, as does carbon dioxide (carbonic acid gas); and, moreover, that the lime-water so clouded effervesced when brought into contact with an acid, just as it does when clouded by the addition of carbon dioxide. The consideration of these facts led him to repeat his experiments, replacing diamond with ordinary carbon; the results were found to be identical, and there was nothing for it but to conclude that the disappearance of diamond was due to combustion. In spite of these apparently conclusive experiments, Lavoisier did not at that time venture to assert that the substance of carbon and of diamond was completely identical.

This was left to be proved by Smithson Tennant, who in 1797 demonstrated that the combustion of a certain weight of diamond resulted in the production of the same amount of carbon dioxide as did the combustion of an equal weight of pure carbon. This observation was confirmed later by other chemists, for example by Sir Humphry Davy, who in 1816 showed in addition that the combustion of diamond was unattended by the formation of even a trace of water. This proved that the conclusions of Arago and Biot, namely, that diamond, on account of its high refractive power, must contain a hydrocarbon, were incorrect. Later, all these results were confirmed by the well-arranged experiments of Dumas and Stas, as well as of Erdmann and Marchand, and others. The combustion of diamond in oxygen gas has now long been an every-day chemical lecture experiment.

These researches have been considered for some time to have finally settled the question as to the constitution of diamond. Recently, however, Krause has suggested that this question should rather be regarded as still an open one, and that the experiments which have been described should be taken to prove simply that the atomic weight of the element of which diamond is composed is identical with that of carbon. He has further suggested that between the two, diamond and ordinary carbon, there might possibly be a relation similar to that existing between the metals nickel and cobalt, which have the same, or very nearly the same, atomic weight and very similar chemical characters. In order to decide this point, Krause allowed the gases produced on the one hand by the combustion of diamond, and on the other by that of pure carbon, to be absorbed by caustic soda. In both cases he obtained crystals: in the one case these were, of course, crystals of sodium carbonate; in the other, the crystals produced agreed so completely with the crystals of sodium carbonate in crystalline form, amount of water of crystallisation, specific gravity, fusibility, solubility, electrical conductivity, &c., that there could be no reasonable doubt of the identity of the two products. This experiment, then, proves definitely and conclusively that the product of combustion of diamond is carbon dioxide, and that the substance of diamond consists, therefore, of pure carbon.

A century previous to the work of Krause, Guyton de Morveau had made experiments with the idea of confirming or overthrowing the results of Lavoisier and Tennant, thinking it inconceivable, as did the majority of his contemporaries, that the rare and costly diamond and a common and widespread substance such as carbon could consist of one and the same chemical element. The method he adopted for the purpose differs from the usual methods of chemical analysis, and is interesting on account of its originality. It depends on the fact that soft bar-iron, when heated with charcoal, takes up a certain amount of carbon and becomes converted into steel. In his experiment, Guyton de Morveau replaced charcoal with diamond, and succeeded in converting the soft iron into steel, the characters of which were identical with those of steel produced by the ordinary process. His experiments thus supplied further proof of the chemical identity of carbon and diamond.

The behaviour of diamond, when raised to a high temperature, varies according to whether it is in contact with air or not. In both cases, however, the stone will be easily cracked or fractured if the rise of temperature is too sudden; such damage to the stone may be avoided by ensuring that both the heating and the subsequent cooling shall be slow and gradual.

In a stream of oxygen gas, a crystal of diamond begins to burn at a low red-heat. It will gradually rise in temperature until it reaches a white-heat, and will then burn uninterruptedly with a pale blue flame, even after the source of heat, such as a gas-flame, applied at first for the purpose of raising the temperature to the point of combustion, has been removed. The crystal gradually decreases in size, and finally disappears, the flame at the last moment often flickering brightly like that of an expiring lamp flame. The combustion of diamond proceeds gradually from the exterior inwards; it is unattended by fusing, or, indeed, by any great alteration in the general form of the crystal, or in the physical characters of its substance, the material of the inner portion remaining unaltered during the combustion of the exterior.

It has been mentioned above that the combustion of diamond in a current of pure oxygen will proceed even when the source of heat is withdrawn; it is otherwise, however, when the diamond is burning in atmospheric air. Should the source of heat in this case be removed combustion will cease, owing to the fact that the oxygen of atmospheric air is largely diluted with nitrogen, a gas which does not support combustion. In the one case the heat evolved during combustion is sufficient to keep the stone above the temperature of ignition, while in the other it is not.

The temperature to which a diamond crystal must be heated in the air before combustion is started, is higher than the **temperature of ignition** in pure oxygen. According to Lavoisier it is a little lower than the melting-point of silver, this being fixed at 916° C. Moissan has recently determined the temperature of ignition of diamond in oxygen at 690° to 840° C. Small crystals are more easily induced to burn than are larger ones; according to Petzholdt, small diamonds placed on platinum foil, heated from below with a blowpipe flame, disappear in a very short time, the whole experiment occupying but a few minutes. Diamond dust burns with greater ease and rapidity the greater its fineness; thus powder of an extreme degree of fineness, when heated on platinum foil over the flame of an ordinary spirit-lamp, burns almost instantaneously with a brilliant glow. Whether in a finely divided condition or not, diamond burns much more easily than does the other crystallised modification of carbon, namely, graphite.

The oxidation of diamond powder, that is, its chemical union with oxygen, takes place with comparative ease if it is mixed with saltpetre, and the mixture then fused. The necessary oxygen is supplied by the decomposition of the saltpetre, and the diamond powder is very quickly burnt up. Diamond powder is also easily oxidised when heated at 180° to 230° C. with a mixture of potassium chromate and sulphuric acid. Diamond resists the action of such powerful chemical reagents as caustic potash, hydrofluoric acid, concentrated sulphuric acid, aqua regia (a mixture of hydrochloric and nitric acids), a mixture of sodium chlorate and nitric acid, iodic acid, and other energetic solvents. Few other substances resist the action of these reagents in the way diamond does ; it will remain in them unaltered even at high temperatures.

On examining a partially burnt crystal of diamond, it will be seen that its edges and corners are more or less rounded, and that its faces are no longer brilliant but dull, rough, and scarred.   On crystals which are bounded by faces of the octahedron, special markings are seen when such faces are examined with a lens or, better still, with a microscope. These markings are regular triangular depressions like inverted pyramids, the bases being equilateral triangles, of which the edges are in all cases parallel to the octahedral edges of the crystal, as is shown in Fig. 31 *r*.   The direction of these triangular pits is the reverse of that of the natural depressions of diamond crystals, shown in Fig. 31 *q*, *n*, and *o*.   Such pits may occur singly on the face of the crystal, or they may be close together and in large numbers ; they are of precisely the same character as etched or corrosion figures, such as may be produced on the faces of other crystals by the action of fused alkalis, or of solvents such as water and acid.   These depressions may indeed in the present case be regarded as corrosion figures, since they are produced by heating the diamond crystal in air or with saltpetre, the etching agent being then hot oxygen gas or fused saltpetre.   The production of etched figures is due to the unequal action of the oxygen over the surfaces of the crystal, the crystal being attacked first at isolated points on the faces, where the material is slowly consumed.

As has been previously mentioned, a diamond heated away from contact with air undergoes no diminution in weight.   The experiment may be performed by packing the diamond in charcoal powder in a closed crucible and heating it in an electric furnace.   The temperature may be as high as the furnace can produce, and may be maintained for any length of time, yet the diamond will still remain unaltered in weight, since in the absence of oxygen there can be no combustion of its substance.   A prolonged exposure to great heat does, however, produce other changes ; the surface of the diamond becomes blackened and soft enough to leave a mark behind it when rubbed on paper.   This is due to the **transformation to graphite** of the surface material ; this change in state of the substance of diamond to the other crystallised modification of carbon, namely graphite, is however, brought about only when a very high temperature is reached.   According to G. Rose, who specially investigated the point, a diamond heated out of contact with air undergoes no change whatever when heated to the temperature at which cast-iron melts, nor even when exposed to the fiercest heat of a porcelain furnace.   At higher temperatures, about that at which bar-steel melts, or in the electric furnace, a superficial blackening and conversion into graphite begins ; if the exposure to the temperature is prolonged sufficiently, this conversion proceeds until the whole of the substance of the diamond is changed into graphite, the original external form of the crystal being, however, still retained.

The behaviour of diamond with respect to its alteration into graphite when heated under other conditions, namely, in the presence of air, has not yet been thoroughly investigated.   While in some experiments no blackening, even at the highest temperatures, has been observed, in others this phenomenon has been seen ; in these cases, however, the blackening may have been due to a sooty coating derived from the burning material which supplies the source of heat and not to the alteration of diamond into graphite.   Many observers, including Lavoisier, have noticed that black spots are formed on the surface of a diamond undergoing combustion.   When combustion is again allowed to proceed these spots may disappear, or they may be still apparent at lower depths as the outer parts of the stone slowly burn away.   According to Rose, no alteration into graphite takes place when a diamond is heated or burnt in a muffle or before the blowpipe, or perhaps even before the oxyhydrogen flame.   When placed in the focus of a concave mirror, however, or when subjected to electric sparking, such a change is observed, but in neither of these cases can the blackening possibly be due to a sooty coating derived from the vapours of the source of heat. It is stated by Jaquet that a diamond placed in the electric arc given by one hundred Bunsen cells softens and becomes converted into a coke-like mass, the specific gravity of which is 2·678, while that of the diamond experimented upon was 3·336. He observed further, that whereas the material of diamond is a bad conductor of electricity, when converted into coke or graphite it becomes a good conductor.

Similar observations have been made by Gassiot, who has stated that before the alteration of diamond into a coke-like mass it softens and has the appearance of a body about to melt. Other statements respecting the melting of diamond, or of appearances referable to this change of state, are to be found in scientific literature. Berzelius reports that he observed a bubble on the surface of a burning diamond, and Clarke saw bubbles on the surface of a diamond when strongly heated in the oxyhydrogen flame. Other observers, on the other hand, under exactly similar conditions, have failed to notice any appearance of the kind, and in the absence of unanimous testimony it is still doubtful whether diamond does really fuse at high temperatures. Observations of the kind quoted are not altogether free from error; the rounding of the edges and corners of a partly-consumed diamond crystal would give it the appearance of having been fused; the rounding is, however, due to the fact that these prominent portions burn more rapidly than do the faces. The Emperor Francis I. sought to obtain a large diamond by fusing together several small ones; the attempt was, however, a complete failure, and the diamonds were burnt.

Probably the highest temperature to which diamond has been artificially subjected was reached in the experiments of Despretz, who employed for the purpose the electric spark given by five hundred to six hundred Bunsen cells. He reported that in the absence of air the usual change into graphite took place, and that if the heating was sufficiently prolonged beads of fused material were formed. Similar beads were also obtained from other varieties of carbon, but it is possible that these consisted of the mineral residue fused into a hard mass. If, however, this were the case, it would appear that the carbon had been volatilised, since combustion could not have taken place. It is very desirable that the researches of Despretz should be confirmed by further observation.

The whole of the substance of a perfectly colourless and transparent diamond is converted, when exposed to sufficient heat, into carbon dioxide, and no residue whatever remains behind. In the case of a deeply coloured or otherwise impure diamond combustion is not so complete, a small amount of incombustible **ash** remaining behind after the diamond has been converted into carbon dioxide. This residue consists of inorganic impurities differing in chemical composition in different stones; these have been enclosed in the diamond during its growth, and are the cause of the colour or cloudiness of the stone.

This residual ash varies very considerably in amount in different stones. In the purest stones it is almost imperceptible, while in less pure stones it varies from $\frac{1}{2000}$ to $\frac{1}{500}$ of the total weight (0·05 to 0·2 per cent.). The largest amount of ash, amounting to as much as 4·2 per cent., is present in carbonado, a peculiar variety of black, porous diamond found in Brazil. These impurities are often evenly distributed throughout the crystal. Occasionally, however, they are collected together at one or more points, which then appear coloured and cloudy, the surrounding portions being colourless and transparent. Such impurities in the diamond are isolated during combustion, but are more or less altered in character by the heat, sometimes being fused into beads, as mentioned above. The uniform distribution of impurities through the whole mass of the crystal is occasionally shown by the incombustible residue remaining behind as a porous mass, and having the form of the original diamond crystal.

The ash of the diamond is of a brownish colour. It contains some yellow flakes, and sometimes a few black grains which are attracted by a magnet. Its precise character depends, of course, upon the nature of the impurity in the diamond. Occasionally a few small, transparent, crystalline grains are present in the ash which have an action on polarised light; these and kindred impurities require a microscope for their detection. Chemical examinations of the ash of diamonds show that silica and iron oxide are invariable constituents, while lime and magnesia appear also in certain cases. An analysis of the ash of carbonado has given : silica 33·1 per cent., iron oxide 53·3, lime 13·2, and a trace of magnesia.

The constituents of the ash of diamond are, as a rule, very finely divided and distributed throughout the mass of the stone, the individual particles of the impurity not being recognisable even under the strongest magnification. Particles of foreign matter are sometimes, however, large enough to be seen with a lens or even with the naked eye ; these bodies, which are referred to as **enclosures,** are isolated grains, splinters, scales, plates, needles, or fibres. They have definite sharp boundaries, and not infrequently are bounded by plane crystal-faces ; they may occur in the diamond singly or in groups.

The nature and character of large enclosures is sometimes definitely known, but more frequently this is not the case. A peculiar and rare occurrence is the enclosure of a small diamond within a larger one, the two sometimes differing from each other both in crystalline form and in colour. In some cases the smaller enclosed diamond is quite free from the larger stone, and when the latter is cleaved open, the small enclosed stone falls out uninjured and perfect. The most commonly occurring enclosures in diamond are small black grains of irregular outline ; they occur in large numbers in diamonds from all localities, and were formerly considered to consist of some carbonaceous substance. This, however, is not always the case. The black enclosures found in a diamond from South Africa by E. Cohen had the characters of hæmatite or ilmenite, and he is inclined to the opinion that all such enclosures consist of one or other of these two minerals. Many of these black grains are incombustible, and therefore inorganic ; but others, according to the observations of Friedel, are consumed with the diamond, and hence must consist of some carbonaceous substance. In a diamond from the Cape a black, viscous, asphalt-like mass has been found, and similar enclosures have been reported in a few Indian crystals. Beside the minerals already named, several others have been determined with more or less certainty to occur as enclosures in diamond ; these include, among others, quartz, topaz, rutile, iron-pyrites, which occur in the form of irregular grains or sometimes as well-developed crystals. Scales of gold have been found, though rarely, as enclosures in diamond crystals from Brazil. Vermiform aggregates of green scales are sometimes observed ; these, however, are differently interpreted by different observers : Des Cloizeaux considers them to be a kind of chlorite, while Cohen regards the green scales in Cape diamonds as some copper compound. Red enclosures are very occasionally met with in Cape diamonds ; they are of unknown nature.

Special mention must be made of enclosures of very fine green needles and fibres interwoven and matted into coil-like masses. In these and in aggregates of a similar kind, the structure of plant cells has sometimes been supposed to have been recognised ; the distinguished botanist, Göppert, indeed, held that such enclosures were undoubtedly of a vegetable nature. It has since been proved, however, that they consist of inorganic material.   There is no single case in which the vegetable nature of a diamond enclosure has been conclusively proved, although the attention of botanists has been more than once directed to this point.

All the minerals which have yet been mentioned as occurring as enclosures in diamond, must have been formed before the diamond commenced its gradual growth around them. There are other foreign bodies, however, which must have been introduced after the formation of the diamond; thus, water containing iron in solution has sometimes penetrated the cracks and fissures of a diamond, and has left a brown deposit of limonite filling up the crack or fissure.

The enclosures of diamond do not invariably consist of solid matter.   Not infrequently there exist cavities in the substance of the diamond which may be vacuous or may contain liquid; these **fluid enclosures** are, however, usually of microscopic size.   The liquid they often contain does not, as a rule, completely fill the cavity, part of the space being occupied by a bubble of gas, which is sometimes fixed in position and at other times movable, thus clearly indicating the fluid nature of the contents of the cavity.   In some cases it can be safely inferred from the behaviour of the liquid when the diamond is heated that it is liquid carbon dioxide; this point will receive further consideration, however, when we come to consider the origin of diamond.   In other cases the properties of the liquid point to its being water or a saline solution.

Other cavities in diamond are quite empty or only filled with gas; like the fluid enclosures these are by no means rare, and, when present at all, occur in large numbers. When observed under the microscope they appear quite black, especially at the visible margins; this is owing to the fact that the rays of light travelling through the stone are almost totally reflected at the surface of separation of the substance of the diamond and the bubble of gas, they therefore fail to reach the eye and the cavity appears dark.   This is a fruitful source of error, for such appearances are liable to be mistaken for solid enclosures of a black colour.   Such mistakes may be avoided, however, by careful observation, for the outline of cavities is usually rounded, while that of solid enclosures is irregular and angular; moreover, most cavities allow the passage of some light, at least in the centre; they will therefore appear to have a bright centre surrounded by a dark border, which would not be the case with solid enclosures.   The presence of these cavities is of practical significance, since to them is due the cloudiness of the diamond, and those faults which have been already considered under the name "silk."   From a theoretical point of view, they will no doubt help to throw light on the obscure question of the origin of diamond.

2. CRYSTALLINE FORM OF DIAMOND.

The diamond is one of the most perfectly crystallised of minerals.   Almost every single stone is bounded by more or less regularly developed faces.   Massive specimens without crystal-faces are scarcely ever found, and when such are met with they are, as a rule, fragments of large crystals or rounded pebbles, of which the original external crystalline form has been destroyed.   As is usually the case with embedded crystals, that is, those which have grown embedded in the mother-rock, most diamonds are bounded on all sides by crystal-faces.   Sometimes, however, irregular areas, by which the crystal might have been attached, can be made out with more or less certainty.

The faces of diamond crystals differ from those of most other crystallised minerals, in that they are, as a rule, much curved and rounded instead of being perfectly plane, as is usually the case.   This curvature is due to the mode of growth of the crystal, and not to subsequent attrition, as might be thought.   It renders the exact determination of a crystal, according to the methods of crystallography, very difficult, and for this reason many questions regarding the crystallisation of diamond are still open to debate. In what follows, the most important general crystallographic relations will be dealt with,

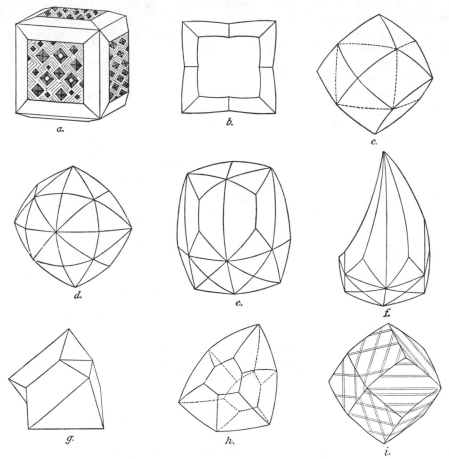

FIG. 31, *a–i.* Crystalline forms of diamond.

while special features peculiar to diamonds from particular localities will be mentioned under the description of these localities.

Observations on the crystalline form of diamond date back to the beginning of the seventeenth century, many diamond crystals having been described by Keppler, Steno, Boyle, and others. Romé de l'Isle and Haüy, the founders of scientific crystallography, were, at the beginning of the nineteenth century, the first to correctly interpret the different forms, and to determine the hemihedral development of the crystals. Great credit is also due to Gustav Rose for his exhaustive study of diamond crystals made at a later date. The results of his investigations were published in 1876 after his death by A. Sadebeck, who added numerous observations of his own.

Crystals of diamond belong to the cubic system, and, according to the views of the majority of mineralogists, to the tetrahedral-hemihedral division of this system. Certain peculiarities, however, render the hemihedrism of the crystals open to question, and some authorities prefer to consider them as holohedral.   All the typical simple forms of the cubic system have been observed in diamond crystals, either alone or in combination with other

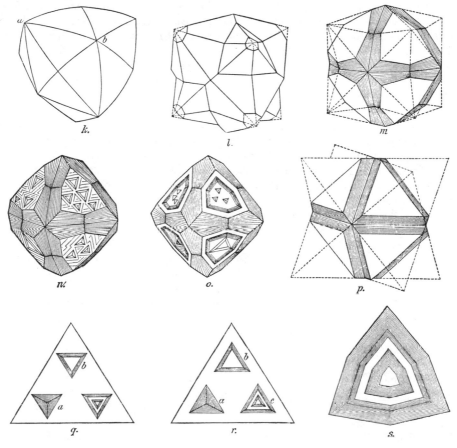

FIG. 31, *k-s*.   Crystalline forms of diamond.

forms.   Some of the more commonly occurring forms, which will be described in some detail, are shown in Fig. 31, *a–p*.

Crystals having the form of a regular cube (Fig. 31 *a*) occur very frequently, but are usually small; this habit is specially characteristic of Brazilian crystals, and is rarely met with in specimens from other localities, especially the Cape.   The faces of the cube are always dull and rough, and show a shallow depression, increasing in depth towards the centre of the face.   The roughness is due to the presence of square-based, pyramidal depressions placed diagonally on the cube face; these are usually small, but may be of fair size.   They occur more or less isolated or closely aggregated together (Fig. 31 *a*). When observed with a lens or, better still, under the microscope, the pyramidal faces bounding the shallow depressions may be distinctly seen; they are marvellously plane and smooth, but just as frequently rough and irregular, and between these two extremes all gradations have been found.   Such a cube of diamond in its matrix is shown in Plate I., Fig. 1.

In most cubes of diamond, however, each edge is replaced by two faces, as shown in Fig. 31 *a ;* the twenty-four faces thus derived, would, if produced or enlarged sufficiently, give rise to the form known as the four-faced cube, or tetrakis-hexahedron.   These faces are, however, as a rule, small ; they are dull and uneven, and are irregularly striated perpendicularly to the cubic edges.   Each face is often divided centrally by a narrow furrow, running perpendicularly to the edge and towards the centre of the cube face ; this is illustrated in Fig. 31 *b*, which shows one cubic face together with the four adjacent faces of the four-faced cube.   A crystal, bounded only by the twenty-four faces of the four-faced cube and with no cube faces, is occasionally met with in diamonds from Brazil and India ; the faces of this form are then bright, but always curved.

The cube is also frequently modified in ways other than by the replacement of its edges.   Not infrequently, for example, its eight corners are truncated by the eight faces of the octahedron.   Moreover, each of the twelve edges of the cube may be replaced by a single plane face ; these twelve truncating faces, if extended, would give the form known as the rhombic dodecahedron, which is of frequent occurrence, and is shown in Fig. 31 *c* and *i*. Its faces are sometimes plane and striated parallel to the longer diagonal (Fig. 31 *i*) ; as a rule, however, they are more or less curved, the lines of intersection of the faces being of course also curved ; in this latter case the faces are not striated but are smooth and bright (Fig. 31 *c*).   These curved faces frequently have a shallow groove running across them in the direction of the shorter diagonal, as indicated by the dotted line in Fig. 31 *c ;* the form is then, strictly speaking, no longer a rhombic dodecahedron, but approaches to that of a tetrakis-hexahedron.   The largest Brazilian diamond yet found, and known as the " Star of the South," is an irregularly developed rhombic dodecahedron ; it is shown in its rough condition in Fig. 48.   This form is frequently to be met with in Brazilian diamonds.

When the faces of the rhombic dodecahedron are grooved in the direction of the longer diagonal as well as in that of the shorter (Fig. 31 *d*), we obtain a form known as a hexakis-octahedron, bounded by forty-eight similar faces, which are always strongly curved, smooth, and bright.   The hexakis-octahedron, which is of extremely frequent occurrence in diamond, approaches, as shown in Fig. 31 *d*, the form of the rhombic dodecahedron ; at other times the same kind of form may approximate to the octahedron, each of the eight octahedral faces being replaced by six faces.   Hexakis-octahedral crystals of the diamond are frequently much distorted by elongation in one direction, as shown in Fig. 31 *e*, a still greater distortion of the same form being represented in Fig. 31 *f*.   Such distorted forms, which appear at a first glance to be quite distinct from that of Fig. 31 *d*, on a closer examination will be seen to be easily derived from that form.

Both the rhombic dodecahedron and the hexakis-octahedron are sometimes, on account of the strong curvature of their faces, almost spherical in shape.   Formerly, when the principal localities for diamond were Brazil and India, the spherical form was known as the *Brazilian type*, and the octahedral form as the *Indian type*.

Occasionally, only half the faces of the hexakis-octahedron are developed, namely, those occupying alternate octants.   The hemihedral form so derived is that of the hexakis-tetrahedron, shown in Fig. 31 *k*.   The faces of this form, which is of rare occurrence, are always curved, smooth, and bright.   If the symmetry of the diamond is really tetrahedral-hemihedral, the complete hexakis-octahedron may be regarded as a combination of two hexakis-tetrahedral forms, but the faces in adjacent octants would then have different surface characters, and this has not hitherto been observed.

A regular or **twin intergrowth** between two hemihedral crystals frequently takes place and results in the production of a holohedral form. The twin intergrowth of two hexakis-tetrahedra gives rise to the twinned crystal, shown in Fig. 31 *l*, in which for the sake of clearness the edges are represented as straight instead of curved lines. The two crystals interpenetrate at right angles, and the sharp corners, *a* (Fig. 31 *k*), of one individual project from the obtuse corners, *b*, of the other, the faces of the two interpenetrating individuals thus forming re-entrant angles. The sharp projecting corners of such a group are not always present, usually being truncated, as is indicated in the figure ; the truncating faces of each individual belong to the tetrahedron, and are never curved but always perfectly plane. The truncation, shown in Fig. 31 *l*, is only slight, while that of Fig. 31 *m*, is more pronounced. These eight truncating faces together complete the octahedron, the faces of which are plane, as is shown in *m* and *n* of Fig. 31, while its edges are replaced by re-entrant grooves, formed, as explained above, by the interpenetration of two hexakis-tetrahedra. On observing the figures it will be seen that these grooves are striated in the direction of their length. The size of the grooves depends on the degree to which the corners of the hexakis-tetrahedra are truncated by the faces of the tetrahedra. When the truncation is a maximum the grooves will be completely absent; but an octahedron of diamond in which such re-entrant grooves are not to be seen is a rarity. An octahedron of diamond with the sharp edges of the geometrical form must be considered to be the same as shown in Fig. 31, *m* and *n*, in which the truncation has quite obliterated the grooves ; in other words, it is the limiting form of such twinned crystals.

This twin intergrowth is simpler when the two individuals are tetrahedra instead of hexakis-tetrahedra, as in the case considered above. Such a twin-crystal is shown in Fig. 31 *p*, where the projecting portions, removed by truncation, of the two interpenetrating tetrahedra are represented by dotted lines. Here the grooves are quite straight, and of the same width throughout, and they do not show the nick in the middle as in the previous twinned form, in which also the grooves widen out away from this nick.

This simpler twinned form, consisting merely of two interpenetrating tetrahedra, is, however, of very rare occurrence in diamond. On the other hand, the form consisting of two interpenetrating hexakis-tetrahedra, as shown in Fig. 31 *m*, is very characteristic of diamond, and is of frequent occurrence. This figure has therefore been drawn again in Fig. 31 *n*, the dotted lines having been omitted and the characteristic markings on the faces inserted. The small faces of the hexakis-tetrahedra, which form the re-entrant grooves due to the twinning, are always somewhat curved and exhibit a delicate striation in the direction of their length. A slightly different form of such an interpenetrating twin of octahedral habit is shown in Fig. 31 *o* ; this also is a frequently observed form of diamond crystal. Here the edges of the octahedron have, in place of grooves, two small planes meeting at a very obtuse angle in a short edge at the middle of, and perpendicular to, the octahedral edge ; and away from this short edge formed by their mutual intersection they gradually widen out. These small planes are curved and finely striated, as shown in the figure, the octahedral planes being as before perfectly plane.

The twinned forms just described (Figs. 31 *m*, *n*, *o*, *p*,) are very characteristic of diamond, and they constitute the octahedral or Indian type. Crystals of this kind, of which one in its matrix is represented in Plate I., Fig. 2, are sometimes known in the trade as " points."

It has already been pointed out above that while the faces of the rhombic dodecahedron and of the hexakis-octahedron show a convex curvature, those of the octahedron are plane and even. The octahedral faces are, however, characterised by the presence of striations and pits, both of which are repeated on the surface with definite regularity and have a definite orientation. The striations are parallel to the symmetrical six-sided outline of the octahedral faces, as shown in Fig. 31 o, for the whole crystal, and in Fig. 31 s, for a single face. They may be either coarse or fine in character, and may be present in small or large numbers. The portions of the face of the octahedron between the striations are often very smooth and bright. These striations are due to the fact that the octahedral face is raised by very low steps towards its centre, each step having the same sharp outline as the margin of the octahedral face itself. It is as if numerous very thin plates, all of the same shape but gradually diminishing in size, had been piled, with their centres exactly superposed, on the octahedral face, so that each layer forms a step, and so a line of the striation.

The triangular pits are regular pyramidal depressions, of which the bases are equilateral triangles. They are usually small and often only to be seen distinctly under the microscope. The pyramidal faces inside the pits are finely striated and may terminate in the apex of the pyramid, as shown at a in Fig. 31 q ; or they may not extend to such a depth into the interior of the crystal, the apex of the pyramid being then truncated by a triangular face parallel to the face of the octahedron, as at b, Fig. 31 q. Sometimes on this inner face there is a smaller pyramidal depression as at c, Fig. 31 q. These depressions are of the same general character as those produced on the octahedral faces of a diamond during its combustion ; but while the corners of the pits of natural origin are adjacent to the octahedral edges (Fig. 31 n, o, q), this position is occupied by the sides of the pits produced by etching (Fig. 31 r) ; thus the two positions are the reverse of each other. The pits occur singly or in large numbers, and the striations may or may not be also present on the same face (Fig. 31 o, n).

Beside, the twin-crystals formed by the interpenetration of two hemihedral crystals, illustrated in Fig. 31, l to p, diamond presents still another type of twin-crystal, which is illustrated in Fig. 31, g to i. Here two octahedral or rhombic dodecahedral crystals are united together along a face of the octahedron. Fig. 31 g, shows two octahedra symmetrically united in this manner, the two individuals having one octahedral face in common. This kind of twin-growth is frequent in diamond but still more so in the mineral spinel, so that the law which governs this kind of twinning is referred to as the spinel twin-law. At the line of junction of the two individuals, three re-entrant angles alternate with the same number of salient angles. These spinel twins of diamond, which are known to the diamond-cutters of Amsterdam as "naadsteenen" (suture-stones), are very frequently flattened in a direction perpendicular to the common octahedral plane; they are, indeed, sometimes reduced to mere thin plates, but the faces and edges always have the surface characters described above.

Very frequently two rhombic dodecahedra or two hexakis-octahedra are twinned according to the same law on a face of the octahedron ; that is, the two individuals have a face in common which occupies the position of an octahedral face, and about which they are symmetrical. These twin-growths also are much compressed in a direction perpendicular to the twin-plane ; this is illustrated in Fig. 31 h, which represents a lenticular or heart-shaped crystal with curved faces. In this crystal only six faces of each of the hexakis-octahedra are developed, and these form low six-sided pyramids with a common base parallel to the six-sided octahedral face shown in the figure.

Fig. 31 i represents another kind of twin growth of rarer occurrence, in which the crystal has the form of a rhombic dodecahedron (Fig. 31 c). Parallel to one or more of the possible faces of the octahedron, which if present would truncate the corners in which three edges meet, are very thin lamellæ in twin positions to the main crystal. Large numbers of these twin-lamellæ may be present, and give rise to striations on the faces of the crystal.

Striations due to the same cause may also be present on the faces of the hexakis-octahedron, where, as before, they are parallel to one or other of the octahedral faces.

All these twin-groupings are quite regular and conform to certain definite crystallographic laws. Other intergrowths of two or more diamond crystals may be met with, in which the grouping is irregular and accidental, and cannot be referred to any general rule, the relative positions of individual crystals being determined by chance. In such intergrowths may be found small crystals growing singly on a larger one, or several crystals of more or less equal size may be united in an irregular group. Such groups are unsuitable for cutting as gems and are usually devoted to technical purposes; the same is true to a certain extent in the case of the twinned crystals above described. Irregular groupings of diamond crystals may, in a crystallographic sense, be referred to as **bort**; in the technical sense, however, the term bort includes all stones which, from some reason or another, are unfit for use as gems; and this term is even applied to simple crystals disfigured by some serious fault, such as imperfect transparency, bad colour, &c.

Bort occurs in a peculiar spherical form, being built up of a large number of small crystals radially arranged, so that the whole group takes the shape of a more or less perfect sphere (Plate I., Fig. 3). Numerous small points project from the surface of the sphere, these being the corners of the individual crystals which form the group. These spheres of bort are found in all diamond mines to the extent of from two to ten per cent. of the total output. Not infrequently only the outer shell of the sphere has the radially fibrous character just described, the central portion being occupied by a large, regularly-formed single crystal, which is usually so loosely attached to the radially crystalline shell that it falls out when the latter is broken.

Massive diamond with a granular crystalline structure and a black colour is known as **carbonado** or "carbonate" (Plate I., Fig. 4). Since it is applied to technical purposes only, it may be regarded in this sense as bort. It is found almost exclusively in the State of Bahia in Brazil; its characters will be further described when the occurrence of diamonds at this locality is under consideration.

Size of diamond crystals.—The size of diamond crystals varies between somewhat wide limits. The smallest which come into the market sometimes measure less than a millimetre in diameter, but still smaller specimens occur in nature. Small stones, measuring not more than one-quarter or one-third of a millimetre along the edge may be separated from a parcel of Brazilian diamonds by sifting with a sieve of fine mesh; the majority of these are octahedra, while cubes and rhombic dodecahedra are but rarely present. The faces of these very small crystals have the same surface characters as those of the larger crystals. By carefully washing for diamonds on the Cape diamond-fields, it is possible to obtain many stones very much smaller than those which usually come into the market, some indeed weighing no more than $\frac{1}{32}$ carat. In the method of washing formerly practised at the Cape and also in Brazil, a large number of the smallest diamonds were lost, their value not being sufficient to justify a special collection of them; the improved washing machinery now in use is, however, capable of saving all the stones however small.

Stones of microscopic dimensions have only recently been observed; previous statements of supposed occurrences, such, for example, as their presence in the xanthophyllite of Zlatoust in the Urals, being based on errors of determination. Microscopic diamonds have now been observed in large numbers in the diamond-bearing rock of the Cape, and there is no reason to doubt that they are present in other diamantiferous deposits.

Smaller diamonds occur in larger number; larger stones are more limited in number; while very large specimens are so extremely rare and valuable that they are known by special and distinctive names, and in most cases form part of the crown jewels of various countries; these famous stones will be described further on in a special section.

The average size of diamonds found in different countries varies very considerably; formerly, when India and Brazil were the only localities at which diamonds were known to exist, stones exceeding twenty carats in weight were of great rarity. During the most productive period of the mines of Brazil, two or three years would elapse before a second stone of this size would be found, while very few stones exceeding one hundred carats in weight were ever found. The largest stone ever found in this locality, that known as the "Star of the South" (Fig. 48), weighed in the rough 254½ carats. The "Braganza," of the Portuguese crown, said to weigh 1680 carats. would rank as the largest diamond ever found

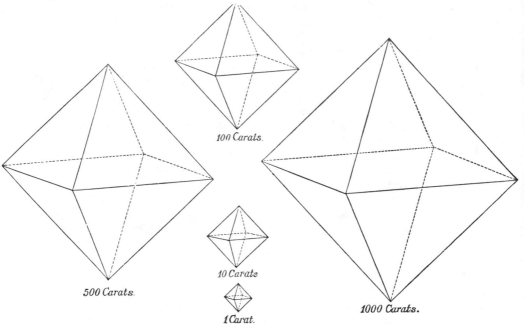

100 Carats.

500 Carats.

10 Carats

1000 Carats.

1 Carat.

FIG. 32.   Actual sizes of octahedral crystals of diamond of 1 to 1000 carats.

in any locality were it indisputably a diamond; the probabilities are, however, that it is a fine piece of colourless topaz.

The chances of obtaining large diamonds in the Indian deposits were more favourable, a considerable number of diamonds exceeding one hundred carats in weight having been found there. Most of the large Indian diamonds are only known in their cut condition so that their original weight can only be estimated. Of large Indian diamonds, known in the rough condition in recent times, the "Regent" in the French crown jewels, is the heaviest; it weighed before cutting 410 carats, and produced a beautiful brilliant of $136\frac{14}{16}$ carats. Other large Indian stones are described below in the section on famous diamonds; they are comparatively few in number. The heaviest of the large diamonds of ancient times is known as the "Great Mogul," which is said to have originally weighed $787\frac{1}{2}$ carats; there is no authentic information, however, either as to its weight or to its present whereabouts. The island of Borneo has produced one or two large stones; the largest reported diamond, weighing 367 carats, is, however, like the "Braganza," almost certainly not diamond, and probably nothing more valuable than a piece of rock-crystal.

Since the discovery of the South African diamond-fields, large diamonds have become less rare; as we shall see later on, stones up to 150 carats in weight have been found there with comparative frequency, while not a few of several hundred carats have been met with. The largest undoubtedly genuine diamond ever discovered, either here or elsewhere, was found at the Cape, in 1893, and weighed 971¾ carats; a more detailed description with a figure (Fig. 51) of this stone will be given later. Probably the largest crystal of diamond to be seen in a public collection is the " Colenso " diamond, presented to the British Museum by Professor John Ruskin; this is a symmetrically developed octahedron weighing 129⅔ carats.

It has been already stated that the size of diamonds, as of all other precious stones, is estimated from their weight expressed in carats. It will be, however, difficult for the general reader to form a correct mental conception of the size of a given stone from its weight in carats alone; hence Fig. 32 is designed to show the actual sizes of diamonds weighing 1, 10, 100, 500, and 1000 carats respectively, each having the form of a regular octahedron, which is the form most frequently presented by crystals of diamond. In the special section devoted to the consideration of the larger and more famous diamonds, figures are given representing the actual sizes of these stones, usually in their cut form (Plates X. and XI.), but in a few cases in their rough form. Plate IX. gives the actual sizes of brilliants varying in weight between one and one hundred carats, and Fig. 44 the actual sizes of rosettes, varying between one and fifty carats.

3. SPECIFIC GRAVITY OF DIAMOND.

The specific gravity of diamond as determined by various observers varies between 3.3 and 3.7. Reliable determinations made on pure stones free from enclosures have, however, in every case yielded values not lower than 3·50 and not much higher than this; the mean value may, therefore, be placed at 3·52. The following are values obtained in particular instances by careful observers using pure material:

3·50 — 3·53	(*Dumas*).
3·524	Brazilian diamond (*Damour*).
3·520 — 3·524	Colourless and yellow diamond from the Cape (*von Baumhauer*).
3·517	Brazilian diamond (*J. N. Fuchs*).
3·529	"Star of the South" from Brazil (*Halphen*).
3·5213	"Florentine" (*Schrauf*).
3·50	Diamond from Burrandong, New South Wales (*Liversidge*).
3·492	Colourless diamond from Borneo (*Grailich*).

The fall of the last value below 3·5, is due to the attachment of a few air bubbles to the stone during the weighing in water.

The small differences in the specific gravity values given in the above table are probably due to the presence of various impurities. Since coloured diamonds always contain a small amount of impurity, the specific gravity will vary with the colour, as is shown in the table below :

		Sp. gr.
Colourless diamond		3·521
Green ,,		3·524
Blue ,,		3·525
Rose ,,		3·531
Orange ,,		3·550

Other values sometimes given are: colourless diamond 3·519, light yellow and green

3·521 ; for colourless Cape diamonds 3·520, and for yellow diamonds from the same locality 3·524.

Determinations which give results much above or much below the mean value of the specific gravity of diamond, namely, 3·52, and specially those which approximate to the extreme values, 3·3 and 3·7, must be regarded either as inaccurate or as having been made on impure material.

Black carbonado has a much lower specific gravity than pure crystals of diamond, values ranging from 3·141 to 3·416 having been determined. This is due to the porous nature of the material, the numerous air spaces enclosed in its substance causing it to be lighter, bulk for bulk, than are crystals of diamond.

4. CLEAVAGE OF DIAMOND.

When a diamond crystal is broken on an anvil by a blow from a hammer, or when it is subjected to sudden changes of temperature, it breaks into a number of fragments which are usually bounded by perfectly plane and bright surfaces. These surfaces of separation, or cleavage, will be found to have always a definite direction in the crystal, being parallel to one or more faces of the octahedron. If a chisel be driven into an octahedron of diamond in a direction parallel to an octahedral face the crystal will be divided into two portions, and the smooth, bright surfaces of separation will be parallel to the same octahedral face. By suitably varying the position of the chisel, the crystal may be divided in the same way into two other portions of which the surfaces of separation will be parallel to any other of the faces of the octahedron. It is not possible, however, to produce a cleavage in a cube of diamond which shall be parallel to the faces of the cube ; in these directions there will be irregular fractured surfaces only. If the chisel is so placed that a corner of the cube is removed, a cleavage surface will be produced, this being as before parallel to a face of the octahedron, the faces of which also truncate the corners of the cube in the natural crystals. The cleavage surfaces of all crystals of diamond, whatever be their outward form, are always parallel to the faces of the octahedron, and in no other direction can plane cleavage surfaces be obtained. Diamond thus possesses an octahedral cleavage only, which is perfect and obtained with the greatest ease ; this gem may indeed be regarded as one of the most perfectly cleavable of minerals.

This perfect cleavage with plane even surfaces throughout can, however, be obtained only in simple crystals. When we are dealing with an intergrowth of two or more crystals twinned according to the spinel-law (Fig. 31, g to i) or irregularly grouped together, the cleavage surfaces will have different directions in each individual, and will not pass uninterruptedly from side to side of the stone as in simple crystals; it is thus impossible to divide such a stone by a single cleavage surface.

From every simple diamond crystal, no matter what may be its external crystalline form, there may be obtained by cleaving parallel to all the octahedral faces a cleavage fragment having the form of an octahedron (Fig. 32). The great importance of this property of cleavage in connection with the faceting of diamonds has already been touched upon, and will be again referred to under the special description of the process of diamond cutting. The property of perfect cleavage is not, however, a desirable one from every point of view, for it is often responsible for the appearance of incipient cracks in the stone, which, if they further develop, seriously diminish its value.

5. HARDNESS OF DIAMOND.

In respect to its hardness, diamond stands alone among all other substances, whether natural or artificial. The hardness possessed by some artificial substances, such, for example, as crystallised boron and carborundum, does, however, approach that of diamond. This gem stands at the head of Mohs' scale, and receives the number 10 as a measure of its hardness. Between this and the next hardest natural substance, namely corundum (ruby and sapphire), there is a wide gap, the difference in hardness between diamond and corundum being far greater than that which exists between corundum and talc, the softest of all minerals. The unique degree of hardness thus possessed by diamond renders it easily recognisable, since it scratches all other substances without exception, and is itself scratched by none.

It is a remarkable fact that degrees of hardness exist in diamonds among themselves, this being shown by the fact that diamonds from one locality are capable of scratching those from other localities. Thus the Australian stones are harder than those from South Africa, which are said to be the softest of all diamonds; and the beautiful black diamonds of Borneo are harder than those of other colours. It is also remarkable that many South African diamonds gradually assume their characteristic hardness only after a more or less prolonged exposure to air.

Diamond forms no exception to the general rule that the hardness of a crystal is not everywhere the same. It has been observed that the powder obtained by rubbing the surface of diamond crystals in the operation of bruting, which will be described in the section on diamond cutting, is more efficacious in the process of grinding than is the powder obtained by pounding up large fragments of diamond. It may naturally be inferred from this, that diamond crystals must be harder on the exterior than in the interior. On the surface itself, however, differences in hardness are distinctly perceptible, some faces of the crystal being more easily scratched than others, while on each face there are certain directions along which scratching may be more readily effected than in others. This being so, it follows that the process of grinding will also be more difficult in certain directions and on certain parts of the stone than in others. This subject will, however, be treated more fully in the section devoted to diamond cutting.

The great degree of hardness possessed by diamond renders it exceptionally suitable for use in personal ornament, since the sharpness of the edges and corners of the cut stone and the lustre of the polished facets are retained in spite of long continued wear.

Several important technical applications of diamond depend on the enormous hardness it possesses; these will be fully discussed in a special section, and we need only mention here the use of diamond powder in the cutting of the harder precious stones and of diamond itself. The hard Australian diamonds, however, are unattacked by the powder of other softer diamonds, they can only be worked by the help of their own powder.

In spite of their enormous hardness, the diamond crystals found in river-sands and gravels often show signs of wear and tear, their edges and corners being rounded and their surfaces dull and roughened. This has been effected by long ages of grinding against the pebbles and quartz-grains, and occasionally the precious stones of river-sands and gravels; thus even diamond itself cannot escape the action of time.

The hardness of diamond has often been confused with its **frangibility** or brittleness. It has been supposed, especially in ancient and mediæval times, that hammer and anvil may be shattered but not the diamond which lies between. This statement was made by Pliny, the great naturalist of ancient days, who was killed in 79 A.D. at the first historic eruption of Vesuvius. He proceeded to say further that the fragmentation of a diamond may be effected by subjecting it to a preliminary immersion in the warm blood of a goat, but that even under these circumstances the hammer and anvil will also be broken! According to Albertus Magnus (1205–1280), the blood is more efficacious if the goat has previously drunk wine or eaten parsley!

Such being the views then held respecting the unbreakable and indestructible character of the diamond, it is easy to understand why the Greek word *adamas*, signifying unconquerable, should have been applied to this stone, although its application to the diamond is singularly inappropriate and inaccurate when its extreme frangibility is considered. Many a doubtful stone has been submitted to the test of the hammer, with the belief that the blow would be resisted only if the stone were a genuine diamond. Probably many beautiful stones have been sacrificed to this old belief. As a matter of fact, diamond is easily fractured, a very moderate blow from a hammer sufficing for the purpose; its perfect cleavage places it among the most brittle of minerals.

6. OPTICAL CHARACTERS OF DIAMOND.

Transparency.—In its pure condition diamond is most beautifully clear and transparent; the presence of enclosures of foreign matter, however, often diminishes the natural transparency of the stone, in some cases causing almost complete opacity. Dark coloured diamonds, especially brown and black specimens, are frequently transparent only at their edges, and black diamonds are often completely opaque. The transparency of a crystal depends also upon the condition of its surface; if this should be roughened, as will be the case after a prolonged rolling about on the bed of a river, the stone will appear dull and cloudy, although its interior may be perfectly transparent, as is evident when the rough surfaces have been removed by cutting.

On the degree of transparency depends largely the quality known as the *water* of a diamond. A stone which is perfectly transparent, colourless, and free from all faults is described as a diamond of the first or purest water. A small degree of cloudiness in a diamond does not entirely unfit it for use as a gem; when, however, the cloudiness exceeds a certain amount the stone can be applied only to technical purposes.

Lustre.—The lustre on the smooth face of a diamond, beside being extraordinarily strong and brilliant, is very peculiar in character and is intermediate between the lustre of glass and that of metal. Being characteristic of diamond, it is known as adamantine lustre, and is shown by very few minerals and by still fewer precious stones. It is therefore possible after a little practice to readily distinguish diamond from other transparent substances, such as glass, rock-crystal, &c., by the character of the lustre alone. As we have already seen, however, there is an artificial glass known as strass, which possesses an adamantine lustre and which is therefore much used in the manufacture of imitation diamonds.

Adamantine lustre is frequently absent from the natural faces of diamond crystals especially after they have become dulled by friction in a river-bed. In such cases, the stone has a peculiar lead-grey metallic appearance, similar to that which is artificially produced by bruting, an operation in the process of diamond cutting which consists of the rubbing together of two diamonds with the object of obtaining an approximation to the form they are finally to assume. Adamantine lustre is seen to perfection on the polished facets of a cut diamond, since here the incident light is reflected quite regularly. The lustre of diamonds which are dark coloured, and therefore have little transparency, approaches that of metals. The same metallic lustre is seen on the facets of a perfectly transparent stone when light falls upon it at a very small inclination, the light being reflected from the facet in such a way that the latter has the appearance of highly polished steel. This phenomenon can be observed by placing the stone with a perfectly smooth facet close to the eye and inclining it towards the light, until, in a certain position, the metallic reflection becomes evident.

Perfect adamantine lustre, in all bodies which possess it, is combined with perfect transparency, very strong refraction, and marked dispersion of light. All substances including diamond itself, which possess adamantine lustre are thus also characterised by strong refraction and dispersion of light; and, conversely, all substances possessing the two latter qualities will be found to exhibit adamantine lustre. Not only the quality but the intensity of the lustre shown by a stone depends upon the strength of its refraction of light; light rays falling obliquely upon the surface of a stone will be the more completely reflected the higher its index of refraction is. Thus diamond, having a higher index of refraction, will reflect more rays of light from its surface, and will therefore show a stronger lustre than will a substance having a lower index of refraction. The qualities of lustre and brilliancy are known collectively as the " fire " of a stone. It will be evident from what has been said that the " fire " of diamond is specially fine.

Refraction of light.—Diamond, like all other substances which crystallise in the cubic system, is singly refracting. A ray of light incident obliquely upon the plane face of a diamond is propagated in the substance of the stone as a single ray, the direction of which, however, differs from that of its path in the surrounding medium. This difference is, in diamond, very considerable, much more than in the majority of other substances; in other words, the index of refraction of diamond is very high.

The power of breaking up white light into its constituent colours, that is, the **dispersion,** possessed by diamond is likewise very marked. The blue rays of light undergo a much greater refraction when passing into diamond than do the red rays; hence the spectrum produced by a prism of diamond is very long, the red and blue ends being widely separated. The various colours into which white light, in passing through a cut diamond, is broken up are widely separated and distinctly perceptible; hence the beautiful play of brilliant, prismatic colours upon which so much of the beauty of diamond depends, and which differentiates it so markedly from other colourless stones, such as rock-crystal, topaz, colourless sapphire, &c., which have a lower dispersion and consequently a less beautiful play of colours. This subject has, however, been fully dealt with above in the section devoted to the consideration of the passage of light through a cut stone.

The action of every diamond upon light is not absolutely identical. A satisfactory explanation of the small differences which exist cannot, however, at present be given. The fact that one stone has a finer appearance than another is probably due to slight differences between them in the refractive index and in dispersive power. In respect of play of prismatic colours Indian diamonds rank highest. Next to these we may place Brazilian stones from the district of Diamantina in the State of Minas Geraes, and from the Canavieiras mines in the State of Bahia. Relatively inferior to these, but yet with a fine play of prismatic colours, are the majority of Cape diamonds. It is a remarkable fact that in many cases diamonds from the Cape and from Canavieiras exhibit a finer play of prismatic colours in artificial light than in daylight, which is the reverse of what is usually the case.

The refractive power and the dispersion of diamond are both given by the values of the refractive indices for different coloured rays. These values give the strength of refraction directly, and the difference between the refractive index for red and that for violet rays is a measure of the dispersion. A comparison of the following determinations by Walter, with similar constants for other precious stones, will show that the refraction and dispersion of diamond are in all cases the greater :

					n
Red light (B line of the spectrum)	.	.	.	.	2·40735
Yellow ,, (D ,, ,,)	.	.	.	.	2·41734
Green ,, (E ,, ,,)	.	.	.	.	2·42694
Violet ,, (H ,, ,,)	.	.	.	.	2·46476

The dispersion co-efficient is thus—

$$2·46476 - 2·40735 = 0·05741.$$

For comparison, the following values of the refractive indices of a particular glass may be given :

									n
Red light	.	.	.	.	.	.	.	.	1·524312
Yellow ,,	.	.	.	.	.	.	.	.	1·527982
Green ,,	.	.	.	.	.	.	.	.	1·531372
Violet ,,	.	.	.	.	.	.	.	.	1·544684

The dispersion co-efficient is here—

$$1·544684 - 1·524312 = 0·020372,$$

less than half as great as that of diamond. The spectrum produced by a prism of this glass will be only about half as long as one produced by a similar prism of diamond under the same conditions.

Anomalous double refraction. — Diamond, being crystallised in the cubic system, should be singly refracting, that is, isotropic. This, however, is only strictly true for such stones as are perfectly colourless, or of a yellowish colour, and are quite free from enclosures of foreign matter, cracks, and other flaws. Such faultless stones when rotated in the dark field of the polariscope remain dark. As previously mentioned, the stone under examination should be immersed in methylene iodide, so as to diminish total reflection as far as possible.

Deeply-coloured stones, and those disfigured by cracks, enclosures, or other faults, when placed in the dark field of the polariscope, allow the passage of light to the eye, but, as a rule, to only a small extent. They have, under these circumstances, a greyish appearance, brilliant polarisation colours being rarely seen. The feeble double refraction possessed by such stones is not an essential character of the substance of the diamond itself, but is due to disturbing influences ; hence it is distinguished as anomalous double refraction. During its rotation in the polariscope it rarely happens that such a stone is uniformly dark or uniformly light over its whole surface ; as a rule, certain areas are dark while others are light, and *vice versâ*. Frequently certain regularly bounded areas or fields behave in a similar manner during rotation, while adjacent fields behave differently. In most cases, however, the areas showing these differences in behaviour have no definite arrangement relative to each other, and areas showing a feeble double refraction are often enclosed in areas which are perfectly isotropic.

The doubly refracting portions of the stone usually surround enclosures or cracks, and it is in the immediate vicinity of these that double refraction is strongest and the polarisation colours most brilliant. As the distance from a flaw of this kind increases the double refraction becomes feebler, and at a certain distance disappears. Sometimes a black cross, the arms of which consist of two dark brushes, is seen when a stone is examined in the polariscope: the arms of the cross are mutually perpendicular and their point of intersection coincides with an enclosure in the diamond. It is clear that such an appearance is due to a strain in the diamond brought about by the presence of the enclosure, and that the strain will be less in portions further removed from the enclosure.

Although the anomalous double refraction of diamond is, as a rule, but feeble, stones exist in which it is comparatively strong, and which show much brighter polarisation colours. This is the case in the " smoky stones " of South Africa, which, because of the great internal strain in their substance, have a tendency to fall to powder for no apparent reason. A parallel case is that of the drops of glass known as " Prince Rupert's drops," which also show strong double refraction as a consequence of internal strain.

There is never the slightest danger of confusing anomalous with true double refraction, for a mineral with true double refraction, such for example as rock-crystal, colourless sapphire or topaz, will appear much more brilliantly illuminated when examined in the polariscope, and, moreover, will be uniformly light or uniformly dark over its whole surface.

Colour.—Diamond is often regarded as the type of what a perfectly clear, colourless, and transparent stone should be. It can by no means, however, be always so regarded, since cloudy and opaque diamonds are actually more common than those which are clear and transparent, while very great variety in colour is found in this mineral. A great number of diamonds are indeed perfectly colourless, and correspond strictly to the popular conception of the stone; this number is, however, only one-fourth of the total number of diamonds found; another quarter show a very light shade of colour, while the remainder, at least one-half of the total, are more or less deeply coloured.

Perfectly colourless diamonds are, at the same time, most free from impurity. Absolutely pure carbon, crystallised in the form of diamond, shows no trace of colour whatever, and stones of this purity are naturally highly prized. A peculiar steel-blue appearance is sometimes observed in stones which combine absence of colour with perfect transparency. With the exception of a few specially beautifully coloured stones of great rarity, these *blue-white* diamonds are the most highly prized of all; they are not of great rarity in India and Brazil, but occur in South Africa with far less frequency.

Any colouring-matter intermixed with the substance of a diamond imparts its colour to the stone, the tone of which will be faint when the pigment is present in small amount and deeper when it is present in greater amount. In all cases the amount of colouring-matter relative to the mass of the stone is extremely small.

Investigations into the precise nature of the various colouring-matters present in diamonds have seldom been undertaken on account of their difficulty and expense. There can be no doubt, however, that the colouring-matter of many diamonds is of an organic nature, possibly some one or other of the hydrocarbons; in other cases the pigment is probably inorganic material in an extremely fine state of division. We have already seen that coloured diamonds contain a small amount of ferruginous material which remains behind as an incombustible ash after the diamond is burnt away, and that with colourless diamonds this is not the case. There seems sufficient grounds here for the inference that in such cases the colour of the stone is due to the inorganic, incombustible, enclosed material, especially as the colour is neither altered nor destroyed after exposure to high temperature, which would be the case were it organic in nature. In the few recorded cases in which a change of colour has been observed on strongly heating the stone, there can be no doubt as to the organic nature of the colouring-matter.

The colouring of many diamonds is so faint that an unpractised observer, unless he is able to compare such a stone with an absolutely colourless diamond or to place it against a background of pure white, will fail to recognise that the stone is coloured at all. The practised eye of the diamond merchant, however, needs no such assistance in recognising the most faintly coloured stones. Such stones are rather lower in value than absolutely colourless specimens of the same clearness and transparency, but the difference in price is not very considerable. The shades of colour which appear most frequently are light yellow, grey, and green. A faintly yellow diamond is not observable as such in any artificial illumination other than the electric light, and then appears to be a colourless stone. Diamonds of a faint bluish tinge are known, but are much less common.

As mentioned above, diamonds showing a pronounced colouration constitute about one-half of the total output. Almost all the colours of the mineral kingdom may be represented in numerous and varied tints, so that the suite of colours of the diamond is very extensive. A magnificent collection of differently coloured diamonds, the most beautiful and the richest in existence, is preserved in the treasury of the royal palace at Vienna. It was brought together by Helmreichen, who spent many years in Brazil, and was so enabled to make the series very complete. The colour which occurs most frequently in diamonds is yellow, in various shades, such as citron-yellow, wine-yellow, brass-yellow, ochre-yellow, and honey-yellow, but sulphur-yellow has not as yet been observed. Most of the Cape diamonds are coloured with one or other of these tints of yellow. After yellow, green is the most commonly occurring colour, especially in Brazilian diamonds. Oil-green or yellowish-green is seen most frequently, then pale green, leek-green, asparagus-green, pistachio-green, olive-green, siskin-green, emerald-green, bluish-green and greyish-green. Brown diamonds are also common at all localities; the different shades are light-brown, coffee-brown, clove-brown, and reddish-brown. Shades of grey, such as pale grey, ash-grey, smoke-grey, are not rare. Black diamonds in well-formed crystals are unusual. The different shades of red, a colour which is rarely met with in diamonds, are lilac-red, rose-red, peach-blossom-red, cherry-red, hyacinth-red. Blue in its two shades, dark blue and pale sapphire-blue, is the rarest of all colours to be met with in diamonds.

The colouring of diamonds is seldom intense, pale colours being much more usual than deeper shades. Diamonds which combine great depth and beauty of colour, with perfect transparency, are objects of unsurpassable beauty; for, in addition to their fine colour, they possess the wonderful lustre and brilliant play of prismatic colours peculiar to the diamond, so that other finely-coloured stones, such as ruby and sapphire, are not to be compared with them. Only a few stones of this description are in existence; they are among the most highly-prized of costly gems.

Of such deeply coloured and perfectly transparent diamonds, bright or deep **yellow** specimens are, since the discovery of the South African diamond-fields, the least rarely met with. The largest of these yellow Cape diamonds is shown in Fig. 52; it is a beautiful orange-yellow brilliant, weighing $125\frac{2}{3}$ carats, and is in the possession of Tiffany and Co., the New York firm of jewellers. A few fine yellow stones dating back to ancient times are preserved in the " Green Vaults " at Dresden.

Diamonds of a fine **green** colour are distinctly rare, only a few examples being known; the same may be said of red diamonds and, even more emphatically, of blue diamonds. The most beautiful green diamond known is a transparent brilliant weighing $48\frac{1}{2}$ carats, preserved in the " Green Vaults " at Dresden; it will be again mentioned in the section devoted to famous diamonds. Another green diamond of the same quality is now in America. Tschudi mentions two beautiful specimens from Brazil, one of an emerald-green and the other of a sea-green colour, while the existence of other Brazilian stones with a colour very similar to that of the yellowish-green of uranium glass, but inclining more to yellow, is mentioned by Boutan.

The ten-carat ruby-red stone, which belonged to Czar Paul I. of Russia, and which is said to be still preserved with the Russian crown jewels, is often mentioned as an example of a **red** diamond; nothing more definite concerning it is, however, known. A better authenticated example is the " Red Halphen " diamond, a ruby-red brilliant weighing one carat; while recently Streeter has reported the discovery of a beautiful red stone in Borneo, and its sale in Paris. Several examples of beautifully transparent rose-red diamonds are known ; such, for instance, as that of the fifteen-carat stone belonging to the Prince of Riccia, a few smaller specimens in the treasury at Dresden, and a thirty-two-carat stone, the most beautiful rose-red known, in the treasury of Vienna. A rose-coloured brilliant, called the " Fleur de pêcher," is among the French crown jewels, and Tschudi mentions a peach-blossom-red stone from the Rio do Bagagem, in Minas Geraes, Brazil.

Blue diamonds are the rarest of all. A magnificent blue brilliant of $44\frac{1}{4}$ carats, the " pearl of coloured diamonds," was formerly in the possession of Mr. Hope, a London banker. It is probably a portion of Tavernier's blue diamond of $67\frac{1}{8}$ carats, stolen in 1792 with the French crown jewels. A small diamond of a deep blue colour, and a pale blue one of forty carats, are preserved in the Munich treasury.

Black diamonds perhaps deserve a brief mention. Crystals of a uniform black colour have been found in Borneo, and also very rarely in South Africa. The opacity of such stones, combined with their high degree of lustre, almost metallic in its character, render them, when cut, of peculiar beauty, and well suited for use in mourning jewellery. These crystals of black diamond must not be confused with the black carbonado of Brazil, to be described later. A few **brown** stones of a delicate and beautiful coffee shade are known ; these also come from Brazil.

As is almost always the case with precious stones and other minerals, the colour of which is due to enclosed foreign matter in an extremely fine state of division, so also in the diamond the distribution of the colouring-matter is not always perfectly uniform throughout its whole mass. The pigment may be collected or accumulated in isolated patches, while the rest of the stone is either colourless or less deeply coloured. In numerous cases, only the thin surface layer of the crystal is coloured, the interior of the stone being colourless ; this occurs often in Brazilian stones, especially those from the Rio Pardo, in the Diamantina district. When the outer layer, which is often pale green in colour, is removed in the process of cutting, a perfectly colourless stone is obtained. Tschudi mentions a fine emerald-green brilliant from Brazil, which before cutting had a sooty black appearance; another specimen of similar appearance retained its black colour almost entirely after cutting, a few facets only appearing white.

Not infrequently the bulk of a rough diamond is colourless, the edges and corners only being coloured; many Brazilian stones are of this description, as well as specimens from South Africa, including some of the " smoky stones " already mentioned. In these the deep smoky-grey colour is sometimes confined to the corners of the stone, the interior being faintly coloured or entirely colourless; a stone so coloured is described as a "glassy stone with smoky corners." Diamonds in which these conditions are reversed also occur, the edges and corners being colourless, and the central portion coloured.

Sometimes, though rarely, a stone shows two differently coloured portions ; the two portions of one mentioned by Mawe were coloured respectively yellow and blue. Stones showing a number of differently coloured sectors, sharply separated from each other, and radiating from a central point, are also of rare occurrence. Thus smoke-grey and colourless rays may have a regular star-like arrangement, or may form a figure like the club of playing cards, on the faces of the octahedron.

Of interest is the fact that diamond sometimes shows a play of colours like that of precious opal. Des Cloizeaux mentioned a few such stones which differed from opal in this respect only in that the colours were less brilliant. Pale blue and yellow stones have been reported by Mawe to show a somewhat similar appearance.

The colour of diamonds is by no means in every case unchanging and unalterable. Some stones are bleached by sunlight; thus a red diamond on exposure to sunlight is reported to have gradually lost its colour and become white. A diamond in the possession of the Parisian jeweller, Halphen, undergoes a peculiar change in colour on exposure to heat. This stone, which is of a faint brownish colour and weighs four grams (about twenty carats), assumes in the fire a beautiful rose-red colour. If kept in darkness this colour is retained for about ten days, after which it returns to its original brownish colour; should the stone be exposed to diffuse daylight, or to the direct rays of the sun, the change to the original colour is much more rapid. The change to rose-red can be produced at will by again exposing the stone to the action of heat. Could means be devised for retaining the rose-red colour of the stone, the possessor would benefit to the extent of many thousand francs, since it is valued when brown at 60,000 francs, and when rose-red at from 150,000 to 200,000 francs. Halphen has also seen a diamond which when rubbed assumed a rose-red colour; this colour, however, was lost again almost immediately.

The colour of diamonds is in some cases affected by exposure to a high temperature. According to Des Cloizeaux, pale green diamonds, after being heated in the oxyhydrogen flame, became light yellow; brown crystals under the same conditions become greyish. Baumhauer also witnessed the colour of one diamond change from green to yellow, and of another from dark green to violet, under the influence of a high temperature. Wöhler caused green diamonds to assume a brown colour by heating them, but found that brown stones remained unaltered in colour. Yellow diamonds, especially those from the Cape, retain their colour at the highest temperatures.

It has already been mentioned that very faintly coloured stones command a somewhat lower price than do those which are perfectly colourless. Hence many attempts have been made to transform faintly coloured stones into the more valuable colourless stones. This is readily effected in those Brazilian diamonds, which have a colourless central portion surrounded by a coloured external layer of no great thickness. The outer coloured layer is in these cases simply burnt away by heating the stone in a crucible with a little saltpetre. The operation is very brief, the coloured external layer disappearing in one or two seconds. This device involves no actual change of colour, but is simply a removal of the coloured portions, which could have been effected just as well by the more lengthy process of grinding.

Attempts have been made, however, to decolourise diamonds in which the undesirable pigment is distributed through the substance of the whole stone. Probably the first to make experiments in this direction was the Emperor Rudolph II.; according to the report of his gem-expert, Boetius de Boot, the Emperor was acquainted with a method by which every diamond could be decolourised, and rendered perfectly colourless. This important secret, however, died with its possessor. At a later date the Parisian jeweller, Barbot, claimed to be able, by the employment of chemical means and a high temperature, to decolourise green, red, and yellow stones, while dark yellow, brown, and black stones only slightly lost their colour. Barbot's method was also preserved as a secret, so that it is impossible to put his assertion to the test; it is probable, however, that he did not possess the power to which he laid claim, in spite of the fact that he described himself on the title-page of one of his books as " Inventeur du procédé de décoloration du diamant." Our present knowledge of the constituents of diamond pigments makes us unwilling to believe too readily in the possibility of their complete destruction ; at all events, no method is at present known which is effectual in all cases.

Although it is impossible actually to destroy the colour of yellow diamonds, such as now so often come into the market, it is easy to disguise their yellow tint and make them appear colourless. This device has often been practised for fraudulent purposes, and was successfully carried out in Paris a few years ago. The yellow stone is placed in a violet liquid, such as the dilute solution of potassium permanganate used for disinfecting purposes. On being taken out and allowed to dry the stone becomes coated with a thin film of the violet substance. The combined effect of yellow and violet colours in certain proportions is to produce in the eye the sensation of white light ; hence, when the violet layer is of a certain thickness, the stone will have the appearance of being perfectly colourless. Should it still appear yellow after the first trial, it may be immersed a second or third time, while if the violet colour is too deep, some of it may be washed off. Instead of a solution of potassium permanganate violet ink may be used for this purpose. A stone treated in this way will, of course, remain apparently colourless only so long as the violet coating remains intact ; directly this is rubbed off, which readily happens, the yellow colour of the stone becomes apparent. A more permanent coating is said, however, to have been recently devised for this purpose. The methods by which yellow stones can be made to appear colourless are very old ; ultramarine is supposed to have been used by the ancient inhabitants of India for this purpose.

According to the investigations of Petzholdt, which are confirmed by every diamond-cutter, the colour of diamond powder or dust varies from grey to black, and is darker the finer its state of division.

Phosphorescence.—Many erroneous statements have been made with regard to the phosphorescence of diamond. Thus it has been stated that diamond phosphoresces in darkness after exposure to the direct rays of the sun, the phosphorescence being specially marked after exposure to blue light, and less so after exposure to red light. It is even said that after being screened by a board or paper from the direct rays of the sun, so that the stone is exposed to diffused daylight only, it will be seen to glow brightly when placed in darkness. Exact researches, however, have proved that only very few diamonds phosphoresce after exposure to sunlight, and that neither the direct rays of the sun nor intense artificial light cause phosphorescence in the majority of stones. Streeter reports that a yellow stone of 115 carats, after exposure to lime-light lit up a dark room ; and Edwards describes a water-clear diamond of 92 carats, which after one hour's exposure to electric light emitted in a dark room a light which lasted for twenty minutes, and was so strong that a sheet of white paper placed near the stone could be distinctly seen. Of 150 diamonds of various forms, sizes and qualities examined by Kunz, only three showed the phenomenon of phosphorescence after exposure to the light of the electric arc.

Although exposure to light has in diamond so small an effect in exciting phosphorescence, the phenomenon is easily produced by rubbing the stone. Kunz observed that all the diamonds he examined, without exception, became self-luminous after being rubbed on wood, leather, woollen or other material. Some stones needed only to be drawn once across the substance, especially if it were of wool, to render them self-luminous ; the most marked phosphorescence was, however, developed in stones by rubbing on wood against the grain. According to other statements, rubbing on metals (iron, steel, copper) is effective.

Whether diamonds are capable, as a rule, of giving out light after being raised to a temperature below red-heat is doubtful, but some specimens, which are unaffected by sunlight and remain quite dark, are induced to glow by exposure to electric sparks. But here, as in all cases, the phosphorescence can be produced only when the stone has not previously been exposed to a strong heat.

The light given out by a phosphorescing diamond is in almost all cases feeble, and much less intense than that emitted by many other phosphorescent substances. The light, which is usually yellow in colour, but may under certain circumstances be blue, green, or red, is strongest when induced by electric sparking. Remarkable observations have been made on the appearance of different faces of a phosphorescing crystal of diamond. Dessaignes (1809) stated that a diamond after exposure to the sun's rays emitted light from the cube faces, but not from the octahedral faces, which remained quite dark. Maskelyne described a diamond crystal which emitted a beautiful apricot-coloured light from the cube faces, a bright yellow from the faces of the rhombic dodecahedron, and a yellow light of another shade from the octahedral faces.

All these appearances are, as a rule, of brief duration. A case is, however, recorded of a diamond which continued to phosphoresce for an hour after the removal of the exciting cause.

The phosphorescence of diamond is said to have been first observed in 1663 by the famous English physicist Robert Boyle.

7. ELECTRICAL AND THERMAL CHARACTERS.

A diamond, whether rough or cut, becomes positively electrified on rubbing; the charge so acquired is quickly lost, never being retained more than half an hour. In contrast to graphite, the other crystallised modification of carbon, which is a good conductor of electricity, the conductivity of diamond is so inappreciable that the stone ranks as a non-conductor.

Diamond being a good conductor of heat appears cold to the touch, and by this means can be distinguished from other substances, as already explained.

B. OCCURRENCE AND DISTRIBUTION OF DIAMOND.

Diamond has been found in all five continents, but not to the same extent in each. It has been longest known in **Asia,** where the famous old Indian deposits have probably been known and worked from the earliest times; now, however, they are almost completely exhausted. In close geographical connection with these are the deposits in Borneo, but the supply from this island, in comparison with the rich treasure of India, has always been limited. Reported discoveries of diamonds in the Malay Peninsula, where, according to one account, the famous " Regent " of the French crown jewels was found, in Pegu and Siam and the islands of Java, Sumatra and Celebes, are for the most part unauthenticated; and the same may be said of the reported occurrence in China (province Shan-tung), Arabia, &c.

In **America** the famous Brazilian diamond-fields were discovered at the beginning of the eighteenth century, and have compensated for the exhaustion of the Indian mines; the richest yield of stones has been given by the mines in the States of Minas Geraes and Bahia. Recent finds have been made in another part of the South American continent, namely in British Guiana. Well authenticated, but of little commercial importance, is the occurrence of diamond in the United States of North America; a small number of stones having been found in the eastern States of Georgia, North Carolina, South Carolina, Kentucky, Virginia, Wisconsin, and in the western States of California and Oregon. Reported occurrences in other parts of the American continent, namely Sierra Madre in Mexico, and the gold mines of Antioquia in Colombia, South America, require confirmation.

The continent of **Africa** is, at the present time, by far the most important source of diamonds, which have been collected here since the late sixties in ever increasing numbers, far surpassing the yield from any other region. The exact locality of the deposits is on the Vaal River, and in the neighbourhood of the town of Kimberley, both these localities being in the division of Griqualand West, in the north of Cape Colony ; also in the adjoining Orange River Colony, which, however, is of far less importance. Compared with the yield of the African fields, all others are insignificant, although in comparatively recent times the markets of the world were supplied from the sources which are now of such minor importance. At the present day the diamond fields of the Cape are the source of 90 per cent. of the stones which come into the market. The reported occurrence of diamonds in the auriferous sands of the river Gumel, in the province of Constantine, in Algeria, is unauthenticated ; three stones were said to have been found here in 1833, but nothing more has been heard of this reputed discovery. The statement of Dr. Cuny, an African traveller, that in the fifties a whole camel-load of diamonds was brought from Western Africa to Darfur seems rather incredible.

In **Europe**, diamonds have been found in Russia, in the Urals in the east, and in Lapland in the west ; the stones are, however, met with only in small numbers, and their importance lies in their mineralogical rarity. The reported occurrence of a few small diamonds in Spain has some degree of probability ; but the supposed discovery of diamonds in a stream in Fermanagh, Ireland, needs authentication. A reputed discovery in Bohemia is certainly false ; in 1869 a single small diamond was noticed in a parcel of garnets at some cutting works at Dlaschkowitz ; the matter was thoroughly investigated by V. von Zepharovich, who proved beyond doubt that the diamond must have been introduced into the parcel at the works, where such stones are used for boring garnets.

Diamonds have been found in recent times in **Australia**, especially in New South Wales, not altogether in inconsiderable numbers ; and Australian stones are at least mentioned in the markets.

Finally we must record the interesting fact that diamond is not only a constituent of the earth's crust, but also of extra-terrestrial bodies, the presence of small stones having been in recent years proved in several **meteorites**.

With regard to the mode of occurrence of diamonds, it is to be noted that in the majority of localities they are found in secondary deposits, such as sands and gravels. These masses of débris produced by the weathering of the original mother-rock of the diamond are usually entirely loose and incoherent ; occasionally, however, as in Brazil and India, they are converted by some cementing materials into firm conglomerates, breccias and sandstones. In Brazil these rock masses, like the loose sands and gravels at other places, lie on the surface, and must therefore be reckoned among the most recent deposits of the earth's crust. In India, and to a certain extent also in Brazil and North America, the diamantiferous fragmentary rocks belong to earlier geological periods, being interbedded with some of the oldest rocks, and thus representing the sands and gravels of very remote ages. When these older fragmentary rocks come to the surface, they are themselves in course of time attacked by weathering agents and supply material for new secondary deposits, from which diamonds are won by the ordinary process of washing.

These relations will be further considered with the description of each special diamantiferous deposit.

The nature and character of the original mother-rock, in the débris of which the diamond is now found, has nowhere been determined with the certainty and clearness as to detail that is desired, although many important steps have been made towards the solution of this important problem. In the following pages we will consider in detail the facts connected with each well-established occurrence of diamond, and endeavour to determine the origin of the stone in each case so far as the available observations permit. In any case it is certain that the original mode of occurrence and the mother-rock are not the same at all localities : in some cases the mother-rock is without doubt one of the older crystalline rocks, such as a gneiss or a crystalline schist, or an eruptive rock, such as granite ; in other cases it is highly probable that diamond originated as a secondary mineral in crevices in the rock known as itacolumite, as will be specially considered when we come to treat of the Brazilian deposits. In the South African diamond-fields, the stones are found for the most part embedded in a green serpentine-like rock, instead of in loose sands as is more usually the case. This mode of occurrence, which is peculiar to this locality, and differs from that of all others, will be considered in detail under its appropriate heading.

The different diamantiferous deposits will be dealt with below in the following order:

1. India.	6. North America.
2. Brazil.	7. British Guiana.
3. South Africa.	8. Urals.
4. Borneo.	9. Lapland.
5. Australia.	10. In meteorites.

1. INDIA.

Diamonds have been known longer in this country than in any other, and the most beautiful, famous and many of the largest stones were found here. A diamond river in India is referred to by Ptolemy ; and the fact that diamonds were known to, and highly prized by, the ancient inhabitants of the country is proved by the rich adornment of the oldest temples of religion with this and other precious stones. The sacred shrines and idols show, moreover, that the art of diamond cutting has been long understood. Until the discovery of the Brazilian deposits in 1728, the supply of the whole world was derived almost entirely from the Indian sources, Borneo being at that time the only other known ocality.

The occurrences of diamond in India are distributed over an extensive area of the country. C. Ritter in his *Erdkunde von Asien* (vol. iv., part 2, p. 343, 1836) collected together the various scattered reports concerning the diamond localities, and was the first to give a detailed and connected account. More recently (1881), Professor V. Ball has given an exhaustive account, in which he has incorporated all the latest information, in the official *Manual of the Geology of India* (Part III. pp. 1-50).

That the occurrence of diamonds in India is almost entirely confined to the eastern side of the Deccan plateau is to be gathered from the finds of the present day and from the reports of earlier times. The southern boundary of the region in which diamonds have been found is the river Penner in latitude 14° N. ; from this river the diamond localities form a frequently interrupted line running northwards on the east side of the Deccan plateau, crossing the Kistna, Godavari and Mahanadi rivers, and reaching the southern tributaries of the Lower Ganges in Bengal, between the rivers Son and Khan, in latitude 25° N. Any other diamond localities outside the area just marked out (see map, Fig. 33) are unimportant, and the reports concerning them are often uncertain. In general, many of the reported localities for diamond are doubtful, there being no exact and reliable information respecting them, or they are simply based on the existence of old mines.

It is often supposed that all Indian diamond mines are of the greatest antiquity. In many cases the date at which the workings were commenced is not known; but the working of the most important deposits known at the present day does not date back to very remote periods, probably in all cases subsequent to the year 1000 A.D. and sometimes much later. Of a few mines it is known exactly when work began, as will be mentioned below.

Diamonds are found in India in compact sandstones and conglomerates, in the loose, incoherent weathered products of these rocks at places where they lie on the surface, and in the sands and gravels of those rivers and streams which have flowed over the diamantiferous strata or their weathered débris and have washed out the stones from their former situations.

The diamantiferous sandstone of India is of very wide distribution. It belongs to the oldest division of the sedimentary formations of the country which usually rest directly upon the still older crystalline rocks, such as granite, gneiss, mica-schist, hornblende-schist, chlorite-schist, talc-schist, and similar rocks. Fossils have not been found in these sandstones, so that it is not possible to determine exactly to which of the European formations they correspond in age; they may however be safely stated to belong to the Palæozoic period, and possibly to the Silurian division of this period.

The oldest bedded rocks known to Indian geologists are included in the Vindhyan formation. Only the lower division of this formation is represented in the Madras Presidency in southern India, and is there known as the Karnul series. In northern India, for instance in Bundelkhand, these lower beds are overlain by the later beds of the Upper Vindhyan formation. As far as is yet known, the diamantiferous sandstones of the whole of India belong to this Vindhyan formation; but while in the southern diamond districts, and probably also in the districts of the Godavari and Mahanadi rivers, they belong to the lower division, namely, the Karnul series, in northern India, for example in Bundelkhand, they belong to the upper part of the Vindhyan formation.

The Lower Vindhyan formation (namely the Karnul series) consists mainly of limestones with interbedded clay-slates, sandstones, conglomerates, and quartzites. In southern India at the base of this series of beds are beds of sandstone and conglomerates which are known as the Banaganapalli group and here constitute the diamantiferous strata. The whole of the Banaganapalli sandstones are on an average ten to twenty feet thick; they are usually coarse grained, sometimes argillaceous, at other times very compact and siliceous, and in places felspathic and ferruginous; they are dark in colour, being red, grey or brown. The pebbles of the interbedded conglomerates have been derived from the denudation of older rocks, and consist for the most part of quartzite, variously coloured hornstones and jasper, as well as compact clay-slates.

The diamonds are found in an earthy bed containing abundant pebbles; this bed is clearly and definitely marked off from other beds and is not repeated at any other horizon. The diamonds, which may themselves be regarded as pebbles since they also show signs of rounding, lie scattered singly among the other pebbles which are of the same materials as those just mentioned. This earthy bed, in which alone the diamonds are found, is of little thickness; in exceptional cases its thickness is stated to be two and a half feet, but it often measures less than a foot, and rarely exceeds this amount.

In Bundelkhand the diamond-bearing stratum belongs to the middle division of the Upper Vindhyan formation, namely the Rewah group, and is situated at the base of this group in the Panna beds. It is usually a red, ferruginous conglomerate, the pebbles of which consist, as in southern India, of quartz, variously coloured jasper, quartz-schists, sandstone, nodules of limonite, &c. The diamonds appear to bear a close relation to certain sandstone pebbles in this bed.

It is often stated, although perhaps further confirmation of the fact is needed, that in Bundelkhand diamond is sometimes found in fragments of a compact, greyish, siliceous sandstone with a peculiar glassy appearance, embedded in the stone in the same way as are the sand grains of which it is composed. These sandstone pebbles in the conglomerates of the Rewah group have been most probably derived from beds of the Lower Vindhyan formation, which, by their denudation, supplied material for the deposition of the later beds of the Upper Vindhyan formation. Thus the diamonds now found in the Upper Vindhyan formation originally belonged to the Lower Vindhyan, where, in southern India, they are still found. With the denudation of these older rocks the diamonds were set free and again deposited with the material of the younger beds, some remaining embedded in fragments of the original rock, while others became isolated and are now found among the pebbles of the conglomerate.

The diamantiferous sandstones and conglomerates are now elevated, and crop out at the surface of the ground, or are covered by younger strata. Where these strata are not so protected, as when a valley cuts across them, they will be exposed to the action of denuding agents, and will be reduced to soft incoherent sands in which the diamonds lie loosely, the whole constituting a diamantiferous sand.

The diamond-bearing strata, together with the sands derived from them by weathering, are everywhere exposed to the action of streams and rivers which transport the material to lower levels. The next resting-place of the diamond is therefore the sands and gravels of the river bed or its alluvial deposits. The most recent of these alluvial deposits lies at the present level of the river ; others are found at higher levels on the sides of the valley, having been formed before the valley had been cut down to its present depth. These diamond-bearing alluvial deposits have a close connection with the diamond-bearing beds from which they have been derived. In any district where diamonds occur in the strata they will also be found in the beds of the streams and rivers, although not always in numbers sufficiently great to repay work on a large scale.

The mining of diamonds is at the present day, just as in former times, almost entirely in the hands of natives of the lower castes. Attempts on the part of Europeans to work the diamond-bearing deposits on a large scale, and according to modern methods, have never been attended with success. The work in many cases is tedious and difficult, and, moreover, the methods used must be altered to suit the conditions in different localities, which vary considerably. The same methods are for the most part now employed as were in use in the oldest times of which records exist ; at any rate, they are identical with those seen and described by Tavernier, the French traveller and dealer in precious stones, in 1665.

The working of the surface layer of sands, that is the loose, weathered product of the sandstone beds, and of the river alluvium, is easy. It consists essentially in removing the larger masses of rock and in washing away the finer earthy material with water. From the sandy residue thus obtained the diamonds are picked out, usually by the women and children of the workers who dig out the gravels.

The working of the sandstone beds is more arduous, and is only attempted where they lie on the surface or at a very small depth beneath it. Where they are overlain by younger beds of any thickness, they are inaccessible to the natives, whose appliances for the sinking of shafts and other mining operations are few and primitive; moreover, in such cases the cost of working would be prohibitive, and the mining of the diamonds can only be effected where the strata crop out on the surface of hill-sides, the workings penetrating to only a very small depth even in these more favourable situations.

Where the diamond-bearing bed lies at a small depth below the surface, a pit or shaft of a few square feet or yards in section is sunk to meet the desired bed, the shaft being usually about 20 feet, rarely 30 feet, and in a few cases 50 feet in depth. The workings at the bottom of the pit extend only to such distances as the stability of the material overhead will permit. The diamantiferous rock so obtained is, when necessary, carefully broken up, and the diamonds obtained from it by washing and sorting in the same manner as from the loose sands and gravels.

The excavation of the hard, solid beds of sandstone which often overlie the diamantiferous stratum is a matter of no small difficulty to the worker whose tools are inadequate for the purpose. In a few districts the difficulty is somewhat lessened by the employment of a device often made use of by the old German miners. A large fire is kindled on the spot at which it is desired to sink a shaft, and when the rock below is strongly heated, it is suddenly cooled by the application of cold water. This causes the rock to crack in many places, and thus the work of excavation is rendered less arduous.

Diamantiferous sandstone, which has been removed from its natural bed, and from which diamonds have been extricated, is often allowed to be exposed to the various atmospheric weathering agencies for some time, and is then again worked over, when a further yield of diamonds may be given, this being sometimes repeated several times. This fact has given rise to a belief among the natives that this second crop of diamonds has originated in the waste rock, or that it is the result of a fusion together of the smaller diamonds originally left behind; similar beliefs are also met with in South Africa. The actual explanation, of course, lies in the fact that during the interval in which the waste rock is exposed to the air, weathering takes place, and any stones which may have been embedded in the larger fragments of rock are thus set free and easily picked out by the searchers. A mass of rock which has once been worked over will naturally be the poorer both in the number and the size of the diamonds it contains. In spite of this, however, the refuse heaps from old diamond mines are in many places at the present time being continually turned over and diamonds as continually found.

C. Ritter arranged the Indian diamond mines known to him in five groups, according to their geographical distribution, and described them in order from south to north. In what follows this grouping will be adopted, the smaller mining districts not mentioned by Ritter being introduced in appropriate places, and information derived from later reports, especially those of V. Ball, incorporated with the matter given by Ritter. A rather different grouping of the mines is given by Ball. The map (Fig. 33) shows the distribution of the diamond-fields in India.

1. The Cuddapah Group on the Penner River.

This group includes the most southerly mines, those furthest to the east are in the neighbourhood of Cuddapah on the river Penner, where numerous mines have been worked for several centuries with varying success. At the present time the majority of the mines in this group—perhaps, at times, the whole of them—are abandoned, but this by no means indicates that the supply of diamonds has been completely exhausted. The spot at which

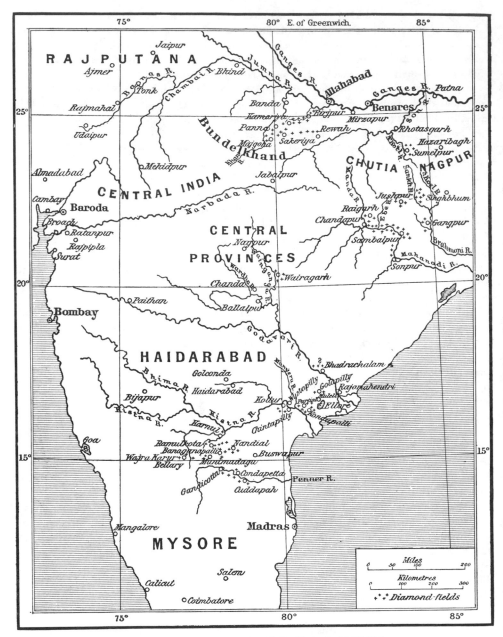

FIG. 33.   The Diamond-fields of India (Scale, 1 : 12,000,000).

diamonds have been most abundantly found is Chennur (Chinon), near Cuddapah, on the right or southern bank of the Penner river.   Westward of this, that is, up the river and on the same bank, mines are situated at Woblapully (Obalumpally).   On the other bank of the river are the mines of Condapetta, referred to by the travellers who formerly visited and described this district, and which probably corresponds to Cunnapurty of the present day. West of Chennur, diamonds have also been found at Lamdur and Pinchetgapadu, and at a few other places, of which Hussanapur (Dupand) may be mentioned as having at the beginning of the nineteenth century yielded many stones. Still higher up the valley of the Penner diamonds were formerly sought at Gandicotta, but with little success.

All these mines are referred to as the **Chennur** mines. At Chennur itself the abandoned mines are in the Banaganapalli sandstone or in the weathered products of this rock. Many stones, some of very fine quality, have been found here. In two particular cases £5000 and £3000 were obtained for single specimens. After a long period of idleness, mining operations were, in 1869, again commenced, but without success. Under the surface soil of this neighbourhood is a stratum 1½ feet thick of sand and gravel with clay, beneath this a tenacious blue or black clay, 4 feet thick, and underlying all, the diamantiferous layer 2 to 2½ feet in thickness, and differing from the clay above only in that it contains many large pebbles and boulders. The pebbles thus included in the diamantiferous clay consist of various minerals; among others there are yellowish transparent quartz, epidote, red, blue and brown jaspery quartz, round nodules of limonite the size of a hazel-nut, and corundum. The boulders are often the size of a man's head, and consist of sandstone, basalt, often of hornstone, as well as fragments of felsite, a rock of which the hills standing 1000 feet above Cuddapah are constituted.

At Condapetta the mines are from 4 to 12 feet deep. Here there is a bed of earthy sand, 3 to 10 feet thick, resting on a bed of pebbles, which vary in size between that of a nut and that of a cobble, and among which the diamonds are found, usually loose, but sometimes cemented to the pebbles. The latter usually consist of ferruginous sandstone or conglomerate, among these being others of quartz, chert, and jasper, the latter being sometimes blue with red veins; also porphyry containing crystals of felspar. The greater number of these pebbles have been derived from the surrounding mountains, but some—for example, those of porphyry—have been transported by water from greater distances. The mines here, as at Chennur, are only worked in the dry season, since in the rainy season they become filled with water, the removal of which would entail too much labour.

The mines at Woblapully were opened somewhere about the year 1750. The diamonds found here are flat and much worn and rounded, so that they show no definite crystalline form. They are specially hard and have a high lustre. In colour they are clear white or clear honey-yellow, also cream-coloured and greyish-white. They are found in alluvial deposits of varying widths which follow the course of the river, and consist chiefly of much-rounded nodules of limonite of about the size of a nut. This district has not been systematically explored, the mines, of which the average depth is 16 feet, are very irregularly scattered about, and have apparently never been of any great importance.

Following up the Penner valley and then turning to the north we reach Munimadagu and Wajra Karur, two important diamond localities in the Bellary district.

The first of these, **Munimadagu**, is sixteen miles west of Banaganapalli and forty-one miles east of Wajra Karur. Here, in a circular area some twenty miles in circumference, are a number of mines which in former times, especially during the period between the beginning of last century and the year 1833, supplied the important market and cutting-works of Bellary with the bulk of their material. The systematic working of the mines on the particular diamantiferous bed has, however, now been given up, although a few stones are occasionally still found in the neighbourhood. The diamond-bearing stratum is of small thickness, and rests upon granite, gneiss, and similar rocks.

Wajra Karur is another locality from which a more abundant yield of diamonds was obtained in former times than is the case at the present day. To emphasise the fact that diamonds are still to be found here, we may mention the stone of $67\frac{3}{8}$ carats discovered in 1881, from which was cut a beautiful brilliant of $24\frac{5}{8}$ carats, valued at £12,000. Some of the largest and most famous of Indian diamonds are said to have been found here. The occurrence of diamonds at this place is peculiar: they lie loosely scattered about on the surface of the ground, and there is no definite diamond-bearing bed. The rocks at the surface are granite and gneiss, and the diamantiferous Banaganapalli sandstone has not been detected in the district. The diamonds are often washed out of the soil by heavy rains, and are then picked up casually, or an organised search for them may be made by the people of the district.

In order to explain the peculiar mode of occurrence of diamond at this locality, it has been supposed that in earlier geological times a diamantiferous bed covered a large area in the neighbourhood of Wajra Karur, and that this has since been entirely removed by denudation, leaving the diamonds behind as an unalterable residue. Although there is nothing impossible about this view, it is supported by no definite facts.

Later investigators have attempted to explain the mode of occurrence of diamond in this district in other ways. To the west of the town of Wajra Karur a pipe of blue rock, very similar in character to a volcanic tuff, was found in the granite or gneiss. This closely resembles the richly diamantiferous rock of Kimberley, in South Africa, and was therefore supposed to be the original mother-rock of the Wajra Karur diamonds. This bluish-green, tuffaceous rock, with interspersed blocks of granite and gneiss, was worked on a large scale by an English company with absolutely no success, not a single diamond having been found.

More recently another solution of the problem has been offered by the French traveller, M. Chaper, who searched the district for diamonds in 1882. This explorer found that the surface rock lying just beneath the soil, which in the neighbourhood of Wajra Karur is gneiss, is penetrated by numerous veins of various igneous rocks. These veins very frequently consist of a coarse-grained, rose-red or salmon-coloured pegmatite containing epidote (pegmatite being a special variety of granite). In the upper, much-weathered portion of such pegmatite veins M. Chaper himself collected two small diamonds, which were accompanied by irregularly bounded grains of blue and red corundum (sapphire and ruby) as well as by other minerals. The two diamond crystals were octahedral in form with perfectly sharp edges, and showed no signs of having been water-worn. Numerous diamonds are said to have been found under the same conditions by the natives. Chaper was convinced that the diamonds he collected had been originally formed in the pegmatite, and had been loosened from it only by the weathering of the matrix. This theory would of course apply equally well to all the other regularly developed crystals of diamond found at the same place.

The Indian geologist, Mr. R. B. Foote, has raised a doubt as to the correctness of Chaper's observation, and specially of the deduction he drew therefrom, suggesting that the French traveller was deceived by his native attendant. A confirmation of Chaper's statement is much to be desired, since it would be of considerable help in elucidating the general problem connected with the identity of the original mother-rock of Indian diamonds. The original matrix of all Indian diamonds may possibly have been similar in character to the rock in the neighbourhood of Wajra Karur, the weathering and breaking down of which has given rise to the sandstones and conglomerates in which the diamonds are now found, but which cannot under any circumstances be regarded as their place of origin. In support of Chaper's view may be mentioned the fact that diamonds in the lower Penner district are sometimes associated with the minerals which Chaper observed at Wajra Karur —namely, ruby, sapphire, and epidote. Foote meets this argument with the statement that ruby and sapphire have never been found at Wajra Karur except with the two specimens found by Chaper, and these, moreover, he considers show signs of workmanship. Were it further confirmed, the reported occurrence of diamond in pegmatitic rocks, both in Lapland and in Brazil (Serra da Chapada, in the State of Bahia), would afford support to Chaper's views.

2. The Nandial Group between the Penner and Kistna Rivers.

This group lies near the town of Banaganapalli, and only about seventeen miles north of the last group. It is situated on the northern margin of the plain, which extends from the western slopes of the Nallamalais as far as the town of Nandial (lat. 15° 30′ N., long. 78° 30′ E.). The mines of this group, which are sometimes referred to—for example, by V. Ball—as the Karnul diamond mines, lie to the east, south-east, and west of Nandial, and are partly in the diamantiferous bed itself and partly in the sands. This group, of which a few only of the more important workings can here be mentioned, includes some of the most famous mines ever worked in India, the majority of which, however, are now abandoned.

The mines at **Banaganapalli,** the village which gives its name to the group of strata containing the diamantiferous sandstone, lie to the north-west of Condapetta and to the south-west of Nandial. According to the observations of Dr. W. King, the sandstones together with the diamond-bearing bed rest unconformably upon the older sedimentary rocks beneath—that is, the lines of bedding of the two series differ in inclination. These older sedimentary beds comprise shales and limestones with interbedded trap-rocks. The diamond-bearing bed and its associated sandstones are from twenty to thirty feet thick. They are penetrated on the hill slopes by pits never exceeding fifteen feet in depth, at the bottom of which the diamond-bearing bed has been removed as far in all directions as the stability of the overlying rock will permit. This bed, which is only from six to eight inches in thickness, is constituted of a coarse sandy or clayey conglomerate or breccia, consisting largely of variously coloured fragments of shales and hornstone. Large diamonds have apparently never been found here. The crystalline forms of most common occurrence are those of the octahedron and the rhombic dodecahedron. Workers of the present day confine themselves for the most part to turning over the refuse-heaps of abandoned mines in search of small stones, but a few mines in the sandstone are being actively worked at the present time.

The mines of **Ramulkota** are situated to the north-west of Banaganapalli and about nineteen miles south-south-west of Karnul. They are in the Banaganapalli sandstone and are worked more deeply and extensively than are those of Chennur, near Cuddapah in the Penner valley. The stones found here are small and not very regular in form ; they may be white (i.e., colourless), grey, yellow or green in colour. The exact output at the present time is not known. The mines in the sandstone are not now worked, but the washing of the neighbouring diamond-bearing sands is carried on to a small extent. Captain Newbold, who visited this district in 1840, saw only twenty men at work here, but in the dry season the number was said to be increased to 500. The rich and famous mines mentioned by Tavernier under the name Raolconda are probably identical with the mines of Ramulkota ; at the time of his visit (1665) these mines had been worked for 200 years and were a source of much wealth. After the working of these mines had ceased, their very situation became completely forgotton ; they were at one time supposed to lie five days' journey west of Golconda, near the junction of the Bhima and Kistna rivers, and eight or nine days' journey from Visapur (now Bijapur) ; the researches of V. Ball have now, however, practically established the identity of these mines with the Ramulkota mines of the present day.

3. The Ellore (or Golconda) Group on the Kistna River.

The mines of this group are situated on the lower portion of the Kistna river and include some of the oldest and most famous of Indian diamond mines, the largest and most beautiful of Indian stones having been derived from these so-called Golconda mines. They derive their name, not from their situation, but from the fact that the diamonds from these mines were sent to the market held near the old fortress of Golconda, not far from Haidarabad, this being also the market for stones from Chennur. At the time of Tavernier's visit to these mines, more than twenty were being worked, most of them being extraordinarily rich. With two or three exceptions, the whole were later deserted, and the situations of many of them, including some which Tavernier described as being most famous, are now forgotten.

The richest of the mines to the east of Golconda were those of **Kollur,** which lies on the right bank of the Kistna, west of Chintapilly and in latitude 16° 42½' N. and longitude 80° 5' E. of Greenwich. This place was referred to by Tavernier under the name Gani Coulour, and now sometimes figures as Gani. This latter is a native word said to signify " mine," while the word Coulour, from which is derived the now common place-name Kollur, is of Persian origin. These mines are not identical, as has often been supposed, with the also far-famed mines of Partial ; the latter, which will be described below, are situated somewhat further east and on the left bank of the Kistna.

The discovery of the diamantiferous deposit at Kollur was made about 100 years before Tavernier's visit, namely, about 1560. A 25-carat stone was first accidentally found, and numerous others soon followed, many weighing from 10 to 40 carats, and some still more. The quality of the stones, however, was not always as satisfactory as their size, cloudy and impure specimens being frequent. Such famous diamonds as the " Koh-i-noor," now in the English crown jewels, and the " Great Mogul," the whereabouts of which, unless it is identical with the " Koh-i-noor," is now unknown, were very probably found in these mines, in addition to some beautiful blue stones, including the " Hope Blue " diamond. Tavernier stated that 60,000 people were engaged in these mines at the time of his visit ; to-day, however, they are completely deserted, as are also numerous other workings situated in the valley of the Kistna, between Kollur and Chintapilly, and between the latter place and Partial. The diamonds here lie in a loose alluvium, which is thus a diamond-sand.

In following the course of the Kistna river, a little beyond where it is joined by the Munyeru river, to the east of Chintapilly, we reach the **Partial** mines, standing on the left bank of the river. These mines also were formerly very rich and probably yielded the " Pitt " or " Regent " diamond, now in the French crown jewels. The workings, which are here in the loose decomposed mass of the diamantiferous bed and in the river alluvium, have been abandoned for a long period, although the diamantiferous bed is probably not exhausted ; in 1850, according to Dr. Walker, only two mines of this group were being worked. Near to Partial, and belonging to the same group, are the old mines of Wustapilly, Codavetty-Kallu, &c. ; the latter is said to have been especially rich, there being a legend to the effect that waggon-loads of diamonds had been taken away. Here again the diamonds occur in sands which are now no longer worked.

Still further east, on the left (north) bank of the Kistna, but at some distance from the river, are the **Muleli** or Malavilly mines, situated between the village of the same name and that of Golapilly, to the north-east of Condapilly, and about six or seven hours' journey west of Ellore. Here pits fifteen to twenty feet deep are excavated in a conglomeratic sandstone or in the surface débris derived from its disintegration. These sandstones rest on gneiss and belong to a somewhat later series of beds than does the Karnul series. The diamantiferous stratum, which according to many observers is overlain by a bed of calcareous travertine, consists mainly of pebbles of sandstone, quartz, jasper, chert, granite, &c., as well as of large fragments of a limestone conglomerate, which show no traces of having been water-worn. All the minerals which accompany diamond at Cuddapah are also present here, with chalcedony and carnelian in addition. These mines have been worked at least as recently as the year 1830, but the yield has since fallen off and they are now abandoned.

In the district in which this group is situated, which lies partly in Haidarabad, the " Hyderabad Company " of English capitalists has acquired working rights. The Company's total output of diamonds in the year 1891 was $862\frac{3}{4}$ carats, valued at 15,530 rupees. The annual output of the whole group of mines is at the present time little greater than this, being perhaps about 1000 carats.

To the north of the district just mentioned, diamonds are said to have been found at Bhadrachalam on the Godavari river. Their occurrence here is, however, doubtful, if not mythical, few if any stones having been found; the whole district is little known, and rendered extremely inaccessible by the thickness of the surrounding forests. Much richer and more important, at least in former times, is the fourth group of diamond mines now to be described.

4. The Sambalpur Group on the Mahanadi River.

This group is situated a good distance to the north-west of the previous group, and lies between latitude 21° and 22° north, in the Central Provinces. The diamonds known to the ancients may have been those of the Mahanadi river, the diamond river mentioned by Ptolemy being supposed to be in this district, and being, in fact, identified by many authors with the Mahanadi river itself. The occurrence of diamond is limited to the neighbourhood of Sambalpur, no other part of the river having given any yield. The mining district extends over a fertile plain, which at the town of Sambalpur stands 451 feet above sea level, and forms the stretch of land between the Mahanadi and Brahmani rivers. The date of the first discovery of stones here is unknown, but Sambalpur has been a familiar diamond locality since very remote times.

The diamonds are found for the most part in the neighbourhood of the confluences of the Mahanadi with some of the tributaries on its left bank. These tributaries, which flow into the river from the north, rise in the Barapahar hills; one of these, which joins the Mahanadi a little above Sambalpur, is the Ebe, and is sometimes considered to be the diamond river of the ancients, but whereas the occurrence of diamonds here has not been proved, there is no doubt as to their occurrence in the Mahanadi valley. In former times the stones were collected in the river-beds after the rainy season. They were found in the Mahanadi river only on the left bank, never on the right, and not higher up than where the Manda tributary enters the main river at Chandapur; according to some accounts, which however are probably incorrect, the mouth of the Ebe is the furthest up-stream limit to the occurrence of diamonds, and this river is therefore often considered to be the one down which the diamonds were transported into the Mahanadi. The whole diamond-bearing stretch of the Mahanadi is about twenty-eight miles long, being limited eastwards by a bend in the river at Sonpur.

One of the most important points on the Mahanadi appears to have been **Hira Khund**, a name which signifies diamond mine; this is an island about four miles long, which lies near the village of Jhunan and divides the river into two branches. Every year about the end of March or later, that is, in the dry season when the river is very low, people flocked in thousands to this place to search for diamonds. The branch of the river on the north side of the island was dammed up, and the diamond-bearing sands and gravels of the river-bed dug out and washed for diamonds by the women. The southern branch of the river was never worked for diamonds, although in the opinion of some experienced persons, they were there to be found, possibly in greater numbers than in the north branch. The damming-up of the south branch would, however, present greater difficulty since the volume of water here is greater and the current stronger than in the north branch.

Diamonds are found near Sambalpur in a tough, reddish mud containing sand and gravel. This material is probably the weathered product of the rocks of the Barapahar hills brought down by the rivers which rise there. The solid rock of this region is not, as far as is known, worked for diamonds, although it is very similar to the rocks which in all parts of southern India yield the precious stone. A certain number of diamonds are found in the small streams which rise in this neighbourhood, near Raigarh, Jushpur, and Gangpur.

Large stones are said to have been found in the Mahanadi with some frequency. The largest was found at the island of Hira Khund in 1809; it weighed 210·6 carats, but ranked only as a stone of the third water, and its subsequent history is unknown. Generally speaking the stones found here were very good in quality, the diamonds of the Mahanadi and of Chutia Nagpur ranking amongst the finest and purest of Indian stones. In the Mahanadi, diamonds are associated with pebbles of beryl, topaz, garnet, carnelian, amethyst, and rock-crystal; these minerals, however, have probably been derived from the granite and gneiss through which the river flows and not from the mother-rock of the diamond. The Mahanadi yields also a fair amount of gold, which is separated from the river sands and gravels by washing at the same time as are the diamonds.

At the present day, diamonds are found in this district only occasionally; systematic work was carried on down to about the year 1850, when, owing to the poorness of the yield, it was discontinued.

The mines of **Wairagarh**, in the Chanda district of the Central Provinces, may be conveniently described with this group. They are about eighty miles south-east of Nagpur, very ancient, and identical with those mentioned by Tavernier under the name Beiragarh; their identity with those of Vena (Wainganga) is uncertain. The remains of these mines are still to be seen on the Sath river, a tributary of the Kophraguri, itself a tributary of the Wainganga. The mines were formerly rich, but have been abandoned since 1827. The stones lie in a red or yellow, sandy, laterite-like earth, but the rock from which this alluvial material was originally derived is unknown. According to Professor V. Ball, this diamond-bearing stratum has a far wider distribution than is generally supposed, and will perhaps at some future time become of importance.

To the north of the Sambalpur district, in the **Chutia Nagpur** (the ancient Kokrah) division of Bengal, diamond mines were formerly worked. These mines are said to have yielded in the sixteenth and seventeenth centuries many large and fine stones, which are stated to have been obtained from one of the rivers of the district. The identity of this river is not exactly known, but it is supposed to be the Sankh, a tributary on the left side of the Brahmani, also a river in which diamonds have been found but at a later date; even such occasional finds are not now, however, to be made.

In Tavernier's time some famous mines, which were described by him, existed at **Sumelpur,** but their exact situation is now not known. According to the account of this traveller, the diamonds were here washed from the sands and gravels of the River Gouel. This river is supposed to be identical with the North Koel river, a tributary of the Son, which in its turn flows into the Ganges, and on the banks of which are the ruins of the ancient town of Semah or Semul, supposed to be identical with Tavernier's Sumelpur (Semelpur). This town must not be confused with Sambalpur, a town on the Mahanadi river which has been mentioned above. The stones found in this district were originally derived from the hills forming the watershed of the rivers North Koel and Sankh. Tavernier states that 8000 people were at work in these mines at the time of his visit, in the dry season at the beginning of February. Many other statements respecting the early finds of diamonds in Chutia Nagpur are now regarded as false, having nothing more substantial than fable as their foundation.

5. The Panna Group in Bundelkhand.

This, the most northerly group of Indian diamond mines, is situated between the Khan and Son rivers in latitude 25° N., and lies on the northern margin of the Bundelkhand plateau where this borders the plain of the Ganges and Jumna. Some of the mines lie in the immediate neighbourhood of Panna (Punnah), to the south-west of Allahabad on the Ganges, others are further away to the west, south and east of this town; all are classed together as the Panna mines. Large stones are not known to occur in this district nor do any appear to have been found in former times, though the number of smaller diamonds of good quality found now as well as formerly is considerable. The form of the crystals is that of the octahedron or of the rhombic dodecahedron. They occur in the special diamantiferous stratum and in the loose surface material derived from the weathering of the same, and have also been transported with river-gravels. The diamond stratum here belongs, as previously remarked, not to the Lower, but to the Upper Vindhyan formation.

In the neighbourhood of Panna, especially to the north and north-east, there are numerous mines; the most important lie close to the town and occupy altogether an area of less than twenty acres. The diamond-bearing stratum is sometimes not more than a span in thickness, and it lies deeper here than at other places where such a stratum is worked, being overlain by a bed of clay of considerable thickness containing pebbles and rock-fragments; these consist usually of sandstone, but at the base of the bed there are numerous fragments of ferruginous laterite. The absence of solid rock above the diamond-bearing stratum makes it impracticable to work the latter for any considerable distance underground; in order to reach this it is therefore necessary to excavate wide and deep pits, measuring about 20 yards across and 10 to 15 yards in depth, a proceeding which involves much labour and time. The diamantiferous stratum consists of a ferruginous clay which contains besides diamonds, fragments of sandstone, quartz, hornstone, red jasper, &c., and deserving of special mention, a green quartz (prase), the abundant occurrence of which is considered a good sign by the diamond seekers. The interior of a diamond mine in this district is illustrated in Plate V. The miners at work in the wide pit are watched by the soldiers of the native ruler. On the left of the drawing are seen the baskets in which the excavated material is hauled up to the surface for subsequent treatment; towards the right is represented a series of earthen bowls, arranged as a chain-pump, for removing water from the pit.

In the mines of Kamariga, north-east of Panna, the diamond-bearing stratum consists of loose, ferruginous earth; it is overlain by a bed 20 feet thick of the firm and coherent Rewah sandstone interbedded with bands of shale.   The solidity of the superimposed rock allows the diamond-bearing stratum to be worked underground from the bottom of the pits for some distance, so that the work is here much lighter than at Panna.   There are also several mines at Babalpur, all of which are now abandoned.

At Birjpur, to the east of Kamariga and near to Babalpur, there are mines standing on the right bank of the upper course of the river Baghin.   The diamond-bearing stratum differs from that at Kamariga, being a firm conglomeratic sandstone, which crops out at the surface and overlies other sandstones; the mining of diamonds is here, therefore, comparatively easy.

At all the mines mentioned above the diamond-bearing stratum itself is worked; the workings in the remaining mines of this group are, however, in the various sands and gravels derived from this stratum.

At Majgoha (Maigama), south-west of Panna and the most westerly point of the district occupied by this group of mines, the mode of occurrence of the diamonds is peculiar. They lie in a green mud, which is penetrated by veins of calcite and is covered by a thick deposit of calcareous travertine or tufa.   This mud is found in a conical depression in the sandstone, about two-thirds of which it fills.   This depression is 100 feet deep and 100 yards wide and being cone-shaped diminishes in diameter as its depth below the surface increases; it may possibly be an old diamond mine filled up by the green mud.   The miners work to a depth of 50 feet and assert that the mud increases in richness as greater depths are reached.   The mine is now apparently abandoned; it is not, however, considered to be exhausted but is reserved for future working.

The mines at Udesna and Sakeriya are of some importance; at the latter place, the diamantiferous gravel is overlain by yellow clay and in part also by laterite.   These mines have been worked until recent times, and possibly may not be altogether abandoned even now.   At Saya Lachmanpur, fourteen miles from Panna, diamonds are found on the top of Bindachul hill.

Finally, we must notice the long stretch of sands in the valley of the Baghin river below Birjpur.   The principal mines are at the lower end of the upper part of the valley, where the pebbly diamantiferous stratum is overlain by about 12 feet of dark brown clayey sand. At the upper end of the valley are two waterfalls, each with a fall of 100 feet, and at the foot of each diamonds are collected at levels which are respectively 700 and 900 feet below that at which the diamond-bearing stratum occurs *in situ*.

The Panna mines are at present the most productive diamond mines in India.   The profits of the workers are, however, greatly diminished by the heavy tribute exacted by the native princes, to whom the land on which the mines are situated (with the exception of Saya Lachmanpur) belongs.   All stones exceeding 6 ratis in weight are appropriated, together with one-fourtn of the value of all other stones found.   In spite of this exaction, more than three-fourths of the inhabitants of Panna and the surrounding villages obtain their livelihood by searching for diamonds.   Owing to the oppressive taxation, dishonesty is rife among the workers, stones being concealed whenever opportunity occurs.

Another place at which diamonds are said to have been found is Simla, on the lower ranges of the Himalayas and to the north of Delhi, this locality being thus quite removed from the districts described above.   Here, about 1870, a few diamonds are reported to have been found after a great storm; this occurrence is by no means an established

PLATE V

DIAMOND MINE AT PANNA, INDIA

fact, but it agrees with an old Indian tradition that diamonds have been found in the Himalayas.

From the mines of these various diamantiferous districts have been derived the enormous number of diamonds, often of large size and great beauty, which, in the course of centuries, have slowly accumulated in the treasuries of Indian princes or have been used in the gorgeous adornments of idols, sacred shrines, and temples. Up to the tenth century almost all the diamonds discovered remained in the country, and it was not till the invasion and plunder of India by other nations that any portion of these treasures was carried away, first into other eastern countries and subsequently into Europe. The first of these occasions was the invasion by the Persians under Mahmud of Ghazni, at the end of the tenth and the beginning of the eleventh century. The magnificence and number of the diamonds amassed in India at that time is related by Ferishtah, the Persian historian of the rise of the Mohammedan power in India. We learn from his account, which was published in 1609, that Mohammed the first, of the Ghuridem dynasty in Persia, who in 1186 founded the Mohammedan rule in India, left at his death 500 muns (= 400 lbs.) of diamonds; all this enormous treasure he had amassed during the thirty-two years of his Indian sway.

Europeans became acquainted with the riches of India mainly through the writings of the Italian traveller Marco Polo, who at the end of the thirteenth century spent many years in Central Asia, China, &c. According to C. W. King, the Portuguese writer Garcias ab Horto was the first to publish, in 1565, any authentic account of Indian diamonds. Towards the end of the seventeenth century the French traveller Tavernier made himself intimately acquainted with the occurrence and mining of diamonds in India, and succeeded in actually viewing the wealth of precious stones amassed by the Great Mogul, Aurungzebe. Tavernier, who in the capacity of a merchant in precious stones spent the years between 1665 and 1669 in India, wrote a detailed description of his journeyings about the country, which is at the present time of the greatest value.

As commercial relations between Europe and the Orient gradually arose and developed, an increasing number of Indian stones found their way into Europe. The principal Indian market for diamonds, and indeed for all precious stones, was, and still is, Madras. At the time of the annexation of India by Britain, a considerable number of Indian stones found their way into English hands; this was the fate of the most famous and beautiful of Indian diamonds, the " Koh-i-noor." Originally the property of the ruler of Lahore, this diamond passed on the dethronement of this prince into the possession of the East India Company, by whom it was presented in 1850 to Queen Victoria.

India has now lost all its former fame as a country rich in diamonds; the most productive mines have long ago been exhausted, and only the poorer deposits still remain. During the devastating wars and native struggles for supremacy, many only partially exhausted mines were abandoned and their very sites forgotten, while from the same cause the demand for diamonds fell off. Moreover, the oppressive and unreasonable tribute demanded by native rulers in former times, and to a certain extent at the present day also, so crippled the industry that many diamond seekers forsook the mines for more lucrative employments, to return perhaps under more favourable conditions.

The chief blow, however, to the diamond mining industry of India was the discovery of the precious stone in Brazil, a country from which diamonds have now been sent to the market since 1728. There could be no competition between these rich and unworked deposits and the Indian mines, whose age can be counted in centuries or even tens of centuries. More recently, the rich yields of the South African diamond-fields have made a profitable mining of the Indian deposits still more impossible. Since in India no new and rich deposits have been discovered to take the place of the old, worked-out mines, as has been the case in Brazil, the time cannot be far distant when India must be excluded from the list of diamond-producing countries.

It has been thought that the diamond-mining industry of India might revive were mining operations to be in the hands of Europeans instead of in those of the natives. Several attempts have been made in this direction, but up to the present have been attended with but little success. The mines most suited for experiments of this kind are those situated in districts directly under English control, namely the Chennur mines in the Penner valley, and those of Karnul and Nandial, Sambalpur, and Chutia Nagpur. Though the economic, social, and legislative conditions even here are none too favourable for the under-taking and carrying out of systematic work, they are less adverse than in districts under the sway of native rulers, such, for example, as those in which the Golconda and the Panna groups of mines are situated, and which are very inaccessible to Europeans. As the geological structure of the country is worked out and becomes better known, it is possible that new occurrences of the diamantiferous beds may be discovered, though it must be said that at present there is no immediate prospect of such discoveries.

The insignificance of the annual output of Indian diamond-mines has already been commented upon ; the proportion of these stones which reaches the European markets is still more insignificant ; indeed, it is doubtful whether any appreciable number leave the country at all. This state of affairs finds its parallel in the times preceding the eleventh century ; now, just as then, the stones are kept in the country to satisfy the passion for gems of the great Indian princes and magnates. Another inducement to dealers to keep the stones in the country is the fact that they will frequently make a higher price there than in the European markets, where they must undergo comparison with the treasures of the whole world and where the price is regulated by the inexorable laws of supply and demand. So brisk is the demand for diamonds in the Indian markets that the native supply is barely sufficient, and many foreign stones are imported, especially from the Cape.

There are not many detailed statements of the mineralogical characters of Indian diamonds ; a few, however, have been collected and are given below.

It is often stated that the usual crystalline form of Indian diamonds is that of the octahedron, while that of Brazilian crystals is more often the rhombic dodecahedron, the two being often distinguished as the Indian and Brazilian types respectively. This view, however, is not in complete agreement with some recent scientific investigations of stones which are certainly known to have been found in India. It appears on the contrary that the octahedral form is seldom seen in India, the more characteristic forms being the tetrakis-hexahedron and the hexakis-octahedron. Of fourteen crystals of diamond in the Museum of the Geological Survey of India at Calcutta, which were examined by Mr. F. R. Mallet, nine show a tetrakis-hexahedron alone, two show this form with subordinate faces of the octahedron, two are octahedra in combination with a tetrakis-hexahedron, and one is an octahedron in combination with the rhombic dodecahedron. A tetrakis-hexahedral form is thus present in thirteen of these fourteen crystals, and on eleven of them it occurs singly or predominates over other forms ; on the other hand, the octahedron is present on five crystals only, and on only three of these does it predominate. Of the fourteen crystals examined, five were from the Karnul district (four tetrakis-hexahedra and one octahedron with tetrakis-hexahedron), one from Sambalpur (tetrakis-hexahedron with octahedron), four from Panna (much distorted tetrakis-hexahedra), the remaining four being said to have come from Simla. Also of thirty-one Indian diamonds in the mineralogical collection at Dresden only six are octahedra, while octahedral faces are present on only two or three more; the majority show the form of a hexakis-octahedron, and a few also that of the rhombic dodecahedron. The crystalline form of the stones found in different districts, when known, has been mentioned above under the special description of each district.

That large diamonds in considerable numbers were formerly found in India has already been stated; a detailed description of the largest and most beautiful will be given in a separate section devoted to the consideration of famous diamonds. The stones found at the present day are usually of small size, so that in this respect also the finds of the present day do not compare favourably with those of earlier times; large stones are, however, occasionally met with even now, as is shown, for example, by the discovery of a stone weighing 67⅜ carats at Wajra Karur in 1881.

With respect to the quality of Indian diamonds not many detailed accounts are available. Though reports dealing with single mines may mention the existence of stones of poor quality, yet, as a general rule, Indian stones rank high in the possession of the most desirable qualities. An Indian stone often shows a combination of lustre, purity of water, strength of fire, and perfect " blue-whiteness " of colour, such as is absent from Brazilian and South African stones. Moreover, India can claim for its own all the finely-coloured stones of blue, green, and red, not however yellow diamonds, which come mainly from South Africa.

2. BRAZIL.

Diamonds were first discovered in Brazil about the year 1725, in the neighbourhood of Tejuco, which is situated in the State of Minas Geraes. According to the usual accounts they were first found during the gold-washing of the auriferous sands of the Rio dos Marinhos, a tributary on the right bank of the Rio Pinheiro. The glittering of the stones attracted the attention of the gold-washers, although they were ignorant of their real nature. The stones were collected and taken occasionally (1728) to Lisbon, where they came under the notice of the Dutch consul, who recognised them to be diamonds of the best quality.

Then began an eager search all over the district, but specially in the water-courses, and it was found that all the streams and rivers there were more or less rich in diamonds. The Portuguese government claimed the stones as crown property, and marked out a definitely bounded diamantiferous district, called the Serro do Frio district, which was to be under its own control, and subject to special laws and regulations preventing the ingress of unlicensed diamond-seekers, while a strict military supervision forbad any dishonesty among the workers.

More extended search showed that diamonds were not confined to the district of Serro do Frio; numerous important discoveries were made in various parts of Minas Geraes and in other States, namely, in São Paulo and Paraná towards the south, in Goyaz and Matto Grosso in the west, and towards the north in Bahia and perhaps also Pernambuco. Discoveries of new and rich deposits have from time to time been made up to a quite recent date, so that it may be safely assumed that further discoveries are in store in the future, such discoveries being the more probable on account of the fact that many of the diamond-fields hitherto worked are situated in districts almost wholly unexplored.

Up to the present time the State of Minas Geraes has maintained its reputation as an important diamond-yielding region in spite of the fact that, owing to long years of mining operations, its present yield is now much reduced, especially when compared with the yield of the years immediately following the first discovery of diamonds. The place of Minas Geraes, as the State from which the richest yields are derived, is now taken by the State of

Bahia, which came into special prominence in the later decades of the nineteenth century. The yield of other States, compared with that of the two above mentioned, is unimportant, but all are as yet too little explored to permit of a final opinion as to their capabilities.

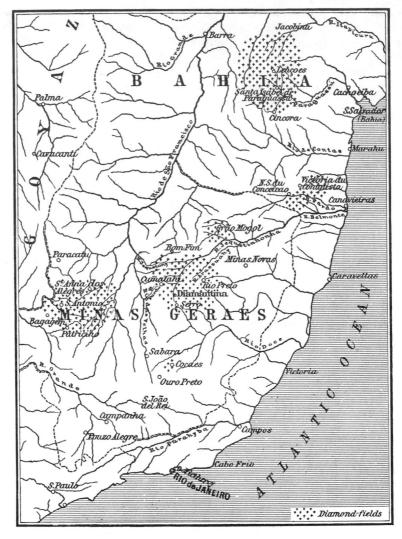

FIG. 34.   Diamond-fields of Brazil.   (Scale, 1 : 10,000,000).

Minas Geraes and Bahia are, however, the only States of commercial importance in respect to the amount and continuity of their production of diamonds, it being doubtful whether stones are found in anything but insignificant numbers in other States.

The important diamond-bearing districts of Minas Geraes and Bahia are shown in the accompanying map (Fig. 34) taken from Boutan.   It is proposed now to deal with Brazilian occurrences, taking the different States in the order of their relative importance, and treating

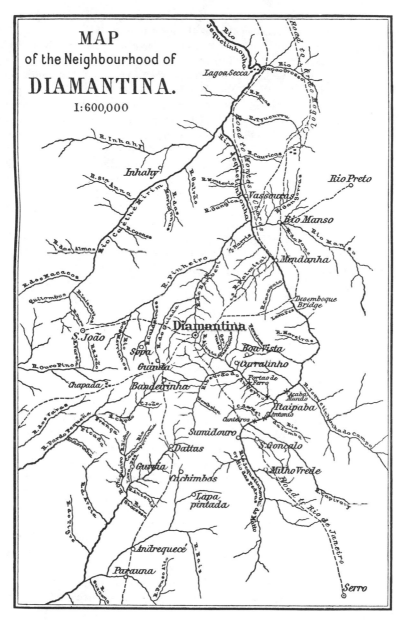

FIG. 35.   Diamond-fields of the Diamantina district, Brazil.   (Scale, 1 : 600,000).

them with more or less fulness according to the greater or less detailed character of published accounts.   We will begiu with the long famous localities of Minas Geraes, of which the majority have been very fully examined, and which, to a certain extent, serve as a type of those which follow.

It is usual to distinguish four diamond-fields in the State of **Minas Geraes**, namely, those of Serro do Frio or Diamantina, Rio Abaété, Bagagem, and Grão Mogol; of these, the first, that of Serro do Frio or Diamantina is the most important.

A sketch-map, after De Bovet, of the district of Serro do Frio or **Diamantina,** is given in Fig. 35. The area is roughly elliptical in outline, the longer axis, stretching from Serro in the south to the Rio Caéthé Mirim in the north, being about fifty miles in length; while the shorter axis, from the Rio Jequetinhonha in the east to a line drawn parallel to the Rio das Velhas through the villages of Dattas and Parauna, is about twenty-five miles. It is a wild mountain district traversed by the northern end of the Serra do Espinhaço; this mountain range runs parallel to the meridian, and forms the watershed separating the Rio de São Francisco and its tributary the Rio das Velhas from the Rio Jequetinhonha and the Rio Doce. The district forms, roughly speaking, a plateau, the rugged margins of which are cut by deep, steep-sided valleys. The principal town, Tejuco, which since the discovery of diamonds has been known as Diamantina, is situated at a height of 4000 feet above sea level and in latitude 18° 10′ S. and longitude 43° 30′ W. of Greenwich; the district of which it is the capital is now usually known by the same name.

The diamonds occur both on the plateau itself and in the valleys which cut into it; the richest and best known of these river valleys is that of the Rio Jequetinhonha, with its two branches Jequetinhonha do Campo and Jequetinhonha do Matto (Rio das Pedras) rising in the Serro do Itambe. The general direction of this river is south-west to north-east, its mouth is in the Atlantic seaboard, near to Belmonte, the latitude of which is about 16° S., and in the lower part of its course it is known as the Rio Belmonte. It yields diamonds from its source down to Mendanha, the stones being found not only in the main river but also in its tributaries. While the tributaries on the right bank, such for instance as the Rio Capivary and Rio Manso, which do not rise on the plateau of Diamantina, are poor in diamonds, the tributaries of the left bank which have their sources on this plateau are rich in diamonds, the Ribeirão do Inferno, Rio Pinheiro, Rio Caéthé Mirim, and to a less degree the Rio Arassuahy, being worthy of special notice. Other important diamond-bearing streams are a few small water-courses flowing westward from the plateau directly or indirectly into the Rio das Velhas, a tributary of the Rio de São Francisco; of these may be mentioned the Rio das Dattas, Rio do Ouro Fino, Rio do Parauna with its tributary the Ribeirão do Coxoeira, and especially the Rio Pardo Pequeña, which has yielded a large number of extremely beautiful stones, and is probably, after the Jequetinhonha, the most important of all.

Next in importance come the deposits of the Rio Jequetahy and the Serra de Cabrol to the north-west of Diamantina; these are separated from the deposits previously mentioned by a zone from which diamonds are absent. With these deposits may be mentioned a small working in the Jequetinhonha valley, about sixty miles below Diamantina. An occurrence which is remarkable in being completely isolated from the other diamond-yielding localities is that at Cocaes, where a few small diamonds have been found; this is situated considerably to the south of Diamantina, and only thirty miles north of Ouro Preto, the capital of the State of Minas Geraes.

Another locality which must be specially noticed is the basin of the Rio Doce, on the east side of the Serra do Espinhaço. The river basin is separated from the rich diamond district of the Rio Jequetinhonha by only a narrow mountain ridge, but in spite of this close proximity, only an insignificant number of stones have been found here, the explanation of which will be given later.

To the west of the Diamantina district is the **Rio Abaété,** a tributary on the left bank of the Rio de São Francisco; this river is fed by the Rio Fulda and Rio Werra, and on the left bank by its tributary the Rio Andrada.   Into the Rio de São Francisco flow also the Rio Indaia, the Bambuy, the Barrachudo, as well as the Paricatú, with its tributaries Santo Antonio, d'Almas, de Somno, de Catinga, de Prata, and others.   The diamonds of this region were discovered by unlicensed searchers (garimpeiros) in 1785, who worked at first without a concession ; in the Rio Abaété they found one of the largest of Brazilian diamonds, which weighed 138½ carats.   Although in 1791 there were as many as 1200 licensed workers at the place, the deposits seem to have been very quickly worked out subsequent to the year 1795, and in 1807 work here practically ceased.

This district embraces a stretch of country on the eastern slope of the Serra da Mata da Corda, about 300 miles long, and it is here that the rivers mentioned above have their sources.   On the western side of the same range, still in Minas Geraes, but near the border of the State of Goyaz, is the district of **Bagagem,** having about the same length as the district of the Rio Abaété, the two districts having together a width of about 250 miles. The whole area embraced by these districts, though it has been only partially explored, has yielded a large number of diamonds, many of which are of considerable size.   These include a stone of 120⅔ carats, and also the famous " Star of the South," the largest of Brazilian diamonds, which was found in 1853, and in its rough condition weighed 254½ carats.

A new diamantiferous deposit, which, however, does not appear to be very rich, has recently been discovered and worked in the district at Agua Suja, about twelve miles south of Bagagem.   The diamonds are here associated with blocks of rock, identical with that which occurs *in situ* not far away, together with much magnetite and also ilmenite, decomposed perofskite, pyrope, and rutile.   Some of these minerals, more especially the perofskite and pyrope, have not hitherto been found associated with diamond at any other Brazilian locality.   This association of minerals recalls the mineral constituents of the " blue ground " of Kimberley in South Africa, which will be described later, as will also the various minerals hitherto found in Brazil associated with diamond.

One other diamantiferous district in the State of Minas Geraes remains to be mentioned, namely, that of **Grão Mogol** (Grão Mogor), which is situated about 180 miles north of Diamantina, in a mountain range to the north-west of, and on the left side of, the Rio Jequetinhonha.   Although this district was first searched in 1813 diamonds were not found here until 1827 ; it is remarkable as being the only locality at which diamonds occur in the solid sandstone, which, at one time, was thought to be their original mother-rock.   Though the yield from this district is now small, it was formerly rather considerable, 2000 people being employed in the industry in 1839.

The **geological relations** of the diamantiferous districts of the State of Minas Geraes, especially that of Diamantina, have been frequently investigated, and, at least in the case of the latter, are fairly well understood, though many doubtful points still await elucidation. The investigations which have been made are due, at the beginning of the nineteenth century, to L. von Eschwege ; a little later to Spix and Martius ; in the fifties, first to Heusser and Claraz and then to Claussen and Helmreichen ; and in recent times to various geologists resident in Brazil, namely, Gorceix, De Bovet, Orville A. Derby, and others.

We learn from their observations that the principal rock in the Serra do Espinhaço is usually a thinly laminated sandstone or quartzite, the laminæ bearing numerous scales of pale green mica on their surface.   Some of the thin laminæ or slabs are so peculiarly constituted that they can be bent without being broken, such specimens being hence described as flexible sandstone.   The increased size of the quartz grains and pebbles renders the rocks in places more coarse-grained in character, so that it resembles a conglomerate rather than a sandstone. This laminated sandstone, which is of great geological antiquity, is usually regarded as a sedimentary rock rather than as belonging to the crystalline schists ; it is very abundant in the Serra Itacolumi, in the southern part of Minas Geraes, and has thus come to be known as itacolumite. Interbedded with it are clay-slates, and various schists such as mica-schist, hornblende-schist, hæmatite schist, &c. The rock is penetrated for short distances by veins which usually contain crystals of quartz. The beds of itacolumite and associated rocks, together with the underlying gneiss, mica-schist, and hornblende-schist, are usually inclined at a steep angle.

On the mountain-tops the itacolumite is overlain unconformably by a younger quartzite, the bedding of which is less steeply inclined than is that of the underlying rock. It is very similar in appearance to the itacolumite, and in places it merges into a conglomerate just as the itacolumite does. From the fact that irregular and angular projections of the lower rock are covered by the younger quartzite, it is evident that the two rocks are perfectly distinct, and probably belong to very different periods.

At some places, and conspicuously so in the basin of the Rio de São Francisco, the beds of itacolumite are associated with slates and limestones, in which fossils of Silurian and Devonian age are found. These slates and limestones have no direct bearing on the occurrence of diamonds, since, as we shall see later, the itacolumite must be regarded as the diamond-bearing rock ; they may, however, serve to determine the geological age, at present unknown, of the itacolumite when its relation to these rocks has been made out, which has not yet been accomplished.

We have already seen that the mode of occurrence of the diamond differs in each of these districts. Three kinds of diamantiferous deposits are distinguished according to their situation, whether on the plateau or in the valley, and, in the latter case, whether above or below the present high-water level. These are known respectively as river-deposits, valley-deposits, or plateau-deposits, according as they are found in the existing water-courses within the present limits of high water, on the sides of the valley above high-water level, or covering more or less large areas on the summits of plateaux.

Both the river-deposits and the valley-deposits are without exception constituted of sands, the plateau-deposits also having in part a similar constitution ; these sands, or alluvial deposits, consist of débris which has been transported by water, and which contains more or less rounded rock-fragments, among which the diamonds occur singly and isolated. The amount of rounding which the rock-fragments have undergone may be regarded as indicating the distance to which they have been transported from their original situation. In places, however, the rock-fragments of the plateau-deposits show no trace of having been water-worn ; when this is the case, such deposits have undoubtedly been formed on the spot they now occupy, and consist usually of much weathered rock-masses, as will be shown in a special description of certain plateau-deposits.

An attentive consideration of the distribution of the three classes of deposits leads to the recognition of a certain connection between them which is of some interest. The various diamond-bearing districts of the plateau are at the same time the collecting grounds of the diamantiferous streams and rivers ; it is a natural conclusion then that the stones now found in the sands of the river valleys have been carried there, together with sand, gravel, and other débris, from their original situation on the plateau by these same rivers and streams. This is especially the case in the neighbourhood of the town of Diamantina, which stands on a plateau, the surface beds of which consist of diamond-bearing rock. The rivers which have their sources in these rocks are in their lower courses rich in diamonds, whereas in other rivers, such, for example, as the Rio Doce and its tributaries, which rise among rocks from which diamonds are absent, no diamonds are to be found.

The connection thus existing between the deposits of the plateau and those of the valley leads to the view that the **mineral associations** of diamond, whether they occur on the hills or in the valleys, are essentially the same over the whole of Minas Geraes. The material of these deposits consists mainly of grains and fragments of the surrounding rocks, from the weathering of which they have been derived, and includes besides the diamonds various minerals which may be in a fresh unaltered condition, or more or less weathered and decomposed. The mineral most frequently and abundantly present everywhere is quartz, of which the transparent and colourless varieties occur, as well as the compact varieties such as hornstone, jasper, &c. All three modifications of titanium dioxide, namely, rutile, anatase, and brookite are met with, the last being represented by the variety known as arkansite; crystals of anatase are sometimes completely altered to rutile, while preserving at the same time their own external form; these pseudomorphs are known in Brazil as " captivos." Other minerals found in the deposits are oxides and hydroxides of iron, especially magnetite, ilmenite (titaniferous iron-ore), hæmatite, hæmatite having the external crystalline form of magnetite (the so-called martite), and limonite; also iron-pyrites, either unchanged or altered into brown hydroxide of iron (göthite), tourmaline, various kinds of garnet, fibrolite, lazulite, psilomelane, talc, mica, yttrotantalite, xenotime and monazite, kyanite, various complex hydrated phosphates (goyazite, &c.), diaspore, staurolite, sphene, and topaz, both white and blue but not yellow. In addition to diamonds, gold is frequently washed for, and is associated with platinum, the latter, however, not in sufficient quantity to be of any commercial importance. Some of the minerals mentioned above are distinguished in the district by local names; thus the black rounded pebbles of tourmaline are known as " feijas " (that is, black beans), and the brown pebbles, consisting of a hydrated phosphate, or of titanium or zirconium oxides, are called " favas " (that is, broad beans).

The minerals mentioned above are not of equally frequent occurrence; the most constant associates of diamond, after the different varieties of quartz, are the oxides of titanium (rutile, anatase, and brookite), hæmatite and martite, pebbles of black tourmaline, and specially xenotime and monazite. Even these, however, occur in varying abundance and frequency at different places; thus the same minerals do not occur associated together in the same way in every river, nor indeed in every part of the same river, this depending on the fact that the lighter minerals are transported at a greater rate than the heavier, and that some are more liable to be altered and reduced to powder than are others.

It should be mentioned here that while in the deposits of Salobro, in Bahia, corundum is found associated with diamond, it is entirely absent from the State of Minas Geraes.

Diamond-diggers are guided to a certain extent in their search for the precious stone by the presence or absence of the minerals usually associated with it, which they refer to as the *formation*. While by reason of their more sparing occurrence and small size diamonds may easily be overlooked, the associated minerals occur usually in larger and more conspicuous crystals and fragments, and are therefore more readily seen. Where the " formation " is absent a search for diamonds is useless, and never undertaken, since they are never found apart from their associates. It by no means follows, however, that diamonds are to be found wherever the " formation " exists; they may be altogether absent, or present in numbers insufficient to pay for the labour of working.

The different constituents of the " formation " are not regarded alike by the diamond-diggers. Those to which a special importance is attached, as being certain indicators of the presence of diamonds, are tourmaline pebbles (" feijas "), the oxides of titanium (especially anatase, less so rutile and brookite), iron oxides (magnetite, ilmenite, hæmatite, and limonite), the phosphates (" favas "), &c.; other minerals, such, for instance, as lazulite, are considered unimportant.   Opinions on this matter naturally vary, and are to a certain extent arbitrary, but it may be taken as a safe rule that the presence of those minerals, which are most constantly associated with diamond, is an indication which must not be disregarded.

We now pass to a more detailed consideration of the three classes of diamond-bearing deposits, namely those of the rivers, valleys, and plateaux, as they occur in the district of Diamantina and elsewhere in Minas Geraes.

The **river-deposits** are the richest of the three ; they are found in the valleys below the existing high-water level, and at the present time are the only deposits of importance, not only in this district but also in the whole of Brazil, and this in spite of the fact that the average size of the stones so found is smaller than that of the diamonds of other deposits.   In connection with the question of size, it is remarkable that stones found in the lower courses of a river are smaller than those in the upper part, and that eventually a point is reached in the river course at which the diamonds disappear altogether.   This is strikingly shown in the Rio Jequetinhonha and other rivers, in which, 60 miles below Diamantina, only very small stones are to be found.   In these rivers the material of the diamantiferous deposit is much rounded, more so than in others ; the edges and corners of the diamonds are also considerably worn.   The fact that the stones diminish in size the farther down the river they are found can easily be explained, when it is considered that material transported by water becomes more and more worn as the distance over which it travels increases, and, moreover, that the smaller the stones become, the more easily are they transported by water and the greater will be the distance they are carried from their original situation.

The diamantiferous débris which lies in the beds of the rivers and the bottom of the valleys consists mainly of rounded fragments of rocks and quartz brought down by the streams and rivers from their sources and upper reaches.   This débris is usually mixed with clay to a greater or less extent, the resulting material in the state in which it is worked for diamonds being known to the diamond-diggers as *cascalho*.   This is usually loose and incoherent, showing no signs of bedding ; at times, however, it has a firmer consistency due to the presence of the clay.   The upper portion of the mass is sometimes bound together, to a greater or less depth, by a ferruginous cementing material so as to form a conglomerate. This conglomerate, consisting largely of rounded quartz grains and pebbles, occurs in extensive beds or in isolated blocks known as " tapanhoacanga," or " canga," which may enclose crystals of diamond.   Such a fragment of conglomerate, with a diamond embedded in it, is represented in Plate I., Fig. 1 ; similar specimens are often exhibited in collections as examples of the occurrence of diamond in its mother-rock, a view which, as we have seen, is incorrect.

The " cascalho " of the diamond-diggers thus contains the diamond with its associated minerals as a finer constituent, and rounded rock-fragments as coarser material, the whole being intermixed with clay or with limonite, which may cement the material together into a more or less firm mass.   This material lies in the beds of the water-courses resting immediately on the solid rocks beneath ; in a few instances, however, the rich diamond-bearing " cascalho," the " cascalho virgem " of the Brazilians, extends upwards to the surface through the whole of the fluviatile deposits.   It has a very variable thickness, and is usually covered by a layer of material from which diamonds are absent, the so-called " barren cascalho " ; this upper layer varies in thickness, from a few centimetres to twenty or thirty metres ; in its lower portion it usually consists of an accumulation of larger rock-fragments.   The " barren cascalho " is constituted of materials similar in character to those of the more deeply situated diamond-bearing layer, to reach which it is necessary to divert the water which flows over the barren layer.

Although the diamantiferous " cascalho " is spread fairly uninterruptedly over long stretches of the beds of streams and rivers, yet its distribution over the whole course of the river is by no means regular. Here it may be accumulated in masses of great thickness, there to only a sparing amount, while in a third place it may be altogether absent. Moreover, the number of diamonds present in the material varies in different rivers, and in different parts of the same river; it is recorded of certain rivers in the Diamantina district, however, that the diamonds were so regularly distributed through the " cascalho " that it was possible to estimate with accuracy the weight in carats of the precious stone which a certain amount of this material would yield; this case is, however, very exceptional.

A large accumulation of specially rich " cascalho " at one particular point is the result of the presence of certain conditions which exist only at that point; such an accumulation is sought for with eagerness. In the beds of the rivers cylindrical holes of greater or less depth are sometimes bored in the solid rock by the action of the running water, such pot-holes or "giants' kettles" being formed in the same way in many other parts of the world. In addition to these, long channels, hollowed out of the bed of the water-course, and following its course for a certain distance, or running obliquely across it, are also to be met with, and are sometimes known as "subterranean cañons." Such hollows in the bed of a stream occur where the water has passed over softer beds, these being worn away to a greater depth than are the surrounding harder rocks. The hollows so formed may be small or of considerable size, and are often filled up with a specially rich " cascalho." In a small hollow in the bed of the Ribeirão do Inferno, which joins the Jequetinhonha near Diamantina, 8000 to 10,000 carats of diamonds were found, the neighbouring part of the river-bed being very poor. Again, in a small pot-hole in the bed of the Rio Pardo, diamonds to the weight of 180 carats were obtained by four negroes in the short space of four days. Again, the three mines in the valley of the Jequetinhonha, which recently have been specially prolific, namely, S. Antonio, with Canteiras above, and Acaba Mundo below, the mouth of the Ribeirão do Inferno, were worked in depressions of the nature of channels or " subterranean cañons " in the river-bed.

The **valley-deposits** (" gupiarras " of the Brazilians) are, as a rule, of small extent; they are formed of the same material as are the river-deposits, and the diamonds are associated with the same minerals. This deposit is also known as " cascalho," and sometimes also as " gurgulho " ; the latter term, however, is more often applied to the material of the plateau-deposits. The valley-deposits also follow the direction of the present water-courses, being situated at the sides of the valley above the present high-water level; they are, as a matter of fact, river-deposits laid down at a time when the bed of the river had not been excavated to the extent it now is. In many cases the successive levels of the former beds of the river are marked out on the sides of the valley by a series of such deposits or river-terraces.

The material of these terraces is much less worn than is that in the bottom of the valley. As a general rule, it is found that the rounding of the rock-fragments is the more pronounced the lower is the level of the terrace in which they occur, and further that the material of any given terrace becomes more worn the further it is deposited from the source of the river. At the bottom of the same valley in which river-terraces are to be seen is the present bed of the river with its deposits, the material of which is more worn and rounded than that of any of the valley-deposits. It is therefore possible by means of this difference for a person acquainted with the region to distinguish a small sample of river-deposit from one of valley-deposit.

The "cascalho" of the valley-deposits rests, as a rule, not directly upon the solid rock, but upon a variously-coloured layer of fine sand mixed with clay; this is not of any great thickness, and is called "barro." It also contains diamonds, and passes gradually, with no sharp line of demarcation, into the "cascalho" above. The "barro," however, is always distinctly bedded, while the true "cascalho," whether on the sides or the bottom of the valley, shows no signs of bedding. It is often, but not invariably, covered by a layer of red muddy earth.

The "cascalho" of the sides of the valley is usually less rich in diamonds than is that found at the bottom; the stones it does contain, however, are less worn and rounded, and relatively larger than those found in the present river-bed.

Plateau-deposits are found at numerous spots on the hills near Diamantina, and in other diamantiferous districts of Minas Geraes. A rich yield of diamonds was obtained from many of these in former times, but at the present day a few only are worked, and these are of less importance than are the river-deposits.

On the hills near Curralinho (Fig. 35), between the Rio Jequetinhonha and the town of Diamantina, lying about due east of the latter, are the rich mines Bom Successo and Boa Vista. On the plateau south-west of Diamantina, and between the basins of the Rio Pinheiro and the Rio Pardo Pequeña, are the mines La Sopa and Guinda, which are now being worked, there being here two diamond-bearing beds of different ages one above the other.

Further on in the same direction, and about twelve miles west of Diamantina, in the district where the Rio Caéthé Mirim and the Rio Pinheiro take their origin, are the specially noteworthy deposits of São João da Chapada, which will be described below. A little to the south of this place, in the neighbourhood of the source of the Rio Ouro Fino, are the diggings of La Chapada, which were formerly very rich, and are now only partially exhausted.

As regards the character of the plateau-deposits, the material of which they consist is very similar to that of the river-deposits, differing from it, however, in the presence of a larger proportion of the heavier of those minerals usually associated with diamond. This is to be expected, seeing that the lighter materials would be the more easily carried away by running water, and the heavier minerals more liable to be left behind. Thus we find the oxides of titanium and of iron present in great abundance, though even here the amount is exceeded by that of the different varieties of quartz. The material of these plateau-deposits is known as *gurgulho*, it occurs usually in horizontal beds, and is built up of coarse blocks of the surrounding rocks, with a more or less red clayey earth. In this material the diamond and its associated minerals are embedded; so indiscriminately, however, is everything coloured by the red clayey earth that it is impossible to distinguish one mineral from another until the material has been washed. In some deposits the earthy material has been removed by a natural process of washing, and here the diamonds and their associated minerals are from the first distinctly seen. The rock and mineral fragments in the "gurgulho" are very slightly, if at all, rounded, and the diamonds themselves still preserve their perfectly sharp edges and corners, and the original natural characters of their faces.

The proportion of diamonds and their associated minerals present in a given weight of the material is rather smaller in "gurgulho" than in other deposits, but the average size of the stones is greater. The distribution of the diamonds is sometimes very irregular, large numbers, aggregating in weight up to 1700 and 2000 carats, being found in a single small nest, while few, if any, in a considerable area of the surrounding " gurgulho."

Under the diamond-bearing " gurgulho," and resting immediately upon the solid rock, there is usually a layer of clay, in which also a few diamonds are to be found. Above the " gurgulho," just as in the valley deposits, is a layer of red clay of varying thickness, from which diamonds are absent. When this layer is absent, as it sometimes is, the " gurgulho " forms the actual surface of the ground and is covered by vegetation. It is said that the observation of diamonds attached to the roots of plants, scratched up to the surface by fowls, or picked up by children at play, has led to the discovery of rich deposits.

Plateau deposits at other places, such as, for example, **São João da Chapada,** on the plateau of Diamantina and about twenty miles west of this town, differ very widely from those just considered. The mines here are situated on the watershed between the Rio Jequetinhonha and the Rio das Velhas, and on the prolongation of the straight line which connects the important deposits of Boa Vista, on the hills near Curralinho (Fig. 35) with those of La Sopa. These deposits were discovered in 1833. Extensive workings were carried on for a long period, but were finally discontinued owing to the exhaustion of the deposit. In spite of this the place remains extremely important from the scientific point of view, for here may be gathered data which afford material help in solving the problem as to the nature of the original mother-rock of diamond in this region.

The diamond occurs here in variously coloured clays, which lie in a trench 40 metres deep, 60 to 80 metres wide, and 500 metres long, somewhat resembling a deep railway cutting. These clays are distinctly bedded, being inclined 50° to the east, and regularly and conformably interbedded with them are beds of itacolumite inclined at the same angle. All the strata, the clays as well as the itacolumite, are penetrated by numerous small veins filled for the greater part with quartz (rock-crystal), rutile, and hæmatite.

The yield of diamonds from these clays was very variable, but on the whole the deposit was considered poor. Tschudi, who visited the place in 1860, reported that in his presence forty-four carats were obtained in two hours, while on another occasion only ten small stones were found in twelve tons of material. The associated minerals are the same as elsewhere, the three just mentioned being specially abundant. It is a noteworthy fact, that where the associated minerals occurred in abundance there diamonds were plentiful, but where, on the other hand, the minerals were present to only a sparing extent, diamonds also were hard to find.

The minerals associated with the diamond are present at this locality in less proportion than in the ordinary " cascalho " or " gurgulho," and the same is true also of the diamond itself. As has already been mentioned, the minerals which occur most frequently are quartz, hæmatite, and rutile, other oxides of iron and of titanium, tourmaline, &c. All are found, like the diamond itself, in perfectly sharp crystals. Even the softest of the minerals found here have preserved intact the sharpness and angularity of their edges and corners. None show any indication of having been transported by running water.

These circumstances have led those who have personally investigated the deposit, namely, Orville A. Derby and Gorceix, to the conclusion that here, in these beds, the diamond is seen in its original home, and that here, in the quartz-veins by which the rocks are penetrated, it slowly took on the shape and form in which we now know it. Although no diamond has ever been found actually in a quartz-vein, yet the minerals associated with it occur in such situations with great frequency, and their constant association with the diamond, not only here but at all other localities of Minas Geraes, seems to point to a common origin for both. The fact that the diamond itself has never been found in a quartz-vein may perhaps be explained by the extreme rarity of its occurrence as compared with that of other minerals. The clays in which the precious stone lies, are decomposition products of the rocks which were originally penetrated by the quartz-veins, and which, like the surrounding schistose rocks, have been reduced to their present state of disintegration by exposure to the action of weathering agencies.

The deposit at **Cocaes,** near Ouro Preto, appears to be very similar to that at São João which we have been considering. The diamonds here occur at a height of 1100 feet above sea-level, on a plateau of itacolumite overlying mica-schist and beneath this granite-gneiss. The minerals here associated with diamond are quartz, ilmenite, anatase, rutile, magnetite, hæmatite, martite, tourmaline, monazite, kyanite, fibrolite, and gold. Of these minerals the first three in the list predominate, and quartz only occurs in rounded fragments. Since the diamond and its associated minerals occur here in a belt running east and west, it is possible that here also they have been derived from a mineral vein similar to many carrying gold and other minerals, which traverse the district of Minas Geraes in an east to west direction.

The occurrence at **Grão Mogol,** in the district of Minas Novas, is of a different type again. This town is situated in the extreme north of the State of Minas Geraes, on the left or northern bank of the Rio Jequetinhonha and about 190 miles north-east of Diamantina. As well as in the normal " gurgulho," diamonds are here found in a solid, compact, conglomeratic sandstone containing much green mica, especially along the planes of bedding. According to some accounts this is to be regarded as a single, isolated, sandstone block of enormous size; others, however, attribute to the diamond-bearing rock an extension of 300 to 400 metres. This deposit was discovered in 1833 and was worked in the thirties and forties, fragments of the sandstone being detached by means of blasting powder. All the fragments of sandstone with embedded crystals of diamond, which are sometimes, though rarely, to be seen in mineralogical collections, have come from this locality. Such specimens are not, however, in all cases genuine, for the crystals of diamond have sometimes been artificially set in the rock.

This diamond-bearing sandstone has been, and is now by some geologists, considered to be itacolumite. Those who hold this view regard the sandstone or itacolumite as the original mother-rock of the diamonds now found in it, it being considered that they are as truly constituents of the rock as are the quartz-grains. Later and more detailed examination has, however, rendered it probable that this sandstone is not itacolumite but the more recent quartzite, such as we have seen to be superimposed unconformably in the Serra do Espinhaço on the beds of itacolumite. This quartzite, although very similar in general appearance to itacolumite, is yet geologically quite distinct and probably of a much later date, having no doubt been formed of material derived from the weathering of the diamantiferous itacolumite. Which of these two views is the more correct has not yet been definitely decided. If the rock be really itacolumite, the origin of diamond at this place will differ from that at São João; if, on the other hand, it should be the later quartzite, which is more probably the case, then its occurrence here is in complete harmony with that at São João, for the later-formed quartzite must of necessity contain not only the constituents of itacolumite but also the minerals, including the diamond, which filled the veins by which it was penetrated, as is in fact the case.

A comparison of the different diamantiferous deposits leads inevitably to the conclusion that each may be regarded as typical of some one stage in the development of a single process.

Thus at São João da Chapada, and elsewhere on the plateau, we see the diamond still at the place and in the rock in which it was originally formed, though the latter has been, in part at least, altered by weathering to a soft clayey mass. In such a case we may distinguish the deposit as original or primary.

Other plateau-deposits, where the rock-fragments are but slightly, if at all, rounded, must have been laid down at an early period when the plateau had been eroded by water-courses to only a slight extent and before the present valleys came into existence. The diamonds and their associated minerals were indeed carried away by water from the disintegrated mother-rock, in which at São João they still remain, but they were re-deposited at spots not very far distant, as is proved by the fact that they are so little water-worn. The diamonds and other minerals were probably re-deposited on the floors of shallow lakes, a hypothesis which would account for the bedding of the material in which they occur. Such deposits are described as secondary or derived.

The slow but never ceasing erosion of the plateau by streams and rivers resulted in the course of ages in the formation of the valleys of the present day. Here, high up on the sides of the valley, the most ancient deposit marks the level of the original river-bed. The material for this deposit was derived partly from the primary or original deposit and partly from the secondary plateau-deposit. Some of the diamonds of the oldest valley-deposits have therefore changed their situation twice, and the material, having undergone a second transportation, is therefore more appreciably rounded. As the rivers carved out for themselves deeper and ever deeper channels, so fresh deposits were laid down, the material of each successively lower level being more and more water-worn, until the wearing down process culminates in the much rounded material of the present river-bed. The older valley-deposits are now to be found forming the terraces high up on the sides of the present valleys, while those in the present river-bed itself constitute the deposits described above as the river-deposits.

What has been said above with regard to the **original mode of occurrence** of diamond in Minas Geraes may be summarised as follows: The home of the diamond is located in those portions of the plateau-deposits in which the diamantiferous rivers take their origin. The diamonds gradually decrease in size and number and finally disappear further down the valleys. The rock, which is *in situ* in the neighbourhood of the plateau-deposits, is everywhere itacolumite, interbedded with schists and covered by the younger quartzite. This itacolumite is evidently the source from which the diamonds and the material in which they are found have been derived. So much was established by L. von Eschwege as early as the beginning of the nineteenth century; this observer noticed that in the Diamantina district diamonds were found only in those rivers, such, for example, as the Rio Jequetinhonha, which flow down the western slopes of the Serra do Espinhaço which are formed of itacolumite, and that, on the other hand, no diamonds are found in those rivers, such, for example, as the Rio Doce and its tributaries, which rise on the eastern side of this range, these slopes being formed of gneiss, mica-schist, &c. As has been already pointed out, those rivers which flow through strata composed of itacolumite are diamantiferous, while in those flowing through districts from which itacolumite is absent no diamonds are found.

The occurrence of the minerals associated with diamond, especially the most important of them, such as quartz (rock-crystal), oxides of iron and of titanium, tourmaline, &c., is also confined to the itacolumite; they do not, however, occur embedded in the rock but in the veins, consisting mainly of quartz, by which it and the interbedded schists are penetrated. The circumstance that the diamond is invariably associated with these minerals, and with them alone, points to the conclusion that it originated in the mineral-veins, as was first insisted upon by Gorceix. This conclusion receives additional support from the fact that Brazilian diamonds, instead of exhibiting a perfect and complete development on all sides such as is characteristic of embedded crystals, frequently show on one side an area by which they seem to have been attached during their growth and development, impressions of quartz-crystals being sometimes seen on such areas of attachment. Moreover, diamonds have been found enclosed in, or attached to, the surface of crystals of quartz, anatase, and hæmatite, and this could scarcely be explained except on the supposition that these minerals have all grown together at the same time and in the same vein. It has been stated by Gorceix that a few diamonds have in places been met with actually in the mineral-veins themselves, and, though in small numbers, have been extracted ; he compares such occurrences with that of the yellow topaz found near Ouro Preto in quartz-veins penetrating decomposed schists. In these districts, then, the diamond is a vein mineral, while in other localities it is an original constituent of the primitive crystalline rocks.

The precise **method of winning** diamonds adopted in Brazil depends more or less upon the nature of the deposit. A diamond-working is known in Brazil as a *serviço*, those in a river-deposit being distinguished as " serviços do rio," while those in valley- and plateau-deposits are distinguished respectively as "serviços do campo" and "serviços da serra." The methods in each case have changed but little during the whole period the deposits have been worked ; the greater part of the labour is performed by negroes working formerly as slaves, but now as freemen.

In the **serviços do rio**, or workings of a river-deposit, the first step is the diversion of the water in order to lay bare the " cascalho " or diamantiferous material. Only a small portion of the river-bed is laid bare at a time, the operation being effected either by cutting a new channel for the river, or by building a dam in the middle of the stream and parallel to its course, so as to confine the water to a bed one-half its previous width, or by conducting the water away in wooden channels. After removing the barren detritus from the surface of the river-bed so laid bare, the diamond-bearing " cascalho " is dug out. This latter is loose and easily worked, but the " canga," or masses of conglomerate, are often so compact that blasting must be resorted to, and thus the working becomes more lengthy and expensive.

The work of excavating the " cascalho " can only be pursued during the dry season, from May to the end of September, when the volume of water in the rivers is at its smallest. During these months as much as possible of the diamond-bearing " cascalho " is excavated and conveyed to a higher level, being deposited, however, as near the stream or river as safety will permit. In the wet season the level of the river rises rapidly and to a marked extent, thus making the excavation of the " cascalho " a matter of impossibility. At this season the material previously excavated is washed and the diamonds it contains collected ; the place at which the operation of washing is conducted is known in Brazil as a *lavra*.

Before the " cascalho " is washed, the larger fragments are separated from the finer material either by hand or by means of a sieve. This fine material is then placed in a shallow, wooden dish of a special kind, known as a batea, and agitated in running water ; the lighter and finer portion is thus carried away, and from the heavier remaining part the diamonds are picked out by hand, the process of washing being all the while proceeding. The washers exhibit a wonderful skill in distinguishing the smallest diamonds, such as might easily be overlooked by even a practised eye, from other mineral fragments.

Plate VI. is a picture of the actual working of a Brazilian diamond washing. The negroes on the left are standing in the stream and washing the " cascalho " in their bateas.

Plate VI

DIAMOND-WASHING IN BRAZIL

Another party of workers on the right is engaged in filling up the bateas with fresh material from the " cascalho " heaped up on the banks of the stream, and carrying them down when filled to the workers standing in the water. The negroes are all the time under the strict supervision of overseers armed with whips, whose duty it is to urge on the workers and to guard against thieving; in order to minimise opportunities for stealing, the clothing of the negroes is of the scantiest description. Each time a stone is found the worker, by raising his hand, signs to the overseer, who takes possession of the treasure. This picture in which the overseers are armed with whips dates back to the days of slavery; since the emancipation of the slaves such coercive measures have, of course, been discontinued, but otherwise the system is unaltered.

The **serviços do campo** on the sides of the valleys above the present high-water level can be worked at all seasons of the year. The water from a neighbouring stream is caused to flow over the deposit to be worked; by this means the surface earth and clay are carried away and the diamond-bearing " cascalho " is laid bare. Since sufficient water for this purpose is only to be obtained in the rainy season this part of the work is usually reserved for that period. The " cascalho " when excavated is washed and the diamonds are picked out of the concentrate, in the same way as in the " serviços do rio."

In the **serviços da serra** the removal of the masses of barren sand and earth covering the " gurgulho " of the plateau-deposits is effected in the same manner, namely, by the agency of running water. As, however, on the plateau there are no natural watercourses having a sufficient head of water, it is necessary to construct artificial reservoirs in which the rain-water may be stored. The water is conducted from the reservoir to the places at which it is required in wooden channels and the diamond-bearing bed thus laid bare as far as possible. As in other cases, the " gurgulho " is first washed and the diamonds then picked out by hand.

In the period immediately following the discovery of diamonds in Brazil, it was the practice of the Portuguese Government to demand in return for the concession of mining rights a certain sum for every slave it was proposed to employ, the total number of slaves so employed to be fixed by agreement. This tax was continually being raised, and so irksome became the conditions imposed on prospective miners that no one could be found to undertake the work. Then from the year 1740 concessions were granted on the payment of a fixed sum, but as the mineral wealth of the country still remained undeveloped, the mining was taken over altogether by the Government from the year 1772 until the separation of Brazil from Portugal. The choicest of the stones found in this period therefore found their way to Lisbon, and were preserved with the Portuguese crown jewels, a collection which comprises many unique and matchless gems. The larger proportion of the Brazilian output was bought by merchants and sent to Europe through Rio de Janeiro and Bahia.

In spite of the laws of almost draconic severity levelled against illicit diamond-mining and trading, there was, besides the Government production, a great deal of surreptitious mining by unlicensed persons (" garimpeiros ") of which of course no records were made. It has been estimated, however, that the contraband production was at least equal in amount to that of the Government. Moreover, a large proportion of the most perfect and beautiful stones fell into the hands of illicit traders, since a Government employee would scarcely risk detection for the sake of a stone of average or poor size and quality. According to other accounts the illicit trade was not of such extent and importance; in any case, however, those engaged in it, when relieved from the necessity of meeting the heavy taxes and high cost of production of the legitimate product, must have found their transactions very remunerative.

In 1834, the year in which the independence of Brazil was established, the Government monopoly of diamond-mining ceased. Since this date the concession of full mining rights has been granted to any petitioner on the payment of a small tax, varying in amount with the area he proposes to work. The landowner is also entitled to demand 25 per cent. of the rough production, and a duty of $\frac{1}{2}$ per cent. is imposed on exported stones.

The negro slaves, by whom the whole of the actual labour connected with the mines was formerly done, were subjected to the strictest supervision. To minimise the temptation to conceal valuable stones, the finders of large diamonds received special rewards; thus, at one time, the fortunate finder of a diamond of $17\frac{1}{2}$ carats received his freedom, but later, when the price of slaves rose, this custom was dropped. Those slaves, on the other hand, who were detected in the act of concealing diamonds, were treated with barbaric severity.

As may be gathered from the scene pictured in Plate VI., the whole of the work was done by hand, the " cascalho " being carried in baskets on the heads of the negroes, and no attempt made to save time and labour by the introduction of machinery or mechanical appliances. Even at the present day the same primitive methods are still in use, the difficulties in the way of the transport of large and heavy pieces of machinery to such inaccessible regions being almost insuperable. It is always more practicable therefore to employ hand labour, especially as, in addition to previous considerations, there is the fact that since any one locality is soon exhausted it would be necessary to move the machinery very often. The occupation of diamond-mining is very lucrative only under exceptionably favourable circumstances; as a rule, the working expenses are very high and the losses by embezzlement considerable.

Brazilian diamonds, when they first appeared in the market, were not favourably received by the diamond-buying public, and were asserted to be either not diamonds at all or inferior stones from India. On this account many Brazilian stones were sent first to Goa, a Portuguese possession in India, and from thence entered the market as Indian stones. When this arrangement came to the ears of the Dutch merchants, they at once entered into a contract by which they secured a monopoly of the trade in Brazilian diamonds, which were subsequently sent direct from Rio de Janeiro and Bahia to Amsterdam. In consequence of a treaty entered into at a later date with the English Government, the whole output was sent to London; in recent times the majority of Brazilian stones are purchased by French houses and put on the Paris market.

In the preceding pages the production and occurrence of diamonds in the State of Minas Geraes, and especially in the district of Diamantina, have been described in some detail, the total output from this State alone having exceeded that of the rest of Brazil. Other States in which diamonds occur have been already named; with the exception of Bahia they are much less known than is the district of Diamantina; their production is also much below that of either Bahia or Minas Geraes and is now probably everywhere at an end; this being so, only a short notice of them will be given.

In the State of **São Paulo,** to the south of Minas Geraes, diamonds have been found in the rivers flowing into the Rio Paraná.

In the State of **Paraná** diamonds have been found, more especially in the basin of the Rio Tibagy. This river runs through the Campos of Guarapuavas and empties itself into the Rio Parapanema, which is a tributary of the Paraná. Diamonds are found also in the tributaries of the Rio Tibagy, especially the Yapo and the Pitangru, and are everywhere associated with a somewhat considerable amount of gold. These rivers are remarkable for the presence in their beds of pot-holes and channel-like depressions, very local in their occurrence, and often containing a large quantity of stones. Just as in the districts previously considered, the diamantiferous deposits may be distinguished as river-, valley-and plateau-deposits. The discovery of diamonds in Paraná was accidental; the stones found were invariably small, rarely exceeding a carat in weight; they were usually, however, of good colour and lustre. Systematic search was undertaken a few years ago, but the yield was small and unremunerative, in spite of the considerable amount of gold present; it was, therefore, soon abandoned. The stones here are supposed to have been washed out of the Devonian sandstone through which the rivers mentioned above flow, and the sandstone itself may have been formed from the weathered débris of itacolumite.

In the State of **Goyaz**, on the western border of Minas Geraes, diamonds have been found in the Rivers Guritas, Quebre-Anzol, S. Marcos, and Paranayba. The upper part of the River Araguaia, bordering on the State of Matto Grosso and its right tributary the Rio Claro (lat. 16° 10′ S., long. 50° 30′ W., of Greenwich), and others in Goyaz are specially rich. The yield from these rivers has been considerable, the diamonds found up to the year 1850 in the Rio Claro alone amounting to an aggregate weight of 252,000 carats valued at £400,000.

In the State of **Matto Grosso** diamonds have been searched for in some of the rivers as far as the Bolivian frontier, and in places a rich yield has been obtained. The majority of the stones have been found in the neighbourhood of Diamantino (not to be confused with Diamantina, formerly Tejuco, in Minas Geraes), in the district of the source of the Paraguay and its tributaries, especially the Rio Cuyabá, a tributary on its right bank (lat. 15° 45′ S., long. 56° W., of Greenwich). The stones from here are usually small and often coloured, some, however, are of the purest water; they are distinguished by the possession of a very brilliant surface, a feature which is usually absent from Brazilian stones. Up to 1850 the State of Matto Grosso had yielded diamonds to the weight of about 1,191,600 carats valued at £1,850,000.

The geology of Goyaz and Matto Grosso is but little known; travellers, however, state that itacolumite is widely distributed; we may therefore assume that the occurrence of diamond in these States agrees in all essential points with that in Minas Geraes.

With respect to productiveness, the State of **Bahia** stands second to Minas Geraes; while the latter, however, is now for the most part exhausted, in the former new and rich deposits have been discovered. Thus the present yearly production of Bahia exceeds that of Minas Geraes, but the reverse is the case when the total production of the two States is compared.

Diamonds had been discovered in Bahia as far back as the year 1755; further search, however, was at that time prohibited by the Government in the fear that the agricultural prosperity of this fertile State might suffer. In spite of this prohibition more and more finds were made, until at the beginning of the nineteenth century the production was quite considerable. It has continued to grow, until now the yearly output exceeds that of Minas Geraes.

The first finds were made on the eastern slopes of the **Serra da Chapada** and, north of this, in the Serra do Assuária, which forms the continuation northwards of the Serra do Espinhaço, a range of mountains stretching across the greater part of Minas Geraes and passing through the district of Diamantina. The stones are found in sands and gravels in the water-courses, and are accompanied by the minerals which constitute the most important associations of diamond at Diamantina, namely, the oxides of titanium and of iron, tourmaline, and quartz (rock-crystal). In addition to these are a few others, which do not occur in Minas Geraes. In a sample of diamond-sand from the Serra da Chapada, Damour determined the following minerals: pebbles of rock-crystal, crystals of zircon, tourmaline, hydro-phosphates, yttrium phosphates (sometimes containing titanic acid), diaspore, rutile, brookite, anatase, ilmenite, magnetite, cassiterite, red felspar, cinnabar, and gold. Garnet and staurolite have also been observed here and recently euclase, but the last only as a rarity. Of these minerals, cassiterite, felspar, and cinnabar have never been found in Minas Geraes in association with diamond. Schrauf argued from the occurrence of these minerals, and especially from the association together of tourmaline, garnet, zircon, staurolite, rutile, &c., that the rocks, from which have been derived the diamond-sands of the Serra da Chapada, were allied to the gneisses and syenites of southern Norway. It has been stated in descriptions of the geological structure of these mountains that they are built up of these particular rocks, but neither in this nor in other diamond districts in Bahia has any thorough geological investigation been made, and since the minerals associated with diamond are the

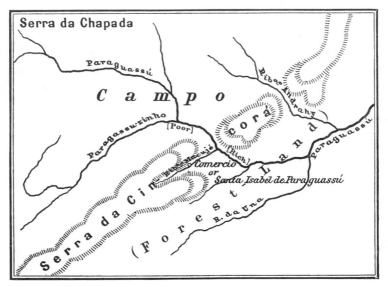

FIG. 36.   Occurrence of diamond in the Serra da Cincorá, Bahia.

same in Bahia and in Minas Geraes, it is probable that the occurrence is also the same, namely, in itacolumite.

Especially rich finds were made in the year 1844 in the **Serra da Cincorá** (Sincorá). This range is situated in longitude about 41° W. of Greenwich, and extends from south-west to north-east between latitudes 13° 15′ and 12° 15′ S. It forms the south-eastern spur of the Serra da Chapada (Fig. 36), with which it is connected at its southern end ; it separates the basin of the Rio de São Francisco from that of the Rio Paraguassú, and constitutes the collecting-ground of these rivers. This range, the Serra da Cincorá, and that of the Serra da Grão Mogol in Minas Geraes, closely resemble each other, both are rugged and inhospitable, and it is highly probable that the Serra da Cincorá consists of itacolumite, although the neighbouring heights are built up of granite and gneiss.

The discovery of diamonds here was due to the observation of a slave, a native of the diamond district of Minas Geraes, who, while engaged in minding cattle, was so struck with the similarity of the soil to that of his home that he began searching for diamonds, and before long collected 700 carats. Scarcely had this find become known, when eager searchers flocked in thousands to the place. According to some accounts, as many as 25,000 people had settled in the neighbourhood in the following year; other estimates, however place the number at from 12,000 to 14,000. Many of these fresh arrivals came from the Serra da Chapada and the Serra do Assuâria, where, in consequence of the stream of emigration, diamond-mining was almost entirely abandoned; the majority, however, were workers from Minas Geraes, where the yield of diamonds had long been gradually diminishing.

The yield of the newly discovered fields was very rich and raised the ever sinking diamond production of Brazil to its former high level. It is said that during the most productive period the daily yield averaged 1450 carats; soon, however, the yield began to decrease and the number of workers fell to 5000 or 6000. Up to the year 1849 the total output of diamonds of this district was 932,400 carats, and this immense production had lowered the prices of the stone fifty per cent. According to the estimates of diamond merchants, Bahia produced in the year 1858 54,000 carats, while from Diamantina came only 36,000 carats.

The occurrence of diamond in the Serra da Cincorá is confined to the alluvial deposits of the rivers. According to J. J. von Tschudi, who quotes the statement of the traveller V. von Helmreichen, the first discovery was made on the banks of the Macujé, a small tributary on the right bank of the Paraguassú. Here, besides a few small villages, there sprang up in consequence of the finds the principal town of the district, Santa Isabel de Paraguassú (also known as Comercio), lying about 190 miles to the west of the town of Bahia. Later, diamonds were discovered at a distance of forty-five miles from Santa Isabel. The principal place to the north of Santa Isabel is Lençoes, in the neighbourhood of which is Monte Vereno, a well-known diamond locality, where the diamond sands consist largely of fragments of itacolumite. Other important localities are Andrahy, Palmeiros, San Antonio, and San Ignacio.

The washings on the west side of the Serra have been poor; a considerable number of diamonds were, however, obtained from the Macujé itself and from those parts of the Paraguassú and Andrahy rivers which cut through the Serra. On the latter river, the principal washings are situated on the small tributary streams of its right bank. In the bed of the Paraguassú river are depressions rich in diamonds similar to those found in the diamond rivers of Diamantina.

Diamonds from the Serra da Cincorá are known as " Cincorá (Sincorá) stones," or as " Bahias," in order to distinguish them from the " Diamantina stones." They are considerably inferior in quality to the latter and command a much lower price. They are usually coloured yellow, green, brown, or red, and almost all have an elongated, irregular form which makes them less suitable for cutting. Diamonds of the purest water are rarer here than elsewhere in Brazil, and in size they are usually small, the large stone of $87\frac{1}{2}$ carats found at the beginning of the fifties being an exception to the general rule.

It is in the diamantiferous district of Cincorá that the peculiar variety of diamond mentioned several times above, namely, the black **carbonado** (" carbonate "), is almost exclusively found. Although found in association with the ordinary diamond, it is so utterly unlike it in appearance that it might be taken for anything rather than diamond.

In contrast to the ordinary diamond, carbonado very rarely exhibits a crystalline form of any regularity, still the octahedron, rhombic dodecahedron, and the cube, with rough faces and rounded edges and corners, have been observed. A crystal of carbonado with the form of a cube is represented in Plate I., Fig. 4. The substance occurs much more frequently, however, in irregular rounded nodules, varying in size from that of a pea to a mass exceeding a pound in weight. The average weight of the nodules is 30 to 40 carats, but specimens weighing from 700 to 800 carats have been occasionally met with. They sometimes have the appearance of being fragments broken from a larger mass, and some show a fine striation similar to that of fibrous coal; this latter feature is believed to be due to friction between several fragments.

The surface lustre of carbonado is dull and sometimes slightly greasy; the interior of the nodule is usually rather brighter and shows numerous brightly shining points. The colour of the exterior always lies between dark grey and black; a fractured surface is a little lighter in colour and shows a tinge of brown, violet, or red.

This substance is but rarely absolutely compact, almost invariably it is more or less markedly porous, so that it is very similar in appearance to coke. When heated in water, numerous air bubbles are expelled from the spaces in the porous material. Its cohesion is usually considerable, but some samples are easily powdered. A microscopic examination of the powdered material shows it to consist of very small octahedra of ordinary diamond, usually semi-transparent, and containing many small opaque inclusions; they are nearly always of a light brown colour and only very rarely water-clear. Carbonado is therefore nothing more than a finely granular, porous to compact aggregate of minute crystals of diamond, and is not, as is sometimes incorrectly stated, amorphous diamond. It differs also from the black diamonds which occur at some localities in regular crystals built up of a uniformly compact substance. Some specimens of carbonado aggregates are penetrated in places by ordinary diamond of a lighter colour, and having the usual strong lustre and non-porous character. Cases are also known of the enclosure in a nodule of carbonado of a small, simple, colourless crystal of diamond, the compact substance of which passes gradually into the porous substance of the carbonado shell, just as do the streaks of paler coloured diamond which sometimes penetrate the dark carbonado. The walls of the cavities in the porous carbonado are sometimes, though rarely, encrusted with minute colourless crystals of diamond.

The largest specimen of carbonado known was found July 15, 1895, in Bahia, in the neighbourhood of the town of Lençoes, between the Rio Rancardor and the stream known as the "Bicas." It is about the size of a man's fist, and when first found weighed 3167 carats; since it was taken from the ground it has gradually lost 19 carats in weight, so that its present weight is 3148 carats, or about 650 grains (nearly $1\frac{1}{2}$ pounds avoirdupois). The heaviest specimen previously known weighed only 1700 carats, and was of inferior quality.

The essential constituent of carbonado, as of ordinary diamond, is carbon; the former, however, contains a much larger amount of impurity than does the latter. After combustion, this impurity remains behind as an incombustible residue, and sometimes forms a skeleton outline of the original fragment or nodule of carbonado. The amount of incombustible ash varies from $\frac{1}{4}$ to over 4 per cent. of the weight of the carbonado burnt. Three specimens examined by Rivot contained 96·84, 99·10, and 99·73 per cent. of carbon, and 2·03, 0·27, and 0·24 per cent. of ash respectively. This ash resembled in appearance a yellow, ferruginous clay, and enclosed microscopic crystals of an undetermined substance. By treating finely-powdered carbonado with aqua regia a portion of the mineral matter constituting the ash may be dissolved out, the solution being found to contain iron and a little calcium, but no aluminium or sulphuric acid. Dana gives as the composition of carbonado: carbon 97, hydrogen 0·5, oxygen 1·5 per cent.; the presence of the last two constituents, however, requires confirmation. The view has been expressed that carbonado is a mixture of crystallised and amorphous carbon, but it is not supported by a microscopical examination of the material.

The hardness of carbonado not only equals that of diamond, but may even exceed it, and its hardness is supposed to be greater the less distinctly it is crystalline. Carbonado cannot therefore be cut by ordinary diamond powder, or at least only with extreme difficulty; it forms a valuable cutting material for ordinary diamonds, and large quantities are used as a grinding material for this and other purposes which require exceptionally hard material. On account of its great hardness, combined with the absence of cleavage (in the mass), carbonado is specially suitable for the rock-drills of boring machinery; moreover, it possesses another advantage over the diamond in that it can be easily shaped into any required form and size, while with ordinary diamond either a natural crystal or a cleavage fragment must be used.

The specific gravity of carbonado is, on account of its porous nature, lower than that of diamond crystals. The values 3·012, 3·141, 3·255, 3·416, &c., have been determined; the last three of these values were determined with the three specimens of which the chemical composition is given above, and in the same order. Carbonado, when reduced to powder, has of course the same specific gravity as ordinary diamond.

That the occurrence of carbonado is almost entirely confined to the district of the Serra da Cincorá has already been mentioned. It was found for the first time in the year 1843 in the "gupiarras" of the river San José, and all the carbonado required for technical purposes is derived from this source. In Minas Geraes carbonado may be said to be completely absent; in South Africa it is present in very small amount; and in India and Australia no trace of it has been met with. In the diamond sands of Borneo it is less rare, and here are to be found nodules of carbonado enclosed by a shell of colourless diamond. In every locality in which it occurs this black, porous variety of diamond is associated with crystals of the usual kind, they are found in the same rocks, and have no doubt a common origin and mode of formation.

The production of carbonado in the Serra da Cincorá, which in former times was considerable, has now appreciably diminished, being scarcely more than 350 grams per month. This, and its ever increasing application for technical purposes, has caused a tremendous rise in price. When first found little use was made of it, and it could be bought for about 2½d. per gram; now, however, for ordinary qualities the same weight costs 32s. and the better qualities 80s., and the price shows a tendency to rise still higher.

Diamonds in considerable numbers have also been found in the southern part of the State of Bahia, near the border of Minas Geraes. This district may be regarded as a continuation in a north-easterly direction through Grão Mogol of the diamond-fields of Diamantina. The stones are found near Salobro (signifying brackish) in the alluvial deposits of the Rio Pardo. This river and the diamantiferous river Jequetinhonha (Rio Belmonte) both empty themselves into the Atlantic Ocean at the foot of the Serra do Mar and near the small haven of **Canavieiras**. The mines are about two days' journey inland from this seaport town, and from it they derive their name of the Canavieiras mines.

The discovery of diamonds here was made in 1881 or 1882 by a forester who had previously searched for diamonds in other districts. Scarcely was the occurrence made known, when the virgin forest, in spite of the unhealthy malarial climate, became peopled by 3000 or more diamond miners. The treasure was obtained at a depth of two feet below a white clay containing decomposing vegetable matter, so that the deposit is a very recent one. The diamantiferous stratum is much more clayey than any in Minas Geraes; it has throughout the character of a plateau-deposit. Diamonds are also found in the rivers Salobro and Salobrinho, tributaries of the Rio Pardo, especially in the "gupiarras" or valley-deposits lying above the present high-water level, just as in the valleys about Diamantina.

The minerals associated with the diamond in these clays are not only less in amount but also differ in kind to a certain extent from those found in Minas Geraes. Monazite in yellowish and reddish broken crystal fragments is present in abundance, also zircon, usually brownish to white in colour, but sometimes violet, and in addition kyanite, staurolite, almandine, hæmatite, ilmenite, magnetite, iron-pyrites, and a somewhat considerable amount of corundum. The occurrence of corundum is remarkable, as hitherto it has been found in no other Brazilian deposit, while all the other minerals mentioned do occur in association with diamond in various parts of Brazil. We may contrast with the occurrence here of corundum the complete absence of certain minerals, which in other parts of Brazil are frequently found with diamond, namely, rutile, anatase, tourmaline, and the hydro-phosphates.

As regards the origin of the diamonds found here, it has been supposed that they are derived from the gneiss, granite, and other ancient crystalline rocks of the neighbouring coast range, the Serra do Mar. In the diamantiferous deposit, however, there is no trace of felspar or mica, two essential constituents of these rocks; moreover, the minerals chrysoberyl, andalusite, tourmaline, beryl, &c., which are frequently present in such rocks in Brazil, are also conspicuous by their absence from these deposits, so that this suggested origin for the diamond seems decidedly doubtful. For a satisfactory determination of the mother-rock of these diamonds, further investigation is required; in any case, it does not seem to be itacolumite, since this rock has not been observed in any part of the surrounding district.

Immediately after their discovery, the yield of these mines was so abundant that other diamond districts became more or less deserted. The stones are distinguished by great purity and freedom from colour, as well as by their very regular octahedral form, which obviates any necessity for preliminary cleaving, and enables them to be cut at once. For a time these mines supplied a large proportion of the total Brazilian output; they may not, however, have been as rich as they appeared, for it has been asserted that many Cape diamonds were sent to Canavieiras in order to be put on the market as Brazilian stones and so command a higher price, just as in former times Brazilian stones were shipped to India to enter the market as Indian stones. At the present time the yield, compared with what it once was, has fallen off considerably, the deposit being now almost exhausted. The same is true to a greater or less extent for all the known diamond-fields of Brazil, and generally speaking it may be said that all are now worked to only a small extent.

Brazilian diamonds, generally considered, show certain characters which are common to all diamonds, but possess other characteristics peculiar to themselves which enable an expert to recognise their Brazilian origin and sometimes even to name the actual district in which they were mined.

In **size** Brazilian diamonds are almost invariably small, being surpassed in this respect by Indian and especially by South African stones, many of which are above the average size. The average weight of Brazilian stones is $\frac{1}{4}$ carat or perhaps less. Large numbers of diamonds smaller in size than the head of an ordinary pin are lost in the process of washing, their size not being sufficient to repay the trouble of collecting. Stones the weight of which lies between $\frac{1}{4}$ and $\frac{1}{2}$ carat are frequent, those varying in weight from 1 to 5 or 6 carats are rare, and the occurrence of still larger stones most unusual. In Diamantina, when the yield was most abundant, only two or three stones of 16 to 20 carats were found yearly, and several years might elapse before the discovery of one of still larger size. Generally speaking, a lot of 10,000 Brazilian stones will contain only one stone of 20 carats, while 8000 of them will weigh but one carat or less each. From 1772 to 1830, the period during which the mines were under Government management, only eighty stones exceeding an oitava (= $17\frac{1}{2}$ carats) were secured by rightful owners; what may have been stolen is, of course, not known.

The largest Brazilian diamond is the "Star of the South," or "Southern Star" (Fig. 48), which was unearthed in the fifties at Bagagem. In its rough condition it weighed $254\frac{1}{2}$ carats, and when cut as a brilliant 125 carats. A stone of $138\frac{1}{2}$ carats was found in the Rio Abaété, and one of $120\frac{3}{4}$ carats in the Caxoeira Rica near Bagagem, while one of 107 carats was reported from Tabacos on the Rio das Velhas. No other stones exceeding 100 carats have been heard of. The famous "Braganza" of the Portuguese crown jewels, a reputed diamond as large as a hen's egg and weighing 1680 carats, is probably only a pebble of transparent, colourless topaz; accurate information on the subject cannot, however, for obvious reasons, be obtained from the Portuguese Government.

The **crystalline form** of Brazilian diamonds is by no means constant, varying in stones from different districts. Moreover, stones from different localities are not equally regular in form, those from the Cincorá district, for example, being more distorted and misshapen than stones from Minas Geraes or Salobro.

Generally speaking, the principal forms for all localities are the rhombic dodecahedron and the hexakis-octahedron, both having rounded faces and often deviating considerably from the ideal form (Fig. 31, c to f.) The octahedron, which is rare, is also frequently distorted, sometimes appearing in the form of triangular plates. The predomination of cube faces (Fig. 31 a) is especially characteristic of Brazilian crystals; such forms are very frequent here, but rare in other countries. The tetrahedron and other hemihedral forms, especially the hexakis-tetrahedron (Fig. 31 k), are only rarely found; twinned rhombic dodecahedra (Fig. 31 h) occur frequently; twinned octahedra (Fig. 31 g) are, on the other hand, rare.

Irregular intergrowths of diamond crystals are frequently met with; indeed the famous "Star of the South" formed part of such an intergrowth, since its rough surface showed several impressions of smaller diamonds. Nodules of bort occur not infrequently; often they are almost perfectly spherical in form (Plate I., Fig. 3), the surface, however, being rough owing to the projection of the corners of the small octahedral crystals which build up the radial aggregate. On the whole, about one-fourth of Brazilian stones are useless as gems; these are also described as "bort" and are applied to technical purposes.

The surface of a rough diamond, that is of the natural crystal, is either smooth and shining, or rough, striated, and dull. Rough stones are usually opaque or translucent, but are sometimes completely transparent; in the latter case they exhibit a fine play of prismatic colours, such as is usually only apparent after cutting. The peculiar surface lustre, characteristic of stones from Matto Grosso, has been previously mentioned; it is found on no other Brazilian diamonds. Diamonds penetrated in all directions by cavities, so that their structure comes to resemble that of pumice-stone, are occasionally met with. Regularly formed depressions may sometimes be seen on the surface of a crystal; very frequently these depressions have the shape of crystals of quartz and must have been formed by the diamond resting during its growth on a quartz crystal. Diamond crystals showing evidences of contact with other minerals have been often described; the "Star of the South" (Fig. 48) is undoubtedly such a crystal, the broad under surface being very probably the area by which it was attached to the parent rock.

The **colour** and the qualities depending on this feature vary considerably, differing in different localities. About 40 per cent. of Brazilian diamonds are completely colourless and of these 25 per cent. are of the purest water and the first quality, the beautiful and highly prized " blue-white " being not of very great rarity.   About 30 per cent. show a slight tinge of colour, and though the remaining 30 per cent. have a pronounced colour, stones of a deep and beautiful shade are rare.   Next to colourless stones, those of a dull whitish or greyish tint occur most frequently.   The lighter tones of colour are, as we have already seen, frequently confined to the surface of the crystal, which may be removed by grinding or by the simpler process of burning, and thus the colourless heart of the crystal obtained. Such stones, and also those in which the colour is confined to the edges and corners, have been found in the district of Diamantina and especially in that of the Rio Pardo and the Serra da Cincorá.   Deep tints of colour usually permeate the whole substance of the stone. Diamonds which are differently coloured in different parts have also been met with.   The enclosure of foreign bodies in diamonds is frequently seen ; these may be dark in colour or black, and sometimes resemble the moss-like markings of a moss-agate.   The colours which have been observed in Brazilian stones are yellow, red, brown, green, grey, and various shades of black ; blue is rare, but a few stones showing a beautiful shade of this colour are said to have been found.

Passing now to the consideration of the general quality of Brazilian stones, it may be stated that this on the whole is good, and surpasses that of Cape diamonds, which, as a rule, have a yellowish tinge.   The quality of Brazilian stones very nearly approaches that of Indian diamonds, the best " blue-white " Brazilian diamonds being in no way inferior to the choicest of Indian stones.

The various diamond localities of Brazil do not, however, produce stones of uniform quality ; the largest, most beautiful, and those most free from colour, have been found at Bagagem.   All the stones mined here do not by any means, however, tally with the above description, many are coloured brown or black, and besides their undesirable colour often exhibit an irregularity of form and numerous other small faults which combine to render them of little value.   The stones from the Canavieiras mines stand next in order of quality to those from Bagagem.   These, though small, possess a perfect whiteness, few faults, and great regularity of form ; by daylight they exhibit a fine lustre and play of prismatic colours ; by artificial light, however, these qualities are less marked and the stones compare unfavourably with Cape diamonds.   Diamantina takes the third place in the quality of the diamonds it produces, and stones from different localities in the district show certain differences among themselves which are well known to the inhabitants ; thus some mines yield white stones exclusively, others yield only coloured stones ; the latter, as a rule, predominate ; the same applies also to the district of Grão Mogol.   Diamonds from the Cincorá district rank lowest of all ; three-fourths of these are coloured, almost all are of irregular forms unfavourable for cutting, and about one-half are fit only to be used as bort.   The colour of diamonds from Bagagem and Canavieiras is confined to the surface, which is usually bright and only very seldom dull.   The surface of stones from Diamantina is not infrequently decidedly rough, it is seldom bright except when the stones have the form of a regularly developed octahedron.

The **production** of Brazilian diamonds has from the time of their discovery, about 1725, been very considerable.   For the eighteenth century and the early decades of the nineteenth century exact official returns were given, but for the years immediately following the first discovery, and also for quite recent years, no absolutely reliable records exist, and the various statements which are met with are based on more or less inaccurate estimates. The official returns account only for stones acquired in a legitimate manner and, of course, leave out of the calculation such as have been surreptitiously mined or obtained by dishonest means.   W. L. von Eschwege, at one time chief mining inspector in Brazil, estimated the contraband product to have been at least as large as the legitimate output, while other estimates place it at one-fifth or one-third of this.

The same authority, W. L. von Eschwege, estimated the yearly production between 1730 and 1740 at 20,000 carats, but for the first twenty years he gives the annual production as 144,000 carats, probably in this estimate making an allowance for smuggled stones. According to the official returns, the total production between 1740 and 1772 was 1,666,569 carats, corresponding to an average yearly production of about 52,000 carats, while between 1772 and 1806 the total of 910,511½ carats, corresponded to a yearly average of about 26,800 carats. For the latter period, 1772 to 1806, F. dos Santos gives the total production as 1,030,305 carats. The production even thus early had therefore considerably fallen off, and was still further diminished during the period between 1811 and 1822, when it stood at 12,000 carats. The total legitimate production of diamonds in Brazil from 1730 to 1822 is estimated by von Eschwege to be 2,983,691⅓ carats. From the first discovery to the year 1850 the total output is given as 10,169,586 carats, or about two tons, and is valued at £15,825,000. Of this, at least 5,844,000 carats, valued at £9,000,000, that is, more than half, has been contributed by the State of Minas Geraes alone.

In 1850 and 1851, in consequence of the discovery of the Cincorá mines, there was a very heavy production, namely, 300,000 carats per annum, but in 1852 it had sunk to 130,000 carats. From 1851 to 1856 the average yearly yield was 196,200 carats; from 1856 to 1861 it was 184,200 carats; and during the following years remained about the same in amount. In 1858 the leading diamond merchant of the country estimated the average annual output for all previous years at about 90,000 carats, of which 36,000 came from Minas Geraes and 54,000 from Bahia. In 1860 and 1861 the yield appears to have again risen.

For more recent times Boutan gives the following totals compiled from information derived from various sources: Diamantina from 1843 to 1885, 1,500,000 carats; other localities in Minas Geraes, together with Goyaz, Matto Grosso, &c., up to 1885, 1,500,000 (?) carats. Chapada in Bahia from 1840 to 1850, 100,000 carats; from 1850 to 1885, 1,500,000 carats. Since diamond-mining ceased to be a Government monopoly, no official records of the production have been kept; the data given above have therefore been compiled from the records of the amounts paid as export duties on diamonds and may be regarded as coming somewhere near the truth, since the number of diamonds exported does not differ widely from the number mined.

The marked fluctuations in the yearly averages, which will be observed on studying the numbers quoted above, are due to the exhaustion of old deposits and the discovery of new ones. Thus the rise in the yield which has recently taken place is due to the discovery of the Canavieiras mines, and it may be reasonably expected that in the future similar new and rich deposits will be discovered, which will have the effect of again raising the total yield. The enormous and steady production of the South African diamond-fields naturally makes the prospector less eager to start in search for new Brazilian deposits. Such, however, may be at any time accidentally discovered, as has, in former times, frequently happened, especially in Bahia.

3. SOUTH AFRICA.

The diamond mines of South Africa are, at the present day, by far the most important and richest in the whole world; at least nine-tenths of the diamonds now marketed being the so-called Cape stones. The diamond markets of the world are now completely controlled by the owners of the South African mines, the output from Brazilian, and especially from Indian, mines being so insignificant in comparison that their effect on the market is inappreciable.

The first exact scientific account of the Cape diamond-fields is due to Professor Emil Cohen, who visited the region in 1872, and his observations are still of great importance. Numerous other inquirers have continued his investigations and have cleared up many details, but no essentially novel theories have been advanced. Comprehensive accounts of these deposits have been published by Moulle, Chaper, Boutan, Reunert, Stelzner, and others, and the details given below are taken from the original works of these and other investigators. A map of the South African diamond-fields is given in Fig. 37.

Diamonds were first found in this region in the year 1867, reported discoveries at dates preceding this—for example, in the eighteenth century—being, for the most part, unfounded. Many versions of the circumstances under which the first discovery was made are in existence. According to one, a traveller of the name of O'Reilly saw a child playing with a bright and shining stone in the house of a Boer by name Jacobs, whose farm, "De Kalb," was situated a little to the south of the Orange River, and not far from Hopetown. This stone the traveller showed to Dr. W. Guybon Atherstone at Grahamstown, who determined it to be a diamond crystal weighing $21\frac{3}{16}$ carats. After being exhibited at the Paris Exhibition of 1867 it was purchased for £500 by Sir Philip E. Wodehouse, then Governor of Cape Colony. O'Reilly obtained from the same Boer a second stone weighing $8\frac{7}{8}$ carats, which had also been accidentally found on his farm ; this also passed into the possession of the Governor of the Colony at a price of £200.

According to another version, the diamond of $21\frac{3}{16}$ carats, the Boer child's plaything, first passed into the hands of Schalk van Niekerk, a Boer who was otherwise connected with the history of the discovery of diamonds in South Africa, since in 1869 he obtained from a Kaffir a stone of $83\frac{1}{2}$ carats, which came into the market under the name of the "Star of South Africa." Schalk van Niekerk is said to have handed over the stone previously obtained to O'Reilly for determination. In any case, it seems to have been the latter who took the initiative in identifying the stones, and thus firmly establishing the occurrence of the diamond in South Africa, so that to him is due all the credit of the discovery.

Scarcely had these events been made known, when the Boers living in the neighbourhood of Hopetown commenced a vigorous search for diamonds. They were rewarded by the discovery of a few scattered stones, but there was no rich, continued yield such as is characteristic of a regular deposit. The searchers soon extended themselves over a wider area, and in the year 1868 the workings on the Vaal River were commenced, and here the yield was much greater. The first actual diamond deposit was met with in 1869, in the neighbourhood of the places now bearing the names of Pniel and Barkly West.

In the years which followed, news of the finds of diamonds gradually spread in Cape Colony, and soon diamond-diggers from the four corners of the earth congregated on the banks of the Orange and Vaal Rivers. Reported rich discoveries attracted miners to the spot in still larger numbers, in spite of the long and toilsome journey across the arid Karoo region, where, in the dry winter season, the region is more than ordinarily barren and inhospitable, and the route is marked out by the bodies of beasts of burden which have perished by the way. Two years after the first discovery of diamonds, namely, in 1869, this previously uninhabited district became peopled by a white population of 1000 souls. These settled on the Vaal River at Pniel and Klipdrift, the latter now known as Barkly West. Here they washed the surface sands of the river for diamonds, and some time elapsed before any systematic digging operations were undertaken.

It was soon discovered, however, that in this region diamonds were by no means confined to the river sands.   In December 1870, a digger from the Vaal observed that the children of a Boer, whose farm, " Vooruitzigt," was situated (Fig. 38) on the plateau

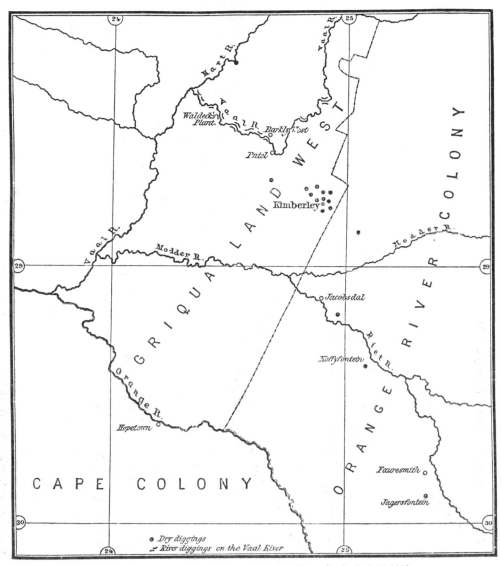

FIG. 37.   Occurrence of diamond in South Africa.   (Scale, 1 : 1,500,000.)

between the Vaal and Modder rivers, and about fifteen miles south of the former, had collected in the neighbourhood a number of small diamonds, of the true nature of which, however, they were ignorant.   According to another story, van Wyk, a Boer who lived at " Du Toit's Pan " farm, situated in the same neighbourhood, discovered diamonds in the walls of his dwelling-house which had been built of mud dug out from a neighbouring pond.   Both stories end in the same way; these accidental finds stimulated further search, which resulted in the discovery of the mine now known as Du Toit's Pan mine (also written Dutoitspan), the first of the four famous mines of Kimberley, the town which sprang up at this spot and became the centre of the diamond-mining industry.

A great influx of people or "rush" to the newly discovered locality at once took place. These newcomers proved a source of great irritation to the Boers in possession of the land, who, seeing that it was impossible to dislodge their unwelcome visitors, sold their valuable possession to an English Company for £6000, a ridiculously low sum considering the discoveries that had been made and were to be expected. The conditions under which the eager searchers for treasure had to work were indeed harassing; exposed to all the intensity of the hot African sun, tormented by storms of dust and insects, deprived of many of the necessaries of life, obliged to fetch drinking water from a great distance, and, for the lack of more permanent dwellings, forced to camp out in the open, their lot was no enviable one, and numbers perished of want and privation. The survivors had no cause, however, for disheartenment in the yield of the deposit; new finds were constantly made, and the conditions of life gradually improved.

Soon another rich deposit, only about half a mile from Du Toit's Pan, was discovered, and became known as the Bultfontein mine, while still another on the farm, " Vooruitzigt," of a Boer named de Beer, who himself commenced mining operations, became the famous " Old de Beer's mine," or, shortly, De Beer's mine (often written De Beers). Finally, on July 21, 1871, a new discovery was made close to the last mentioned mine; this was at first known as " Old de Beer's New Rush," or as the " Colesberg Kopje." Later, however, it became known as the Kimberley mine, and proved to be the richest of the whole group. These four mines, which still form the nucleus of the diamond-producing area, are all situated close to the town of Kimberley, which was founded by the diamond miners, and which has now a white population of 30,000. Two miles to the south-west of Kimberley is the suburb of Beaconsfield with 10,000 to 11,000 inhabitants. The situation of the mines is shown in Fig. 38; they all lie in a circular area not more than three miles in diameter, and besides the four important deposits there are half a dozen others too insignificant to be worked to any extent.

After the first accidental discovery had drawn attention to the occurrence of diamond here, the four important mines mentioned above were all discovered in the course of six months. Very soon following these discoveries, other but less important deposits were found to the south of Kimberley, namely, the Jagersfontein mine near Fauresmith and the Koffyfontein mine on the Riet River between Jacobsdal and Fauresmith, both in the Orange River Colony. The Jagersfontein mine was discovered almost simultaneously with the Kimberley mine, from which it is situated about eighty miles distant. Practically the whole of the present enormous production of diamonds in South Africa is derived from these six mines and from the washings on the Vaal River.

The method of winning diamonds in these mines furnishes a great contrast to the work of collecting them from the river sands. In the latter case the sands are washed in a manner similar to that employed in Brazil and India. In early days on the arid and waterless plateau of Kimberley, the stones were picked out of the dry fragmented rock, such workings being known as " dry diggings," to distinguish them from the " river diggings " on the Vaal. These terms are still in use, but the former is now somewhat inappropriate, as at the " dry diggings " water is now also used to separate the diamonds from their matrix.

All the diamond mines of South Africa, which are of any importance, lie to the north of the Orange River, and are confined to a comparatively limited area, as may be seen by reference to the map (Fig. 37). They are situated in the stretch of country lying between the line of longitude 26° E. of Greenwich and the fork of the Orange and the Vaal rivers, the two principal water-courses of South Africa. The north or right-hand bank of the Vaal must, however, be also included, and the very first discovery of diamonds in South Africa was made a few miles to the south of the Orange River. All the known mines and washings lie in a quadrangle bounded by parallels of latitude 28° and 30° S., and by meridians of longitude 24° and 26° E. The town of Kimberley lies very near the centre of

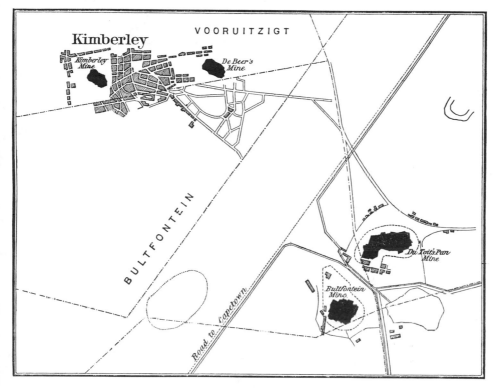

Fig. 38. Diamond mines at Kimberley. (Scale, 1 : 40,000.)

this quadrangle, and the boundary between Cape Colony and the Orange River Colony very nearly coincides with the north-east and south-west diagonal. The Kimberley mines are not only central in position but also in importance, for they supply 90 per cent. of the total output of South African diamonds.

The diamond localities of this district (with the exception of the washings on the Vaal River) are situated on an almost straight line, 125 miles in length, running north-north-west and south-south-east, from the confluence of the Hart River with the Vaal, to beyond Fauresmith in the Orange River Colony. On this line, about fifteen miles from the Vaal, Kimberley is situated in latitude 28° 42′ 54″ S., and longitude 24° 50′ 15″ E., of Greenwich, at a height of 4050 feet (1230 metres) above sea-level. Koffyfontein is about forty miles, and Jagersfontein about double this distance, from Kimberley. A stone of 70 carats was once picked up at Mamusa, on the far side of Jagersfontein, but the find has remained an isolated one.

Outside this district no diamonds have been found ; within the district they are confined to a few isolated points, some of which have not yet been properly investigated, since, the yield being poor, they were abandoned almost as soon as they were discovered.

False assertions as to the occurrence of rich deposits in certain localities have sometimes been made with the sole object of attracting diamond miners, thus promoting the sale of food and spirituous liquors, and incidentally enriching the vendors thereof ; a flocking together of miners attracted by such assertions is known as a " canteen rush."

The mines mentioned above, from which rich yields are at the present time derived, were all known as far back as 1872. Since that time other districts have been vigorously prospected, but without success ; still the region is extensive enough to warrant the belief that fresh discoveries may yet be made, especially as in 1891 a new deposit was found in the Kimberley district, one mile east of Du Toit's Pan, on the farm " Benauwdbeidfontein," of J. J. Wessels, senior. This deposit lies under a thick layer of calcareous tufa, and the mine known as the Wesselton or Premier mine promises to become of importance.

Before the discovery of diamonds the whole of this now important stretch of country was almost valueless, and was peopled by only a few hunters and Boers, who derived a meagre living from its scanty vegetation, and whose lot no one was inclined to envy. There were thus in this region no rigidly defined spheres of influence, and when it suddenly acquired an enormous value and importance complications arose in the shape of rival claims, various portions being asserted to be the property of the Orange Free State, the Transvaal, or of native chiefs. In 1870 the diamond-fields near Pniel, on the Vaal River, were proclaimed as British territory, on behalf of a native chief who had ceded his rights to Great Britain, and on November 17, 1871, the British flag was hoisted at Kimberley. The matter was formally settled in July 1876, by the London Convention, according to which the Government of the Orange Free State agreed to give up its claim to the diamond-fields in consideration of a payment of £90,000 from the British Government. Griqualand West, the division in which Kimberley is situated, remained a Crown Colony until October 1880, when it was formally incorporated in Cape Colony. In it are situated all the rich diamond mines of South Africa, with the exception of the Koffyfontein and Jagersfontein mines, which are in the Orange River Colony, but which yield only about 6 to 7 per cent. of the total South African output.

Not only has the discovery of the precious gem enormously increased the importance of South Africa as a country, but it has also so raised the value of the comparatively small plots of land on which the mines stand, as to make a comparison between their present and their former values of interest. Thus the farm " Vooruitzigt," on which now stand the De Beer's and Kimberley mines, was bought from its owner in 1871, the time of the discovery of diamonds there, for £6000, while only four years later £100,000 was paid for it by the Cape Government, the transfer being made with the object of putting an end to the frequent and ever arising disputes between the mine-owners and the miners as to the dues to be paid by the latter to the former.

It is now intended to consider the different deposits in more detail, commencing not with the most important but with those first discovered, namely, the river diggings.

River Diggings.

The richest of these deposits lie on both banks of that portion of the Vaal River flowing between the mission stations, Pniel and Barkly West (formerly Klipdrift), to the east, and Delport's Hope, at the junction of the Vaal with the Hart River, to the west, Barkly West being at the present day the centre of the diamond-washing industry. In addition, diamonds have been found in small numbers further up the river at Hebron, and even as far as Bloemhof and Christiana in the Transvaal ; also in the opposite direction, at the junction of the Vaal with the Orange River. A few diamonds have also been found in the Orange River, between its confluence with the Vaal and Hopetown, as well as in some of the tributaries of the Vaal, notably the Modder and the Vet. The yield in all these places was, however, so poor that the workings were soon all abandoned except the portion of the Vaal River mentioned above, and a stretch of its valley, parallel to the same portion of the river, and measuring fifty miles in a straight line, or seventy-two following the windings of the river. At the present day whole series of mines even in this region are practically deserted, the workers having left the river for the far richer dry diggings of Kimberley. The production of the river diggings up to 1871 was of some importance, but is now quite insignificant ; in spite of the poorness of the yield, and the miserable conditions under which they have to work, a small number of a certain peculiar class of diamond-miners still cling tenaciously to their holdings in the hope no doubt of better days coming. Counting both black and white men, their number for many years probably did not exceed two or three hundred ; they work singly or in twos or threes, not in large companies, and are most frequently to be seen in the neighbourhood of New Gong-Gong, Waldeck's Plant, and Newkerke. The amalgamation of the "dry diggings" to form the De Beers Consolidated Mines, has had the effect of increasing the number of river diggers, it being estimated that there are now 1000 of them, exclusive of native workers. Companies have been formed with the object of working the deeper beds of river sand, but have met with little success. The river-diggings, on account of their poor yield, are known as "poor men's diggings."

The bed of the Vaal is strewn with blocks of basalt, often amygdaloidal in character, and with other rocks which are probably of metamorphic origin. These blocks are usually of considerable size, and have been washed down from the sides of the valley and from the surrounding hills ; between them lies a loose material consisting of gravel, sand, and mud, and it is in this that the diamonds are found. The whole deposit, which varies in thickness up to 40 feet, rests on basalt, this rock being *in situ*, and is here and there scooped out to form deep hollows, known as pot-holes or "giant's-kettles," similar to those found in the beds of the diamond-bearing rivers of Brazil, which have been worn out by the continued whirling of pebbles in the eddies of the stream. The diamond-bearing débris accumulates in these depressions, which often yield a rich harvest to the finder.

The search for diamonds was at first confined to the bed of the river, but it was soon discovered that the sands and gravels of the river-terraces were as rich or richer than the river-bed, so that these also came to be worked. The terraces and their workings are usually only a few yards above the present high-water level, but one or two are 200 feet above this level. The workings in the river-terraces are easier to manage and more secure than those in the river-bed, since the latter are liable to be flooded, and thus considerably damaged ; it has, therefore, been proposed in recent times that the stream should be diverted into another channel, but this scheme has never yet been carried out.

The diamonds found in this sandy clay are, as a rule, distinctly water-worn, though not of course to the extent of the other pebbles and sand grains which accompany them. These accompanying pebbles consist of various minerals, the different varieties of quartz (agate, jasper, silicified wood, &c.), which have all travelled down from the upper courses of the river, being especially abundant. Pebbles of the rocks which occur *in situ* in the neighbourhood are present in large numbers in these alluvial deposits. The minerals which are associated with the diamond in the dry diggings are less abundant, but small fragments of garnet, ilmenite, vaalite, &c., are met with. It is among these pebbles that the diamond is to be found ; its distribution is, however, extremely irregular, a miner who hits on a favourable spot may make his fortune in a very short time, while his comrades toil on month after month unrewarded by the smallest success.

The method of work does not differ essentially from that followed in the diamond-washings of other countries or in the gold-washings of the same country. The sand and clay in which the diamonds and other pebbles are embedded, must first be excavated ; this, when the diamantiferous material is overlain by blocks of basalt, &c., of considerable size, is no light task. This material is placed in a cradle, and the clay and fine particles washed away by rocking the cradle under a stream of running water ; what remains after this process is put through a sieve, and the coarse residue, which contains the diamonds, is spread out on a sorting table and the diamonds picked out by hand. This final operation is easily performed, for the peculiar lustre of diamonds enables a practised sorter at once to distinguish them from other pebbles.

The yield is not very great, only on an average about 15,000 or 20,000 carats a year ; in 1890, however, 28,122$\frac{2}{3}$ carats, valued at £79,231, were obtained ; a production of 30,000 carats (about 13 lbs. avoirdupois), is seldom reached, and never exceeded.

The quality of the yield in part compensates for its small quantity, stones from the river diggings being on the average far superior to those from the dry diggings. The average value of the former is in consequence much higher than that of the latter ; for example, in the eighties a river-stone weighing one carat was worth 56s., while a carat stone from Kimberley only fetched on an average 22s. 9d.

A few specially large stones have been secured in the river diggings, such, for example, as the " Star of South Africa," a diamond of the purest water, weighing, in its rough condition, 83$\frac{1}{2}$ carats ; also the slightly yellow " Stewart," weighing 288$\frac{3}{8}$ carats, which was found at Waldeck's Plant on the Vaal River.

The sands and gravels in which the diamonds are found in the river diggings are secondary deposits. It has been suggested that these sands and gravels have been derived from a deposit similar to that in which diamonds are now found in the dry diggings, and situated somewhere in the neighbourhood of the source of the Vaal River. The denudation of such a deposit would supply the diamantiferous débris carried down by the river. That the diamonds have been transported some distance is shown by their distinctly water-worn character, and in all probability the original deposit was situated somewhere below Bloemhof in the Transvaal, since no diamonds have been found above this town. The fact that very few of the minerals associated with the diamond in the dry diggings occur in the Vaal River is easily explained when we consider that these minerals are not very hard and would be reduced to powder before they had been transported any great distance by the running water ; whereas the harder minerals, found in the basin of the upper part of the Vaal, and now associated in the river deposit with diamonds, would resist the action of the water for a longer period and would be transported over greater distances. Furthermore it is possible that the characteristic minerals of the dry diggings now known may have been of sparing occurrence in these original deposits, if not indeed absent altogether. The higher quality of the river stones as compared with those from the dry diggings does not militate against the truth of this theory as to their origin, since the quality of stones found in the Jagersfontein dry diggings is well above the average ; it simply leads to the conclusion that the deposit from which the river sands and gravels were derived was also above the average in quality.

Dry Diggings.

The nature of these deposits was not at first known, and they were supposed to be similar to the river deposit and to consist merely of superficial layers of alluvium.   It was soon recognised, however, that this was by no means the case, and that the deposits were absolutely unique in character.   The geographical position of these deposits has been already described, they are situated on a high plateau, far removed from any water-courses and formed of rocks belonging to the Karoo formation.   This formation, which has a total thickness of about 10,000 feet, consists of sandstones and shales with numerous intruded dykes and bosses of igneous rocks, variously referred to, according to the form of the mass and the character of the rock itself, as trap, dolerite, melaphyre, basalt, diabase, &c.   The age of the sedimentary rocks is not exactly known as yet, but in any case they are later than the Carboniferous, the lower beds probably corresponding with the Permian, and the upper beds with the Trias of Europe.   In this upper and younger part occur the deposits of diamonds in Griqualand West which we have now to consider.

The account which follows deals mainly with the half-dozen mines having the richest yield, and specially with the four best known Kimberley mines, others being passed over as insignificant or not completely examined.   The main features of all are identical, and as the individual deposits differ only in unessential points, it is unnecessary to consider each one singly in any great detail.

The diamond-bearing material is contained in pipes or funnel-shaped depressions which penetrate the Upper Karoo beds in a vertical direction to an unknown depth.   The outline of a cross-section of one of these depressions may be circular, elliptical, kidney-shaped, or more or less irregular.   The rock which fills these pipes differs entirely from the surrounding beds of the Karoo formation, the so-called " reef," and is sharply separated from them.   The occurrence of diamonds is confined exclusively to the material filling the pipes; nowhere in the surrounding reef of sandstone and shale, or elsewhere in the Karoo beds, has a single stone been found, although enormous quantities of these rocks have been removed in the course of the mining operations.

The upper extremities of the pipes are elevated above the surface to the height of a few yards each, thus forming a small kopje; in the case of the Wesselton mine, however, there was a slight depression.   The pipes vary in diameter from 20 to 750 yards, the usual diameter being from 200 to 300 yards.   In 1892 the diamond-bearing material had been excavated in the Kimberley mine, which is the deepest of all, to a depth of 1261 feet, and, as in the other mines, with no sign of exhaustion; the rock is therefore continued to an unknown depth.

The cross-sections of different pipes taken at the earth's surface differ widely both in shape and area, as will be seen from the following data: Du Toit's Pan (Dutoitspan), 192,000 square yards in area, of a flat horse-shoe shape, 750 yards long and 200 yards broad; Bultfontein, 118,000 square yards in area, almost circular in outline with a diameter of 363 yards; De Beer's (De Beers), 66,000 square yards in area, elliptical in shape, measuring 320 yards from east to west and 210 from north to south; Kimberley, 49,000 square yards in area, oval in shape, 290 yards long and 220 yards broad, with a small projection measuring 37 yards towards the east.   The size of the pipe of the Jagersfontein mine is not exactly known, its cross-section is between 100,000 and 110,000 square yards; exact details respecting the Koffyfontein mine are also wanting, but in any case it is smaller than the mine last mentioned.   A peculiar feature of the Kimberley mine is the gradual contraction of the pipe in sectional area as greater depths are reached; thus at a depth of about 300 feet the two diameters are reduced to 260 and 160 yards respectively, and the contraction is continued as still lower depths are reached. A diagrammatic section of the Kimberley mine is given in Fig. 39, an explanation of which is given below.

The **rocks composing the reef** are, on the whole, much the same everywhere, still, in the various mines, certain differences do exist.

The neighbourhood for a considerable distance round Kimberley is covered with a layer of red clay, 1 to 5 feet thick; underlying this is a bed from 5 to 20 feet thick of calcareous tufa, also of wide distribution. This tufa is of recent origin and has no genetic connection with either the reef or the diamond-bearing pipes, since it covers both indiscriminately, and to a certain extent penetrates cracks and crevices in them. Beneath this tufa lie the rocks of the Karoo formation which constitute the reef.

The uppermost part of the reef in the Kimberley mine consists of a series of bedded shales, 40 to 50 feet thick, greenish-grey in the upper part and yellowish or greyish in the lower; they are of varying hardness, and at different levels in the mine are interbedded with a fine-grained to compact olivine-basalt. Beneath these pale shales are about 270 feet of black bituminous shales, very similar in character to the shales of the English coal-measures; certain of these beds are impregnated with iron-pyrites, and they often contain nodules of clay-iron-stone, small bands of calcite, and thin layers of coal, while interbedded with them near their base is a sheet of basalt one foot thick. Beneath the black shales is a hard grey or green amygdaloidal diabase (melaphyre), the base of which is not exposed in the open workings, but is seen in the underground shafts at a depth of 440 feet below the upper surface of the mass.

Beneath this igneous rock the shafts penetrate a bed of quartzite of about the same thickness, and under this again black shales, both of which are penetrated in places by dykes of eruptive rock (dolerite). The deepest shaft of the mine has not yet penetrated to the base of the black shales, so that the total thickness of these beds is unknown. Probably at still greater depths, as yet untouched by mining operations, there are deep-seated rocks, such as granite, gneiss, or olivine-rocks, but this question will be discussed later.

In the De Beer's mine, a sheet of basalt 47 to 61 feet in thickness is met with in the upper part of the reef, otherwise the beds are the same as in the Kimberley mine. A similar sheet of basalt is present in Du Toit's Pan mine, but is absent from the Bultfontein mine. The walls of the pipe consist here, as far as they have been laid bare, only of shales, which are much displaced, sometimes having an inclination of at least 15° to the horizon; this is also the case to a certain extent in the De Beer's mine, while in other places the beds are horizontal. In the Du Toit's Pan and Bultfontein mines the shales have not yet been penetrated to their base, and their thickness appears to be greater here than in the Kimberley and De Beer's mines, which lie a little further to the north.

The **material filling the pipes**, like that of the surrounding rock, is essentially the same in every mine, and in every part of each mine, but in all mines the upper portions of the pipes to a fairly considerable depth have suffered the effects of weathering. Observable differences do exist, however, and an experienced miner can sometimes recognise not only from which mine, but also from what part of a particular mine, any given specimen of material has been taken. These small differences are usually connected with variations in colour, hardness, and composition, the nature of the enclosed minerals and fragments of foreign rock, &c., and are, as a rule, unimportant.

The different kinds of rocks constituting the material which fills the pipes are separated by no trace of bedding planes, but masses of rock, slightly different in character have been recently observed to be separated from each other in quite another way.

Vertical, or nearly vertical crevices, not more than three-eighths of an inch across and filled with a mineral substance resembling talc in character, penetrate the material down to the lowest depth to which the mines have been worked.   These divide the whole contents of each pipe into a number of vertical or nearly vertical columns, each differing slightly from the others in composition, but showing no difference in its own mass.

These small variations in the material filling the pipes, as well as its character as a whole, do not depend in any way upon the nature of the various rocks of the surrounding reef.   It was formerly contended that the character of the reef had a more or less marked influence on the richness in diamonds of the material filling the pipes ; thus it was feared that when the base of the black shale in the Kimberley mine had been reached, the yield of diamonds would cease, since the formation of diamonds was supposed to have been dependent on some way on the presence of carbon in these shales.   In consequence of this belief, the value of mining claims for a time fell; but the yield of diamonds at lower levels, where the pipe is surrounded by melaphyre, turned out to be just as good as at the higher levels in the shale.

Between the material filling the pipes and the enclosing rocks or reef there is always a sharp line of demarcation and never a gradual transition.

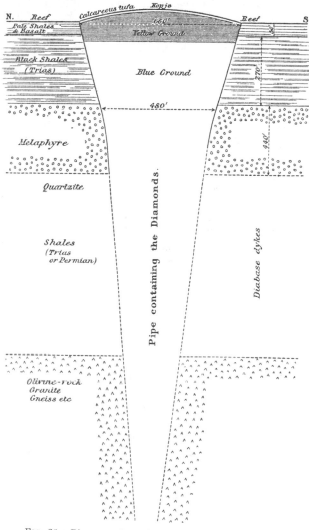

Fig. 30).   Diagrammatic section through the Kimberley mine.
(Scale, 1 : 4,000.)

Usually the two sets of rocks are in immediate contact, but not infrequently they are separated by a space, sometimes of considerable width, into which project beautiful crystals of calcite.   Other secondary minerals are also found in the numerous crevices by which, in addition to the vertical cracks, the rock is penetrated.

The actual diamond-bearing rock itself which fills the pipes must now be considered. In the upper portion of the pipes it consists of a light yellow, soft, sandy or friable material known to the diamond miners as " yellow ground " or " yellow stuff." This upper portion, which has a thickness of from 50 to 60 feet, has now, in the Kimberley mine, been completely removed in the course of mining operations ; in the other mines, however, a little still remains to be seen. The rock at a greater depth has the character of a volcanic tuff or breccia, it is of a green or bluish-green colour, and is known as the " **blue ground** " or " blue stuff." Throughout the whole depth to which it has been worked it shows no deviation from these characters.

The passage of the " yellow ground " into the " blue ground " is as a rule abrupt, and the line of division is never quite horizontal, but inclined from 5° to 15° to the horizon. Sometimes there is an intermediate reddish layer, known as " rusty ground " which passes upwards into the " yellow ground " and below into the " blue ground." Neither the " yellow ground " nor the " rusty ground " is anything more than the weathered upper portions of the " blue ground ; " the latter originally filled the pipes up to the surface, but the portion exposed to atmospheric influences became altered and transformed into what is now known as " yellow ground." Similarly the " rusty ground " is a layer in which the alteration has not proceeded as far as in the " yellow ground "; the uppermost layer, therefore, of " blue ground " marks the level below which the weathering process has not yet commenced. In the early history of the mines, this change in the colour of the diamantiferous material also had the effect of diminishing the value of claims, since it was feared that the " blue ground " might be deficient in yield. Experience of course showed that these apprehensions were groundless, for the rock at greater depths proved as rich, if not richer, than the upper levels.

The " blue ground," which thus fills the pipes, and from which the uppermost " yellow ground " has been derived, has the appearance of dried mud, and consists of a green, or dark bluish-green ground-mass, which gives its colour to the whole rock. It binds together numerous fragments, larger or smaller in size, and with sharp, or in some cases rounded corners, of a green or bluish-black serpentine rock. The actual material of the mud-like ground-mass, and of the blocks which it cements together, is identical, the one being in a finely divided condition, and the other in compact masses. These are the chief constituents of the " blue ground," but it contains also numerous mineral grains as well as fragments of foreign rocks in large numbers. A piece of " blue ground " of its natural colour, and containing a crystal of diamond embedded in it, is depicted in Plate I., Fig. 2.

Although the ground-mass is not very hard, it has a certain amount of toughness which renders it difficult to work with a pick-axe ; it readily yields, however, to the chisel. It can be scratched with the finger-nail and is somewhat greasy to the touch. The qualitative chemical composition of the " blue ground " is almost identical throughout the whole mass, but certain differences in the quantitative composition of different portions are detected in analysis. All analyses which have been made of this material record the presence of silica and magnesia in varying amounts, some ferrous oxide, usually only a little lime, some water and carbonic acid, and little or no alumina. The material is thus essentially a mixture of hydrated magnesium silicate and calcium carbonate.

The following is a quantitative analysis of a specimen of " blue ground " from the Kimberley mine given by Professor Maskelyne and Dr. Flight :

	Per cent.
Silica (SiO_2)	39·732
Alumina (Al_2O_3)	2·309
Ferrous oxide (FeO)	9·690
Magnesia (MgO)	24·419
Lime (CaO)	10·162
Carbon dioxide (CO_2)	6·556
Water (H_2O)	7·547
	100·415

The carbon dioxide (carbonic acid) is present in nearly sufficient amount to combine with the whole of the lime to form calcium carbonate; deducting this, the remainder, consisting of hydrated magnesium silicate, with some of the magnesium replaced by ferrous oxide, has approximately the composition of the mineral serpentine. It has thus become customary to speak of the whole rock as a serpentine breccia, and this term, or that of volcanic tuff or agglomerate, will be used in referring to the diamantiferous material.

The blocks of foreign rock embedded in the breccia, which are often known as boulders, have usually perfectly sharp edges and corners, though occasionally these may be rounded. The size of these rock fragments varies from that of a small splinter to that of a block several thousand cubic yards in dimensions. In the pipe of De Beer's mine there is a block of olivine-basalt called "the island," which has a sectional area of 330 square yards, and has been traced to a depth of 237 yards. Large masses of similar rock occur commonly in all the mines, they are referred to as "floating reef," in contradistinction to the "main reef" which surrounds the pipe. This "floating reef" is more frequently met with in the upper than in the lower levels of the pipes. Smaller fragments of the same rock have, however, been met with at the greatest depths to which the mine shafts have been sunk, and here as elsewhere they form a large proportion of the material filling the pipes, through which they are distributed with the greatest irregularity.

Some of the rock fragments agree completely in character with the rocks of the main reef, frequently consisting of amygdaloidal basalt (melaphyre), shales, &c. In some places, the highly bituminous and carbonaceous shales are present in such large amounts that sometimes the presence of the fire-damp characteristic of coal-mines has been observed. It has been asserted that diamonds occur only in those portions of the agglomerate in which bituminous shales are present in large amount, and it has been argued from this that the diamonds were actually formed from the carbonaceous matter present in these shales. There is, however, reason to believe, as will be shown below, that the diamonds were formed not in the pipes themselves, but at far greater depths in the interior of the earth from which they have been brought up by the action of volcanic forces.

Beside the blocks, which have evidently been detached from the reef surrounding the pipes, there are others which have not been found in situ in the neighbourhood, and which, therefore, must necessarily have been brought up from below. In the Kimberley mine at depths below 230 feet, there are found large blocks, several cubic yards in extent, of grey or greyish-white sandstone, the grains of which are bound together by a calcareo-argillaceous cement. They are of much the same character as the sandstone which, in other localities, forms part of the Middle Karoo formation, and which from geological considerations must form part of the reef at the Kimberley diamond mines at a great depth below the surface. There also occur, though not so frequently, fragments of quartzite, mica-schist, talc-schist, eclogite, and granite. The last named rock is rarely found, and when met with is so decomposed as to be only doubtfully recognisable as granite. It was found in numerous large blocks, and in smaller fragments in the upper portion of a small mine known as Doyl's Rush about a mile from Kimberley. Such rocks crop out at the surface some distance north of the diamond-fields; it is therefore probable that their southern extension lies at a great depth below Kimberley, and forms the base of the reef. Rock fragments of materials not found in the reef enclosing the pipes, which in all probability have been brought up from below, are called "exotic fragments."

The minerals embedded in the agglomerate are usually distributed through it with some regularity, but very sparingly; they constitute only about $\frac{1}{4000}$ of the total mass of the rock and are therefore rather inconspicuous. A complete collection can only be made from the residue left after the process of diamond-washing.

Among these minerals the most important, but not the most frequently occurring, is the diamond : it is found in crystals developed regularly on all sides, and also in fragments, such as would result from the breakage of larger crystals. It is remarkable, however, that different portions of the same crystal are never found lying close together. The edges and corners of the crystals are always perfectly sharp, not even the faintest trace of rounding can be detected, so the stones of the dry diggings are easily distinguished from those of the river diggings. A more detailed description of the special characteristics of Cape diamonds will be given later, here we are concerned only with their mode of occurrence.

The diamond is a constituent part of the agglomerate in which it is embedded, and its mode of occurrence in no way differs from that of other minerals contained in the rock. Each crystal or fragment of a crystal occurs alone, firmly embedded in the agglomerate, from which it can be extracted only with difficulty ; its surface is usually clean, but in some cases is coated with a layer of limonite (iron hydroxide) or with a calcareous film, both of which are easily removable. Until recently, no diamond had ever been observed attached to another mineral in such a way as to suggest that the two grew side by side at the same time. The discovery, however, of a diamond crystal attached in this way to a garnet shows that such a growth does take place, though rarely.

Diamonds are to be found at the surface, and downwards through the "yellow ground," the "rusty ground," and the "blue ground," as far as the deepest mines have yet penetrated ; they do not occur, however, in equal number in all mines, nor in different portions of the same mine ; numerical details will be given later. In the Kimberley mine, which is unique in this respect, the richness of the yield increases rapidly as lower levels are reached. The different columnar divisions of each pipe vary in the number of diamonds contained, some being so poor that the working of them is unprofitable, others on the contrary being just the reverse. The total number of diamonds contained in a given mass of any particular column is so constant that it is quite possible to calculate beforehand how many carats of stones a certain amount of "blue ground" will yield.

The presence of diamonds in the "blue ground" is of enormous economic importance ; regarded as a rock constituent, however, they are quite insignificant, being present in such small amount that, had they been less highly prized, and of less general interest, they would probably have been scarcely mentioned in a petrographical description of the rock. A striking illustration of their sparing occurrence is furnished by the fact that in the richest part of the richest mine, namely the Kimberley mine, they constitute only one part in two millions, or 0·00005 per cent. of the "blue ground." In other mines the proportion is still lower, namely one part in forty millions, a yield which corresponds to five carats per cubic yard of rock, and which can be profitably worked. When the absolute amount of diamond present is so small, slight variations in this amount in different parts of the rock, though of great economic importance, are of little scientific significance.

In respect to the **associated minerals** of the diamond in the agglomerate, certain differences exist between the various mines and between different parts of the same mine. These minerals occur either in homogeneous grains of the same kind, or in small groups consisting of minerals of various kinds. Those of most frequent occurrence are red garnet, green enstatite, and vaalite (an altered mica), others are less widely distributed, some indeed being regarded as great rarities. The most important of the minerals associated with the diamond in these deposits will be now considered in order.

The garnet is of constant occurrence, and always in relatively considerable amounts.

It is found in the form of rounded or angular grains, crystals, or even indications of crystalline form, being never observed.   The grains are usually in a fresh and unaltered state, and therefore appear bright and transparent, some, however, are cloudy and opaque, and of a reddish-brown colour, in consequence of a process of decomposition having commenced.   The colour of the unaltered garnets is variable, a deep wine- or hyacinth-red is most frequent, red tinged with violet is less common, while light or dark brownish-yellow, and a beautiful ruby-red are colours which are seen but rarely.   Garnets of this ruby-red colour are cut for gems and enter the market under the name of " Cape rubies."   In size the garnet grains vary from mere dust up to the size of a walnut.   All specimens yet examined contain a little chromium, and have the chemical composition of pyrope, itself well known and much used in jewellery under the name Bohemian garnet.

The members of the pyroxene group most frequently met with are enstatite (and bronzite) and chrome-diopside.   The enstatite has the usual composition, but not the usual appearance of this mineral.   It occurs generally in fragments about the size of a hazel-nut.   It is transparent, with the colour of green bottle-glass, and has a distinct cleavage and a conchoidal fracture.   It closely resembles olivine in appearance, and is frequently confused with this mineral.   It is often found intergrown with garnet in such a way that single grains of garnet are enclosed by a shell of enstatite.   This variety of enstatite is of more common occurrence than is the garnet.   Another variety of enstatite (bronzite) also occurs, but more rarely.   It is brown in colour and less unlike the ordinary mineral in appearance ; moreover, it has a distinct plane of separation in one direction.

Chrome-diopside, sometimes referred to as chromiferous diallage, though less common than garnet, is yet very frequently met with.   It occurs in irregular polyhedral grains of about the same size as the garnet grains and with no trace of crystal-faces.   It is emerald-green in colour, translucent in mass, but transparent in thin splinters, and is usually distinctly cleavable in one direction.   Wollastonite, another mineral of the pyroxene group, is also said to occur in the " blue ground."

The amphibole group of minerals is represented by the green smaragdite, which, however, is of rare occurrence.   It has possibly been derived by the alteration of chrome-diopside.   The occurrence of tremolite and asbestos has also been reported.

An altered magnesium mica occurs everywhere in small shining scales of a greenish or brownish colour or completely bleached.   These thin plates or prisms frequently have a regular six-sided outline, and show the characteristic cleavage of mica ; optically they are almost uniaxial.   This altered mica, which is distinguished as vaalite, sometimes occurs aggregated into brown balls the size of a hen's egg, and in some places forms the chief constituent of the " blue ground."   The glittering scales of mica embedded in the " blue ground " are sometimes mistaken at first sight by the unpractised eye for diamonds.

Ilmenite (titanic iron ore) is another mineral frequently associated here with the diamond.   It occurs in shining black rounded grains with no indication of crystal-faces. It contains some magnesia and is not magnetic.   Formerly the diamond miners imagined this mineral to be the black variety of diamond known as carbonado, and at present found almost exclusively in Brazil.   They were not easily convinced of their error, and the name they gave it, carbonado, still remains.   True carbonado occurs only very sparingly at the Cape.   Magnetite (magnetic iron ore) in grains and of the usual character is said to be of frequent occurrence.   Chromite (chromic iron ore), found in brilliant shining black grains up to the size of a pea, and with a conchoidal fracture, is also fairly common.   Zircon, known to the Kimberley miners as "Dutch bort," occurs very rarely in transparent to translucent grains of a very pale flesh colour, and about the size of a lentil or pea.   Other minerals associated with diamond in the "blue ground" are iron-pyrites, sapphire, kyanite topaz, and on very rare occasions colourless olivine.   Apatite has been detected by chemical tests, and gold was once found enclosed in eclogite at the Jagersfontein mine.   Under the microscope, graphite, tourmaline, rutile, and perofskite, among other minerals, have been detected.   The common mineral quartz, on the contrary, has never yet been observed.

The majority of the minerals mentioned above occur in all the mines, but some are confined to particular pipes.   Thus gold has been met with only in the Jagersfontein mine, and up to the present the occurrence of sapphire also is confined to the same mine.

The parti-coloured residue left by washing the "blue ground" after sorting out the large rock-fragments consists largely of grains of red garnet and zircon, the green minerals of the pyroxene and amphibole groups, and black ilmenite and magnetite mixed with small fragments of diabase.   The absence from this residue of the other minerals mentioned above is explained either by their rarity or by their having been lost in the washing process.   Diamonds are of course present, and are picked out by hand.

All the minerals mentioned above are original constituents of the "blue ground," and were already formed at the time it first filled the pipes.   There are others, however, which are of secondary formation, owing their origin to the weathering processes undergone by the upper layers of the "blue ground."   Such a secondary mineral is calcite, a not unimportant constituent of the rock-mass, and occurring also in veins and crevices and as crystals encrusting the walls of cavities in the rock.   Other secondary minerals are zeolites, especially mesolite and natrolite, sometimes found in beautiful groups of acicular crystals; also in places rough fragments of a bluish hornstone.   Barytes, which is of rare occurrence, is also probably a later-formed mineral.   All these minerals of secondary origin, but particularly the zeolites, are found most abundantly in the upper part of the pipes, which is more exposed to the action of atmospheric agencies.   At successively lower levels they diminish in amount and finally disappear.

Stanislas Meunier has described a total of eighty different species from the "blue ground," but the existence of some of these as distinct mineral species probably requires confirmation.

One other rock found in the "blue ground" of De Beer's mine remains still to be mentioned, but is of no great importance.   It penetrates the "blue ground" as a dyke five to seven feet thick, and on account of its tortuous path is known locally as "the snake."   It is a compact greenish-black rock of much the same composition and consisting of essentially the same minerals as the "blue ground," but it contains no diamonds.

The **manner in which the pipes have been filled** with the material we have been considering has been explained in many and various ways.   The first investigator to formulate a theory in accordance with all the observed facts was Emil Cohen, and the views he propounded in 1873 have never been seriously controverted by any one of the numbers of observers who have followed him in this field of inquiry.

He regards the pipes as volcanic vents or chimneys comparable with those, also extinct, of the Eifel, and considers that the serpentine breccia now filling the pipes was brought up from below by the action of volcanic forces, but at what period of geological history this took place neither he nor later authorities can say.   To quote Cohen's own words:

"I consider," he says, "that the diamantiferous ground is a product of volcanic action, and was probably erupted at a comparatively low temperature in the form of an ash saturated with water and comparable to the material ejected by a mud volcano.   Sub-sequently new minerals were formed in the mass, consequent on alterations induced in the upper part by exposure to atmospheric agencies, and in the lower by the presence of water. Each of the crater-like basins, or perhaps more correctly funnels, in which alone diamonds are now found, was at one time the outlet of an active volcano which became filled up, partly with the products of eruption and partly with ejected material which fell back from the sides of the crater intermingled with various foreign substances, such as small pebbles and organic remains of local origin, all of which became embedded in the volcanic tuff. The substance of the tuff was probably mainly derived from deep-seated crystalline rocks, of which isolated remains are now to be found, and which are similar to those which now crop out at the surface, only at a considerable distance from the diamond-fields. These crystalline rocks, in which the diamonds probably took their origin, were pulverised and forced up into the pipes by the action of volcanic forces, and, embedded in this erupted material, these same diamonds, either in perfect crystals or in broken fragments, are now found. Analogous cases of the simultaneous ejection of broken and of perfect crystals are afforded by some of the active volcanoes of the present day, and moreover, in many other localities, the mother-rock of the diamond is probably to be found in the older crystalline rocks. At any rate, these rocks contain, as a rule, just those minerals which are most frequently associated with diamond. The beds of shale and sandstone interbedded with sheets of diabase were broken through and fractured by the force of the eruption, and so large blocks (floating reefs) and small fragments of these rocks became embedded in the tuff. Since in well-borings in the neighbourhood of the mines bands of coal are often met with interbedded with the shales, the coal, which is occasionally found in the diamond-bearing ground, and which has been incorrectly thought to have some genetic relation with the diamond, must have been derived from the seams of coal interbedded with the shales."

The fact that there is no genetic relation between the coal found in the tuff and the diamonds, or in other words that the diamonds have not been formed in the pipes from fragments of coal, is clearly shown by the frequent occurrence of diamond crystals in broken fragments. Had the diamonds been actually formed in the " blue ground," it would be difficult to find any explanation of the occurrence of so many broken crystals. If, on the other hand, we suppose them to have been formed in a deep-seated crystalline rock, which by the action of powerful volcanic forces was pulverised and forced up into the pipe or funnel, the fragmentation of many of the crystals follows as a matter of course.

That the material filling the pipes was not washed into them by flowing water is proved by the absence of any trace of wear in the minerals and rock-fragments enclosed in the tuff, all of which preserve intact the sharpness of their edges and corners. Had the soft and fragile materials, such as the abundantly occurring shales and mica, been transported by water over even the shortest distance, they would inevitably show some sign of their journeying.

A volcanic origin for the diamantiferous deposit thus appears to be the only possible conclusion which can be drawn from the observed facts ; it should be noted, however, that according to this theory the diamond itself did not originate in the same way, but was formed in a deep-seated rock before the eruption took place. Cohen's theory is so closely in agreement with the observed facts, that it has been very generally accepted, and up to the present has only required modification in a single particular. In this theory it is assumed that each pipe was formed and filled up by a single manifestation of volcanic activity and that, excluding of course the effects of subsequent weathering and alteration, the pipe as we now see it is the product of this single eruption. A consideration of the vertical columns into which the pipes of " blue ground " are divided, and which differ from each other in such characters as colour, composition, contained minerals and richness in diamonds, has led

Chaper to the conclusion that each of these vertical columns is the product of a distinct eruption. Since the columns are similar in general character and differ only in minute details, he considers that they have been formed by a series of eruptions of the same type ; in short, that each diamantiferous deposit or pipe is the product of a long-continued period of volcanic activity. Thus, according to Chaper's view, the pipe of the Kimberley mine, in which fifteen columns have been observed, is the result of fifteen successive eruptions. Further observations in this direction are, however, desirable.

From the considerations brought forward above, it seems very probable that the South African diamonds were formed in a deep-seated crystalline rock which became fragmented and erupted to the surface by the action of volcanic forces ; and moreover, that the greater part of this ejected, fragmentary material fell back again into, and filled up the vent or crater produced by the eruption. From the nature of the minerals which accompany the diamonds in the volcanic tuff, it is perhaps possible to draw some conclusions as to the character of the rock in which the diamonds were formed. Almost all the minerals, which are constituents of the rocks generally known as olivine-rocks, and which are widely distributed in the earth's crust, are found amongst the minerals associated with the diamond in " blue ground." It is therefore highly probable that the original mother-rock of Cape diamonds was an olivine-rock, situated at a great depth below the earth's surface and containing as constituents biotite (represented by the altered mica, vaalite), enstatite (bronzite), garnet, and all the other minerals already mentioned, including of course the diamond. Such a rock, which would be similar to a lherzolite in composition, has indeed, though in a somewhat different sense, been named *kimberlite*, and from this, the " blue ground " filling the pipes has been referred to as a kimberlite-breccia or a kimberlite-tuff. This kimberlite-breccia or tuff, at least as far down as it has been reached by mining operations, has undergone great alteration, its originally predominant constituent olivine being almost completely altered to serpentine, so that very little of it is now to be seen. This more or less complete alteration of olivine to serpentine in an olivine-rock is not at all unusual, being a matter of common observation in all parts of the world. The other constituents of the original rock have undergone less alteration and are in a more or less fresh condition. At greater depths the decomposition of the olivine has been less complete, and here may be found traces of kimberlite still unaltered, in which the olivine retains more or less completely its original character.

The foregoing pages have been devoted to the consideration of the manner in which the diamantiferous rock-mass actually occurs in the pipes. The question which naturally follows, namely how the diamond itself was formed in its original mother-rock, the kimberlite, will be treated generally below. We must first, however, notice certain other theories as to the formation of the pipes, which are more or less opposed to that of Cohen, as set forth above. According to the theory first promulgated by the late Professor H. Carvill Lewis, and which has some substantial support, the "blue ground" is not fragmentary material or tuff, but was forced up from below into the pipe as a molten mass which consolidated on cooling. According to this view, therefore, the " blue ground " is an ordinary igneous rock, which solidified in the situation in which it is now found, and it was with this supposed origin in his mind that the name kimberlite was proposed for the diamantiferous rock by Carvill Lewis. This rock, which was originally an olivine-rock, is supposed to have subsequently undergone the same alteration processes as described above.

There appears to be a similar, though very sparing occurrence, of diamonds near the village of Carratraca in the province of Malaga in Spain. According to the statement of

A. Wilkens, a mine-owner resident there, a few diamonds were found in the seventies along with pebbles of serpentine in a stream, in the neighbourhood of which serpentine with nickel ores occurs *in situ*. It is therefore not impossible that these diamonds had been derived from the serpentine. Similar relations between diamonds and serpentine rocks have also been reported from Australia and the western part of North America.

Although the association of diamond with serpentine in South Africa renders it very probable that the former has been derived from an olivine-rock, yet it is a noteworthy fact that the only mineral found actually inter-grown with and firmly attached to the diamond is garnet. Professor Bonney has recently (1899 and 1901) described the occurrence of colourless octahedra of diamond as a constituent of rounded boulders of eclogite. These boulders of eclogite, which is an igneous rock composed of garnet and green diopside, came from the " blue ground " of the Newland's Diamond Mine in Griqualand West, about forty-two miles north-west of Kimberley. The same observer describes the occurrence, also in this mine, of rocks rich in olivine, such, for example, as saxonite and lherzolite, in which, however, diamonds have not yet been observed.

At all other known diamond localities, especially those of India, Brazil and Lapland, olivine or serpentine as a mineral associated with diamond is conspicuous by its absence. In such localities, therefore, the mother-rock of the diamond cannot be an olivine-rock. On the other hand, diamonds have occasionally been found in meteoric stones, of which olivine is an invariable constituent, and the association of these two minerals in extra-terrestrial matter is a fact of considerable interest and importance.

The **mining operations** for obtaining the diamonds at the dry diggings were commenced at the end of the year 1870 ; by 1872 the industry was in a flourishing condition, and since this date it has steadily developed. At first the deposits were worked, regardless of future inconvenience, in an irregular and haphazard fashion, the aim of the miners being to amass the greatest possible amount of treasure with the least possible immediate expenditure of labour and money. Thus much valuable ground was covered with débris, which subsequent workers were forced to remove at a great sacrifice of time, labour and capital. In the deposits more recently discovered, the authorities have profited by former mistakes and mining operations have from the first been carried out in a systematic manner, with due regard to future necessities.

Each diamantiferous area was at first divided into square lots or claims, as was the custom in the gold-fields of California and Australia and also at the river diggings of South Africa. These claims in the Kimberley and De Beer's mines were 31 feet, and in the Du Toit's Pan mine 30 feet square ; each claim therefore had an area of 100 square yards or a little more. In the Kimberley mine there were 331 such claims, in the De Beer's mine 591, in the Bultfontein mine 886 and in Du Toit's Pan, 1430. In the three last-named mines the claims were laid out in such a way that there was no means of access to those in the centre except over the surrounding lots; this inconvenient arrangement materially increased the difficulty of mining and transporting material. When the Kimberley mine was opened, the Government Inspector of Mines in what was then the Orange Free State, profiting by past experience, arranged that every claim should be directly accessible by the construction of fourteen or fifteen road-ways, each 15 feet wide and all running in a north and south direction across the narrowest part of the mine. By this regulation every possessor of a claim lost $7\frac{1}{2}$ feet of ground, and until the advantage of the arrangement was realised it was bitterly opposed. In Plate VII. is shown the Kimberley mine as it appeared in 1872, with these road-ways.

Up to 1877, no single individual was permitted to possess more than two claims, the only exception to this rule being that of the discoverer of the mine, who was allowed three. Every intending digger had the choice of any of the claims which happened to be vacant, and each tenant of a claim paid the owner of the land 10s. per month in return for a licence permitting him to work the claim.   Until 1873, the penalty enforced for leaving a claim unworked during a period of seven days was forfeiture of the claim, which could then be transferred to another digger.

To keep the whole of a claim constantly worked proved somewhat too heavy a tax on the energy and resources of a single individual, the claims therefore came to be divided up, one man making himself responsible for a half, a quarter or even one-sixteenth of a claim.

More important than the sub-division of the claims was the amalgamation of several under one management, a system which began to be adopted in 1877, after the regulation preventing it had been rescinded.   Companies were formed to buy up a number of claims, and these, being under one central control, could be worked more expeditiously and economically ; before very long, there were very few claims, or portions of claims, worked by single individuals.   Thus in the middle of the eighties, almost all the claims into which the Kimberley mine was divided were in the possession of one or two large companies, while ten years before, these same claims were the separate property of about 1600 persons ; the same change also took place in the management of the other mines.   The number of claims in the possession of each company, of course, varied considerably, some having as few as four and others as many as seventy.   Many of these companies were formed with perfectly legitimate objects, others however were nothing more than swindles, and the claims they had acquired were usually very soon abandoned.

The immediate result of the formation of these companies was a large increase in the production of diamonds ; thus while the total output of diamonds in 1879 was about two million carats, in 1880 and 1881 it suddenly rose to over three millions.   This large increase was rightly ascribed to the advantages resulting from the partial amalgamation of claims which had taken place, and it was strongly urged that this policy should be pursued to its logical conclusion, and that the whole of each mine should be placed under one management and one system of working.   This proposition met at first with great opposition, but was effected in 1887 by the managers of the De Beer's mine, a company which, formed in 1880, gradually acquired claim after claim, until in 1887, with a capital of £2,332,170, it came into possession of the whole of the De Beer's mine.   The formation and development of this company, which since 1887 has been known as the " De Beers Consolidated Mines, Limited," had the effect of greatly reducing the working expenses of the enterprise, which from 1882 to 1887 had amounted to 40 per cent. of the value of the output.   In 1882 the production of the diamonds cost the company 16s. 6d. per carat, while in 1887 this amount had been reduced to 7s. 2d. ; at the same time the deposit had increased in richness as greater depths were reached, an increase in the production of about 40 per cent. having taken place. This increase in the richness of the deposit, combined with a diminished cost of production, naturally affected the dividends of the company, which rose from 12 per cent. in 1886 to 16 per cent. in 1887 and 25 per cent. in 1888.

In spite of the success which has attended this effort in the direction of amalgamation, the management of the other mines has not yet been altogether unified, though they are more or less under the control of the powerful and heavily capitalised De Beers Company. This company now has possession of the whole of the Kimberley mine, and the new Wesselton mine, as well as of parts of the Du Toit's Pan and Bultfontein mines.   Neither of these two latter mines, however, is now worked, since the open workings have become

Plate VII.

Kimberley Diamond Mine in 1872.

covered by falls of reef and the underground workings are not yet organised. An idea of the importance and influence of the De Beers Company may be gained from an inspection of the returns dealing with the production of diamonds. Thus from April 1, 1890, to March 31, 1891, this company alone produced 2,195,112 carats of diamonds, valued at £3,287,728, this being more than 90 per cent. of the total yield (2,415,655 carats) of the four mines at Kimberley, or indeed of the whole of South Africa. This result is of course due in some measure to the large amount of capital, £3,950,000, at the command of the company.

The inception of this company and its pursuit of a policy of buying up all available claims resulted in a considerable rise in the price of the latter, £10,000 or even £15,000 being asked for single claims, and proportionate prices for portions. The claims thus acquired a definite market value, which depended on the richness of the deposit at that particular place, usually known fairly accurately, and which of course varied at different times. Thus the claims in the Kimberley mine in 1875 were worth from £200 to £2500 each, in 1878 from £50 to £6000, and in 1882 from £150 to £15,000; the value of the whole mine being in these years £525,000, £1,300,000 and £4,150,000 respectively. At one time the value of the shares in the Kimberley mine amounted to £8,000,000.

In the other mines the deposit was poorer and the price of claims correspondingly lower. In the year 1880 the values of the richest claims in the Kimberley, De Beer's, Du Toit's Pan, Bultfontein, Jagersfontein and Koffyfontein mines were in the proportion of $10 : 5 : 2 : 1 : \frac{1}{10} : \frac{1}{15}$. In other words, the richest claim in the Kimberley mine was 150 times more valuable than the richest claim in the Koffyfontein mine; for the former £15,000 would be demanded and paid, while the latter would cost at the highest from £30 to £100.

For a short time after the opening of the mines each owner of a claim worked alone on his own piece of ground. It was found, however, that comparatively cheap labour could be obtained by employing the native Kaffirs, and these were soon engaged in large numbers. It is stated that in the seventies 10,000 to 12,000 Kaffirs were employed in the Kimberley mine alone, and by some authorities this number is doubled. The diamantiferous rock was excavated by the help of pickaxes or blasted with gunpowder, the latter agent being replaced later on by dynamite. The excavated material was then either loaded into carts or simply carried away from the mine. The whole mine was thus honey-combed with square pits which varied in depth in different claims, those which had been vigorously worked being very deep and enclosed by high vertical walls, and others having the appearance of rectangular columns, so high that they sometimes fell over and buried neighbouring claims with débris. The road-ways by which the claims in the Kimberley mine were separated, soon came to be mere walls, the surfaces of which rose high above the floors of surrounding claims and gave the whole mine a peculiarly striking appearance, as may be gathered from Plate VII.

Owing to the ease and rapidity with which the tuff composing these walls became weathered, they formed anything but stable boundaries, and as early as 1872 they had to be removed. After their removal, the mine had the appearance of one gigantic pit; the rock subsequently excavated could not then be removed in the same manner as before, and other means had to be devised. The mine was surrounded by high, wooden stagings provided with ropes and winding machinery, by the aid of which the diamond-bearing material was hauled up in sacks or buckets of hide. The owner of every claim, or part of a claim, had his own hauling rope, so that at this time, about 1874, the total number of these ropes was very large and gave the mine, which is pictured in the upper figure of Plate VIII., the appearance of a huge cobweb. The winding was first effected by hand windlasses, then by horse-power, and finally by steam, the delay in the adoption of the latter being caused by the cost of importing machinery and coal.   In spite of this difficulty, there were in 1880 no less than 150 steam-engines employed at the Kimberley mines, and in 1882 this number had been increased to 386 with a total horse-power of 4000, and this was further supplemented by the use of 1500 horses and mules.

The continual increase in depth of the claims was attended by increasing difficulty in excavating the tuff and by frequent accidents, due to falls of loosened material.   These difficulties were still further complicated by the fact that falls of reef also began to take place.   Often masses of rock would fall sufficient to bury, wholly or in part, many of the surrounding claims ; and in such claims no further excavation of " blue ground " was possible until the overlying mass of reef had been removed.   In September of 1882, in the

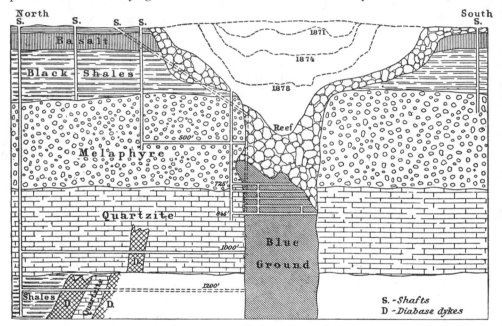

Fig. 40.   Section through the Kimberley mine.   (Scale, 1 : 4800.)

Kimberley mine, there was a fall of reef, estimated at about 350,000 tons, which buried no less than 64 claims ; in 1878 one-quarter of the total area of the mine was strewn with fragments of reef.   In 1879 and 1880, £300,000 was expended in removing this fallen material, and in 1882, £500,000 more was spent for the same purpose, and even then this obstacle to progress was not entirely removed.   From the Kimberley mine alone a total of about four million cubic yards of reef have been removed, at a cost of £2,000,000.   To what an extent the difficulties occasioned by a fall of reef influence the production of stones can be seen from the fact that the yield of the Kimberley mine, during the 18 months which preceded the catastrophe mentioned above, was 1,429,728 carats, but in the following 18 months only 850,396 carats.   The frequent falls of masses of reef and the removal of other masses which threatened to fall, resulted in a great increase in the surface area of the mine.   Thus in the middle of the eighties, the Kimberley mine, a representation of which is shown in the lower figure of Plate VIII., was a crater-like pit 385 yards long, 330 wide and 400 feet deep.

The appearance of water in the mine still further added to the embarrassment of the workers, and constituted a difficulty which was quite insuperable so long as the owners of the claims worked independently. The necessity for co-operation was met in 1874 by the institution of the Kimberley Mining Board, a body which undertook all work of public benefit, such as the removal of water, of fallen reef, and of reef about to fall, the expense incurred being shared equally by the owners of the claims. It was about this time that the formation of companies began to take place, although at first this form of co-operation was strongly condemned by individual miners, yet as time went on it became more and more apparent that the increasing difficulties and expense of working could only be overcome in this way. The larger capital at the disposal of the companies enabled them to employ the best machinery and to adopt all the improved modern methods of working, and thus to decrease the working expenses, and at the same time to increase the production.

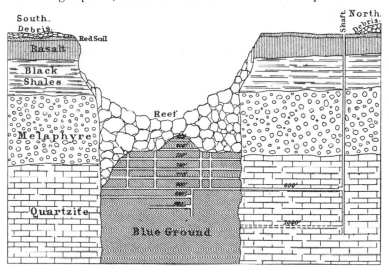

FIG. 41.   Section through the De Beer's mine.   (Scale, 1 : 1800.)

Although the amalgamation of individuals and capital rendered it possible to prolong for a time the system of open workings, yet, as time went on, it became very evident that this system could not be continued indefinitely, and that the open workings would have to be replaced by systematic underground workings, if the treasure hidden away in the depths of the mine was ever to be reached. A very successful beginning was made at the Kimberley mine in 1885, and in 1891, at this same mine, a shaft was driven into the reef to a depth of 1261 feet, from which horizontal galleries or tunnels were excavated to meet the diamond-bearing rock. In the section of the underground workings of the Kimberley mine shown in Fig. 40 may be seen these tunnels or galleries, situated partly in the diamantiferous pipe itself and partly in the surrounding reef. The lowest depth at which material could be excavated in the open workings was about 400 feet, so that the construction of the underground workings made accessible large quantities of fresh material. Moreover, the new system did away with the liability of the workers to injury from falling reef, and many of the earlier regulations dealing with this danger, became then unnecessary. The same system was also introduced in the De Beer's mine, a section through which is given in Fig. 41, although here the falls of reef had been less troublesome.

The earlier methods of **extracting diamonds from the rock** when excavated, were as primitive as were those first adopted for mining the rock. It has been mentioned above that the dry diggings are situated on an arid plateau, and at the time of their first discovery the water required for every purpose had to be fetched from the Vaal River, many miles away. This necessity forbad the washing of the diamantiferous material as was practised at the river diggings. The mass had therefore to be coarsely broken up with wooden pestles and the coarse and fine material separated by sieves ; the material of medium grain was then, as in the river diggings, spread out in a thin layer on a sorting-table, and any diamonds it contained picked out by hand.

By this method all the stones which passed through the fine sieve, the mesh of which was about $\frac{1}{8}$ to $\frac{3}{16}$ inch, were of course lost, these smaller stones being not then considered worth the time and trouble involved in their collection. The larger rock-fragments separated by the use of the coarse sieve with a mesh of $\frac{3}{8}$ to $\frac{5}{8}$ inch were thrown aside, though many would of course contain diamonds. It is estimated that during the period in which these methods were practised, at least as many diamonds were overlooked as were found, and in 1873 the débris was reworked and yielded a rich harvest. The material taken from the richest part of the mine has been worked over even a third time, and thanks to the use of improved methods, the result amply repaid the workers for their trouble. Hundreds of poorer miners, who were not fortunate enough to possess a claim, gained a living by working over the material of old mine heaps, as is still done at some places in India.

The lack of water was not felt for long ; very soon a main 18 miles long, bringing water from the Vaal, was constructed, and this supply was further supplemented by the numerous springs in the district, and by the water pumped out from the mines themselves. Thus it soon became possible to treat the diamantiferous material in the same way as at the river diggings, and so the term dry diggings came to be a misnomer. At first, the " blue ground " was reduced to fragments and then washed by the aid of the same simple appliances as had been in use at the river diggings. Improved methods, however, were gradually introduced ; thus in 1874, a washing machine worked by hand was employed for the first time, and this in 1876 and 1877 was itself replaced by a machine driven by steam. The construction of this was so much improved that it was capable of dealing with almost two thousand times as much material as could formerly be treated by washing, and of collecting stones which, on account of their small size, had formerly been lost. The diamonds are picked out by hand from the heavy residue, and are finally freed from any foreign matter which may adhere to them, by treatment with a mixture of sulphuric and nitric acids, after which they are ready for the market.

The " blue ground " excavated from the deeper parts of the mine is too hard and compact to be washed without previous treatment. It is therefore spread out in thin layers on the hardened ground of large fenced-in spaces, known as depositing floors. Here, exposed to frost, rain and sunshine, it gradually weathers, becoming friable and crumbling, when it is fit to undergo the process of washing. The weathering of the material, which is accompanied by a change from the normal colour of " blue ground " to that of " yellow ground," takes from one to nine months, according to the character of the weather to which it has been exposed, and to the mine from which the material was taken. " Blue ground " from the Kimberley mine weathers in about half the time required for the same process by the material from De Beer's mine ; the latter sometimes requires several years for the completion of the process, while a few months is usually all that is necessary for material from the Kimberley mine.

PLATE VIII

KIMBERLEY DIAMOND MINE IN 1874

KIMBERLEY DIAMOND MINE (WEST SIDE) IN 1885

The longer the period required for the weathering process, the more will the profit derived from the yield be diminished. For during this period there are many expenses and losses incurred, such for example as ground rent, which is very high, wages of labourers and watchmen, losses due to thieving, &c. Any means whereby the slow natural process could be hastened would therefore be welcomed, but up to the present no such means have been devised.

A factor which has largely contributed to the hardships of the South African diamond fields is the high price of the ordinary necessities of everyday life. This scantily-peopled region, in which only the barest necessaries could at one time be obtained, became inhabited with comparative suddenness by a population of at least 30,000 white people. The many and various articles necessary to their existence on these barren arid wastes had all to be conveyed from Capetown, Port Elizabeth, or some other seaport town. The transport was effected in waggons, drawn by horses, mules or oxen, and the long and difficult journey to Kimberley occupied several weeks. The rates for transport from Port Elizabeth to Kimberley, a distance of 500 miles and requiring about four weeks, were from 10s. to 30s. per 100 lbs of goods, and from Capetown to Kimberley, a journey of 650 miles, occupying about six weeks, they were still higher. Other prices were of course correspondingly high: thus, Cohen relates that in the year 1872 a bottle of beer cost 3s. 6d., and good Rhenish wine 18s. a bottle; a cabbage could never be obtained for less than 3s., potatoes were as high as 1s. per lb., and eggs were 6s. a dozen. At some seasons of the year a day's supply of fodder for a horse cost 15s., English coal fetched £16 10s. per ton, and a waggon load of wood of about 4½ tons was worth £30. The same authority states that £8000 was paid for a steam-engine of 100-horse power delivered in Kimberley. It is not surprising then, that with these prices for coal and machinery, steam power was so long in coming into general use, especially as it was not at first known that the diamantiferous deposits were so extensive. The cheapest food available was antelope flesh, a whole animal the size of a deer costing only from 3s. to 8s.; meat, therefore, was the staple article of food, and every drop of water had to be bought.

The wages paid to overseers and miners had of course to correspond with these high prices. The overseers and officials, who were all white men, were paid up to £2000 per annum. White miners, of whom in 1882 and 1883 there were about 1500, received from £4 to £8 per week, while the native workers, about 11,000 in number, were paid 22s. to 30s. per week.

All these details apply to the time when Kimberley was still unconnected by railway with the coast towns. Since 1885, it has been joined to Capetown by a line 647¼ miles long, and to Port Elizabeth by one of 485⅓ miles. The construction of these railways considerably diminished the cost of transport, and, in consequence, the price of many of the necessaries of life fell; moreover, it became possible to make more extensive use of coal, which was brought both from England and also from the South African mines at Stromberg in the Indwe district of Cape Colony, a place which has also been connected by rail with Kimberley.

For comparison with the prices quoted above, a few of more recent date may be given. In 1891 a ton of English coal cost at Kimberley £8 10s., and 100 lbs. of wood fetched 2s. The transport of goods from Port Elizabeth to Kimberley costs from £6 to £8 per ton, and the journey occupies only about thirty hours, instead of four weeks, as was formerly the case. The reduction in the cost of living is of course accompanied by a corresponding fall in wages; from £3 to £6 10s. per week is paid to white men, while Kaffirs earn at most 24s. per week, exclusive of housing, wood, water and medical attendance.

The climate of Kimberley cannot be considered anything but healthy; in winter it is mild and pleasant; in summer, however, from September to March, it is often very hot, in spite of its elevation of 4012 feet above sea level. There is often no rain for months together, and the whole of the rainfall usually takes place in a few heavy downpours. Since the erection of suitable dwellings for the miners and the improvement in their mode of living, the deadly camp-fever has been almost unknown, and the district can no longer be considered an unhealthy one, a consideration which has an important bearing on the output of diamonds.

Year.	Weight in carats.	Value per carat.	Total value per year.	Total value for 5 years.
		£ s. d.	£	£
1867 } 1868 }	200	—	650	
1869	16,550	1 10 0	24,813	
1870	102,500	1 10 0	153,460	
1871	269,000	1 10 0	403,349	
1872	1,080,000	1 10 0	1,618,076	
1873	1,100,000	1 10 0	1,648,451	2,200,348
1874	1,313,500	1 0 0	1,313,334	
1875	1,380,000	1 2 6	1,548,634	
1876	1,513,000	1 0 0	1,513,107	
1877	1,765,000	19 6	1,723,145	
1878	1,920,000	1 2 6	2,159,298	7,746,671
1879	2,110,000	1 4 6	2,579,859	
1880	3,140,000	1 1 6	3,367,897	
1881	3,090,000	1 7 0	4,176,202	
1882	2,660,000	1 10 0	3,592,502	
1883	2,410,000	1 2 9	2,742,470	16,275,758
1884	2,263,734	1 4 9	2,807,329	
1885	2,439,631	1 0 5	2,489,659	
1886	3,135,061	1 2 4	3,504,756	
1887	3,598,930	1 3 7	4,242,470	
1888	3,841,937	1 1 0	4,022,379	15,786,684
1889	2,961,978	1 9 3	4,325,137	
1890	2,504,726	1 13 3	4,162,010	
1891	3,255,545	1 5 8	4,174,208	
1892	3,039,062	1 5 8	3,906,992	
				20,590,726
Total	50,910,354	Mean 1 4 8	62,600,187	

In spite of the many and varied difficulties which have been encountered, the development of mining operations at Kimberley has been so extensive that, although the stones are of relatively sparing occurrence in the "blue ground," an enormous number must have been found. An idea of the extent of the **output** may be derived from an inspection of the above table, which is copied from Reunert. In this table is given the yearly export of diamonds from South Africa since 1867, the total value of this export, and the mean value per carat, the whole of the information having been derived from the most reliable sources available. The yearly export, though not exactly identical with the production, approaches it very nearly, and is sufficiently close for all practical purposes. The numbers quoted in the table may differ slightly from other returns, but are accurate enough to convey a correct idea of the gigantic scale of the output.

It should be remarked that the numbers given in this table for the years 1867 to 1882 are based only on estimates. Exact statistical records have only been kept since the establishment of the " Board for the Protection of Mining Interests " in 1882. It may be difficult, from the numbers given in the table, to form a correct conception of the enormous quantity of diamonds which have been exported from South Africa ; a few concrete examples are therefore appended as an aid to the imagination. The total weight of stones exported amounts to almost 51,000,000 carats, which is equal to 10,500 kilograms, or nearly $10\frac{1}{3}$ tons. These stones would fill a box five feet square and six feet high ; they would also form a pyramid having a base nine feet square and a height of six feet.

An exact record of the yield of each of the Kimberley mines has also been kept since September 1, 1882. The unknown, but probably very considerable, number of diamonds stolen by the workers from their legitimate owners, cannot of course be included in these records ; the value of the diamonds misappropriated every year is variously estimated at from £500,000 to £1,000,000. During the three years between September 1, 1882, and September 1, 1885, the four Kimberley mines, from which, as has already been mentioned, over ninety per cent. of the total South African output is derived, have yielded, according to official returns, the following amounts in carats :

	Sept. 1, 1882 to March 1, 1884.	March 1, 1884 to Sept. 1, 1885.	Total for the 3 years.
Kimberley mine	1,429,726$\frac{7}{8}$	850,396$\frac{1}{4}$	2,280,123$\frac{1}{8}$ carats
De Beer's mine	656,427	790,908$\frac{3}{4}$	1,447,335$\frac{3}{4}$ „
Du Toit's Pan mine . . .	709,877$\frac{1}{8}$	773,306$\frac{2}{4}$	1,483,183$\frac{7}{8}$ „
Bultfontein mine	738,230$\frac{1}{4}$	877,647$\frac{1}{2}$	1,615,877$\frac{3}{4}$ „

This gives the yearly average for the four mines together at 2,372,809$\frac{3}{8}$ carats, valued at £2,628,289 3s. 7d.

In the year 1886, the production of these four mines, and of a few others of less importance, as well as of the river diggings, amounted to :

	Carats.	Value. £. s. d.
Kimberley mine	889,864	836,767 17 7
De Beer's mine	795,895	739,937 2 8
Du Toit's Pan mine	700,302$\frac{1}{4}$	909,023 11 5
Bultfontein mine	661,339$\frac{1}{4}$	623,339 17 3
St. Augustine's mine	239$\frac{1}{4}$	317 19 4
River diggings	38,672$\frac{7}{8}$	181,156 9 2
Orange River Colony	73,303$\frac{3}{4}$	121,654 15 1
	3,159,617$\frac{3}{8}$	3,412,197 12 6

The figures in the three tables given above are in all probability too low, since there are many sources the supply from which it is difficult or impossible to estimate. Thus the total production of the river diggings, and of the Jagersfontein and Koffyfontein mines in the Orange River Colony, is not exactly known, the numbers in the table referring not to the total production, but to the diamonds sent from these mines and diggings to the central market at Kimberley. The first of the above three tables gives the total yearly production for the whole of South Africa, from the discovery of diamonds in 1867 up to the year 1892. Details connected with the production in certain years of individual mines are collected together in the second and third tables.

Formerly, when the claims were in the possession of single individuals or small companies, the aim of the owners was to obtain and sell as many stones as possible. Now, however, the output is controlled by the " De Beers Consolidated Mines," and only a certain number of stones are placed on the market, in order to keep up the price. Experience has shown that the annual demand for diamonds, both for jewellery and for technical purposes, amounts in value to about £4,000,000. Taking the average value of rough stones at 21s. per carat, this corresponds to a weight of rather over 3,800,000 carats, or about $15\frac{1}{3}$ cwts.; the yearly production of Cape diamonds is rather over 3,000,000 carats, and the difference between these weights represents the amount supplied by Brazil, India, Australia, and Borneo.

Although the total weight of stones collected is so enormous, yet the **relative amount of diamond in the "blue ground"** is extremely small. The proportion which this constituent bears to the whole mass varies both in different mines and in different parts of the same mine. Thus in some mines the " blue ground " excavated from successively lower depths has shown a marked increase in richness. It would be naturally of great importance to those interested, were it possible to foretell whether such an improvement in yield would be maintained or not; this, however, is not possible, and it is equally impossible to give any explanation of the variation.

The Kimberley mine, since its discovery in July 1871, up to the present day, has been the richest of all the deposits. Many of the original miners made their fortunes in less than a month, and one case is quoted in which diamonds to the value of over £10,000 were found in the short space of a fortnight. A cubic yard of " blue ground " from the Kimberley mine contains more diamonds than a mass of equal size from any other district; this being so, mining operations were here prosecuted with great vigour, all the other mines being at one time forsaken. Owing to the want of reliable statistical reports dealing with early times, it is not possible to determine with certainty whether the deposit increases in richness as greater depths are reached, or not. Probably it does not, and if this be so, the Kimberley mine differs in this respect from the others.

Boutan has collected statistics dealing with the content of diamonds in the " blue ground " excavated from the Kimberley mine by certain companies. From the year 1881 to 1884, this varied between 3·04 and 7·17 carats per cubic metre (about 1·3 cubic yards), an amount corresponding to two- to five-millionths of a per cent. This yield, however, applies only to the richest parts of the " blue ground," the mean yield for that part of the deposit which is worked, taking rich and poor together, is about 4·55 carats per cubic metre (3·5 carats per cubic yard). If, however, the material of the western side of the mine, which on account of its poverty is not worked, is also included, the mean yield for the whole deposit will be 4·20 carats per cubic metre of " blue ground," or on the average three-millionths of a per cent.

The dimensions of the portion of the deposit which is worked being known, it can be calculated that the excavation of " blue ground " to a depth of one metre would result in the production of 88,000 carats of diamonds, which at £1 per carat would be of the aggregate value of £88,000. From the same part of the deposit, a cubic metre of worked material would contain on an average £4 11s. worth of diamonds.

The De Beer's mine is $1\frac{1}{5}$ times the size of the Kimberley mine; the central point of the former is situated 1771 yards to the east of the central point of the latter. The deposit ranks next to that of the Kimberley mine in richness, although it was at first very poor, only yielding ⅓ carat to the cubic metre of rock. The yield, however, increased rapidly as greater depths were reached ; at a depth of 300 to 400 feet it had increased tenfold, so that 3½ carats of stones were then obtained from a cubic metre of rock. From the year 1882 to 1884 the yield obtained by some companies varied between 1·28 and 3·52 carats ; the mean yield for the whole mine was estimated at 3·15 carats per cubic metre of rock, worth, at the higher price commanded by stones from this mine, £3 9s.

The richness of this mine varies not only with the depth, but also at different places on the same level. The best part of the deposit is not surpassed in richness by any part of the Kimberley mine, while other parts are so poor that they are not worked at all. The central portion is very rich, and extensions of this stretch out especially towards the north and south, forming a great contrast to the western third of the mine, which is extremely poor. A beautiful yellow octahedron, weighing 302 carats, was found in the eastern side of the deposit, on March 27, 1884.

The increased richness of the lower lying parts of the deposit led to an attempt being made, soon after the opening of the mine, to excavate the deeper portions by means of underground workings, leaving the poorer portions standing above. The attempt was very successful as far as yield was concerned, but, owing to the imperfect methods adopted, was attended by so many accidents that it had to be abandoned. Since 1885, however, when, as we have already seen, the construction of underground workings began to be more skilfully engineered, the excavation of the deeper lying material has been resumed.

The Bultfontein mine is situated 4840 yards to the south-east of the Kimberley mine. A cubic metre of the surface material yielded only a small fraction of a carat, but here also a rapid and regular improvement in the deposit as lower depths were reached was manifested, the yield at a depth of 200 feet having increased threefold. The increase in the richness of the deposit took place in this mine with almost mathematical regularity, and was attended both by an improvement in the quality of the stones, and by a diminution in the number of fractured crystals. In 1887 a depth of 460 feet had been reached, a depth which has never been exceeded in open workings. In the period 1881 to 1884 the yield varied between 0·56 and 1·27 carats per cubic metre of " blue ground," the mean yield being about 1·05 carats, so that a cubic metre of rock contained diamonds to the value of about 23s. Underground workings have not yet been established in this mine so that from it may be derived an idea of the appearance presented by the other mines in earlier days.

Du Toit's Pan mine is 1320 yards distant from Bultfontein mine, and 3542 from De Beer's mine. In 1874 the mine was almost deserted, the yield being so small, and it has only been systematically worked since 1880. Here again the surface rock was poor, and yielded at the best of times only ¼ carat per cubic metre of material ; here also the deposit improved at greater depths, but not so rapidly as in the De Beer's mine. At a depth of 175 feet, at which the yield had doubled itself, a peculiarity not hitherto noticed in any other mine was observed : the richness of the deposit was found to be absolutely identical at all points on the same level, so that no variation in the yield was experienced. From this point downwards the yield rapidly increased, approaching that of the Kimberley and De Beer's mines. From 1881 to 1885 the yield varied between 0·31 and 1·11 carats, the mean yield being 0·77 carat per cubic metre of " blue ground," having the average value of 22s. In this mine, as in the Bultfontein mine, the increase in richness of the deeper lying deposit was accompanied by an improvement in the quality of the stones, and by a diminution in the number of broken crystals. The open workings were here excavated to a depth of over 400 feet, and underground workings have been scarcely as yet commenced.

The Jagersfontein mine, near Fauresmith, in Orange River Colony, contains only 0·10 to 0·35 carat of diamond per cubic metre of "blue ground." The poorness of the yield is, however, in some measure compensated for by the singular beauty and size of the stones. For the year ending March 31, 1891, the average value per carat of stones from this mine was 37s., stones from the Kimberley mines being worth only 25s. 6d. per carat. The largest diamond known was found in this mine in 1893; it weighed 971¾ carats and will be figured (Fig. 51) and described later on. A very fine stone of 655 carats was found here also at the end of 1895. The mine was opened in 1880, was abandoned for a time about 1885, but was subsequently re-opened.

The Koffyfontein mine, also in the Orange River Colony, gives a smaller yield still, amounting to only about two-thirds that of the Jagersfontein. The stones, however, are of good quality and are worth about 30s. per carat; from December 1887 to April 1891, 9912 carats of diamonds, valued at £14,640, were mined here.

The relative importance of the different mines is also shown to a certain extent by a comparison of the number of workers employed in each. Thus in the year 1890 the numbers were as follows·

	Whites.	Blacks.
De Beer's mine	682	2780
Kimberley mine	495	1800
Du Toit's Pan mine	67	400
Bultfontein mine	37	300
	1281	5280

The number of persons employed in the mines at Kimberley in the year 1892 is given in the following table. This includes two mines not before specially mentioned, namely Otto's Kopje and St. Augustine's, the former being situated a couple of miles to the west of Kimberley, and the latter in the town itself.

Name of Mine	Above ground.		Below ground.		Total.		Grand Total.
	Whites.	Blacks.	Whites.	Blacks.	Whites.	Blacks.	
Kimberley . .	372	982	133	822	505	1804	2309
De Beer's . .	693	2098	229	1812	922	3910	4832
Du Toit's Pan .	—	—	—	—	96	654	750
Bultfontein . .	—	—	—	—	186	933	1119
St. Augustine's .	9	15	10	13	19	28	47
Otto's Kopje . .	3	54	—	—	3	54	57
	1077	3149	372	2647	1731	7383	9114

We turn now to the consideration of the characters and the quality of Cape diamonds, that is to say, the form and condition of crystallisation, colour, size, &c., peculiar to stones from this region.

These diamonds usually occur as distinct crystals, symmetrically developed in all directions and with perfectly sharp edges and corners; but fragments of larger crystals, bounded by cleavage surfaces, and which are therefore cleavage fragments, also occur with considerable frequency. These cleavage fragments are sometimes of fair size, the original crystals, of which these fragments are part, probably varied in size from 3 to 500 carats; large cleavage fragments are known as "cleavages," while fragments weighing less than a carat are referred to as "splints." It is a remarkable fact that these cleavage fragments are nearly always white, that is colourless, or at least very faintly coloured; fragments of a dark colour, or of a decided yellow, are extremely rare, so that we must conclude that such stones offered greater resistance to fracture than did the colourless diamonds.

The **crystalline form** is on the whole very regular, and the edges and corners never show signs of having been water-worn except of course in stones from the river diggings. Octahedra with curved and grooved edges (Fig. 31, n and o) are very frequent, while rhombic dodecahedra with curved faces (Fig. 31 c), and with singly or doubly nicked faces (Fig. 31 d) are rare. Crystals of this kind, when not unduly distorted, are greatly prized, especially the octahedra, for it is possible to give such stones the desired brilliant form with very little preliminary cleaving and loss of material. Cubes (Fig. 31 a), which are specially characteristic of Brazil, are practically non-existent at the Cape, extremely few diamonds with this form having been found. While hemihedral forms (Fig. 31 k) are of rare occurrence, twinned crystals, on the other hand, are very abundant; the twinning takes place according to the usual law, with a face of the octahedron as the twin-plane, and the individuals of the twin being two octahedra (Fig. 31 g), two rhombic dodecahedra, or two hexakis-octahedra (Fig. 31 h), which are much flattened in the direction of the twin-axis. The external form of twin-crystals varies with the development of the individuals, and may be tabular, lenticular, heart-shaped, &c. On account of their small thickness they are not very suitable for cutting as brilliants, and are generally used for rosettes; they are for this reason less highly prized than are other forms, such as the octahedron, and command a lower price. When the junction of the two individuals in flattened twin-crystals is distinctly to be seen, the stones are known in the trade as "twins," while those in which the junction is less conspicuous are referred to as "macles" (mackel).

Besides the occurrence of twinned crystals consisting of two individuals which have grown together in a symmetrical manner, there occur groups consisting of two or more individuals irregularly intergrown. An example of this irregular intergrowth is furnished by the spheres of bort, which have been already described and which occur very frequently both here and in Brazil. The surface of these spheres is seldom quite smooth, the projecting corners of the numerous small octahedra, of which the sphere is built up (Plate I., Fig. 3), being the cause of the irregularities of its surface. The size of these peculiar crystal aggregates is sometimes quite considerable, spheres weighing as much as 100, or even 200 carats, having been found. Occasionally, the centre of a sphere of bort is occupied by a single large, colourless crystal, which falls out of the rough, grey shell, when the latter is broken.

The **size** of the Cape diamonds is extremely variable, and ranges from that of the largest to that of the smallest stones yet found in any country.

When the operation of washing is performed with sufficient care, it is possible to collect numerous stones weighing no more than $\frac{1}{32}$ carat (about 7 milligrams). The improved washing machinery now in use is capable of collecting stones of this small size just as easily as larger specimens. Formerly these small stones were lost in the process of washing, and this gave rise to the belief that diamonds less than $\frac{1}{4}$ carat either did not occur at all, or only very rarely, at the Cape. The existence, hitherto unsuspected, of large numbers of microscopically small diamonds, together with particles of carbonado and of graphite, in the "blue ground" has been recently demonstrated; the occurrence together of diamond and graphite is worthy of special remark.

The most salient feature of the South African diamond-fields, as compared with those of other countries, is the prevalence of stones of large size.   It will be remembered that in Brazil the discovery of a stone of 17 carats was such an event that its finder, if a slave, was rewarded with his freedom.   In South Africa stones of this size occur in hundreds and in thousands; and the discovery of a stone of 100 carats causes less excitement than did the finding of a 20-carat stone in Brazil.   Stones of 80 to 150 carats are of common occurrence, scarcely a day passes in which a stone between 50 and 100 carats in weight is not brought to light.   Since the year 1867, when the South African deposits were discovered, the number of large stones, which have been found there, far exceeds not only the number unearthed in India in the course of a thousand years and in Brazil during a period of 170 years, but also the total production of large stones in these two countries added together.   Diamonds which weigh after cutting upwards of 75 carats have occurred at the Cape in greater numbers than in any other known locality.   While the mean size of Brazilian diamonds is scarcely one carat, the majority of South African stones are of this or larger size, excepting, of course, those stones which are rejected as being unsuitable for cutting.

We have already mentioned that the largest diamond known, the "Excelsior," was found at the Jagersfontein mine in 1893.   It is a stone of the first water, weighing $971\frac{3}{4}$ carats, and will be described and figured (Fig. 51) in the section dealing with large diamonds. The next largest Cape diamond is the one of 655 carats found in the same mine in 1895, which is stated to be of unusually fine quality.   Another stone of 600 carats, but of poor quality, is said to have been found in this mine.   A stone of $457\frac{1}{2}$ carats was found in one of the mines, but which one is not recorded; one almost as large, weighing $428\frac{1}{2}$ carats, was obtained from the De Beer's mine, while in 1892 the Kimberley mine yielded a diamond of 474 carats, from which was cut a brilliant weighing 200 carats.   "The Julius Pam," a stone of $241\frac{1}{2}$ carats, which gave a brilliant of 120 carats, came from the Jagersfontein mine.   A few large stones have also been contributed by the river diggings, the largest being the "Stewart" of $288\frac{3}{8}$ carats, which was cut as a brilliant weighing 120 carats.

Although the diamond-fields of South Africa are unique as concerns the number and size of the stones found there, the same can by no means be said of the **quality** of diamonds from this region.   Not only are the stones frequently so dark and unpleasing in colour that they can only be applied as bort for technical purposes, but they are also very often disfigured by "clouds" and cracks, the so-called "feathers."   Moreover, these cracks, especially in stones from the Du Toit's Pan mine and from the diggings on the Vaal River, are often rendered still more conspicuous by being filled with films of limonite. The presence of enclosures of foreign matter is also common; these are usually black and resemble particles of coal, but are probably hæmatite or ilmenite.   There are also green enclosures of a pecular vermiform character, which, according to Cohen, are probably some compound of copper, and red enclosures of unknown nature.   It is stated by Streeter that on an average only 20 per cent. of Cape stones are of the first water, 15 per cent. of the second, and 30 per cent. of the third, the remaining 35 per cent. being bort. According to Kunz, however, only 8 per cent. are of the first quality, 25 per cent. of the second, and 20 per cent of the third quality, the remainder being bort.

Cape diamonds show a great range of **colour.**   Perfectly colourless or pure white to deep yellow, light to dark brown, green, blue, orange, and red specimens have all been found.   At the same time the stones may be transparent and clear or cloudy and opaque.

Pure white, absolutely colourless stones are very rare, still the finest blue-white diamonds, such as are found in India and Brazil, are not altogether absent.   Only about 2 per cent. of the total number of stones found reach the standard of absolute perfection, among such the most general form is that of a symmetrically developed octahedron. The " Porter Rhodes," found in the Kimberley mine on February 12, 1880, is one of the finest of Cape diamonds; it is said to weigh 150 or 160 carats, and is a stone of singular beauty. The largest diamond known, that of $971\frac{3}{4}$ carats, as well as those of 655 and $209\frac{1}{4}$ carats respectively, all from the Jagersfontein mine, are also of this high quality. As a rule, large stones are patchy and impure or coloured yellow, often a deep shade of yellow, which greatly diminishes their value. The otherwise poor deposits of Jagersfontein and the river diggings are remarkable for the purity and beauty of the stones found there, especially in the latter.

The majority of what are usually regarded as white Cape diamonds are in reality more or less tinged with yellow ; this, though not apparent to an unpractised eye, is at once remarked by an experienced diamond merchant. Stones of this tint are described as being " Cape white," while others, in which the faint yellow tint is replaced by an equally faint greenish tinge, rank as " first by-water." Although the yellowish and greenish tinge is so slight, yet it manifestly exercises a considerable influence on the lustre and refractive power of the stone. Such a stone scarcely attains to the fire and play of colour of a perfectly colourless Indian or Brazilian stone ; moreover, even though cut in the best brilliant form, it will appear dusky when compared with the latter and will therefore be less highly prized.

Stones of a distinct, though pale, yellow colour are specially common ; they vary in shade from a canary- or straw-yellow to a light coffee-brown. They form the majority of those Cape stones which are suitable for cutting, and are naturally less prized than the Cape whites or others already mentioned. As a rule, these stones, the different shades of which are distinguished by the terms second by-water or off-coloured stones, pale yellow and dark yellow, are less disfigured by faults than are the colourless stones. The abundance of these pale yellow stones is a feature peculiar to the South African diamond-fields ; nowhere else are they found in such numbers. Before the discovery of these deposits, stones of this colour were extremely rare and were sought after as much as are now the stones of a fine red, blue, or green colour, which are still rare. Such diamonds are referred to as " fancy stones," and are perhaps more rare at the Cape than at other localities ; a representative of such " fancy stones " from the Cape is a rose-violet diamond of 16 carats. Diamonds of these beautiful colours, even when found, are invariably small. Transparent stones of a dark brown or black colour are very rare ; though the qualities most highly prized in colourless diamonds are absent in such stones, yet, on account of their rarity and their application in mourning jewellery, they command a high price. Very darkly coloured or impure stones, as well as those which are cloudy and opaque, are unsuitable for cutting and are used as bort.

Another unique feature of the Cape diamond-fields is the occurrence of the peculiar " smoky stones," which have been already mentioned. These occur for the most part at Kimberley and are scarcely known elsewhere ; they are distinguished by their very regular octahedral form and by the possession of a peculiar smoky-grey colour, which is either distributed uniformly or accumulated at the edges and corners of the stone, which, in the latter case, is known as a " glassy stone with smoky corners." In these diamonds there is a liability, as has been already mentioned, to fall to powder with no apparent external cause ; this is certain to happen sooner or later when such a stone is once taken out of the ground, and many and various are the devices adopted by the unfortunate possessor to postpone the catastrophe, at any rate, until he has prevailed on some inexperienced buyer to take the stone. Thus, immediately after it is taken from the rock, the finder will perhaps place it in his mouth, or smear it with grease ; and when it must be sent on a journey it will often be placed inside a potato, this being considered the safest method of packing such stones, probably because they are thereby protected from contact with other diamonds or hard objects, the slightest scratch being sufficient to bring about the bursting of the stone. The singular behaviour of these stones is due to the existence of intense internal strains in their substance ; the same phenomenon being also the cause, as we have seen, of the strong anomalous double refraction possessed by some diamonds.

The collective characters of the stones found in each mine and in each part of a mine are distinctive, but single stones of every quality occur in all mines. Thus, though it may be impossible to state the particular mine in which a single stone was found, yet an experienced Kimberley diamond merchant would have no difficulty in naming the mine, or portion of a mine, from which a parcel of stones had come, provided that the parcel formed a fair sample of the yield of that particular deposit.

The stones found in the rich Kimberley mine are usually poor in quality; and broken fragments, the latter invariably uncoloured but containing many black enclosures, are present in great abundance. A large percentage of the material yielded by this mine, especially from the north side, is unsuitable for cutting and only applicable as bort, 90 per cent. of South African bort being furnished by this mine. Broken fragments are confined to a certain extent to the middle and south side of the deposit ; while the north-east corner and the west side of the mine have yielded brown octahedra and " smoky stones " in great abundance; yellow diamonds, so numerous everywhere else, are here almost wholly absent. The stones found in the east and south-east portions of the deposit closely resemble those from Du Toit's Pan mine.

From the De Beer's mine are obtained crystals of every kind and colour, the surface of which is almost invariably finely granular, glimmering, and somewhat greasy, surface characters which are met with in crystals from no other deposit. Bort is rare, but broken fragments containing black specks are abundant. Large yellow rhombic dodecahedra are very frequent, the De Beer's stones being, on the whole, remarkable for their large size, while stones from the Kimberley mine are conspicuous for their whiteness.

The diamonds found in the Du Toit's Pan mine are usually well crystallised and of considerable size, yellow octahedra being often specially large. Bort, very small stones and " smoky stones," are practically absent, and crystals disfigured by black specks are seldom met with. The colour of stones from this mine is often rather dark, but the proportion of Cape white and yellow stones found is greater than elsewhere in this region. On the whole, the diamonds yielded by this mine are more beautiful than those of any other deposit in the neighbourhood of Kimberley.

The Bultfontein mine yields principally small white octahedra, much modified on their edges and usually full of faults and spots. Large stones, broken fragments, bort, and deeply coloured stones are here practically absent.

The average value of stones from the different mines of course varies in correspondence with the variations in quality we have just been noticing. In the first column of the following table, compiled from the estimates given by Moulle, will be found the average price per carat of rough stones from the different mines during the period between September 1, 1882, and the end of March 1884. The corresponding prices for 1887, given in the second column, are somewhat lower, but the proportion existing between them is about the same and has indeed remained practically unaltered up to the present day.

	1882–4.	1887.	
	s. d.	s. d.	
River stones	54 11	46 7	per carat
Du Toit's Pan mine . . .	28 1	24 3	,, ,,
Bultfontein mine	21 0	17 11	,, ,,
De Beer's mine	20 11	17 5	,, ,,
Kimberley mine	19 2	17 2	,, ,,

These four mines yield, on an average, respectively, 0·77, 1·05, 3·15, and 4·55 carats of diamonds per cubic metre of " blue ground," so that the stones found in mines of which the yield is poor, surpass in quality those found in richer mines.

In the Jagersfontein mine, as has been already mentioned, are found the whitest, largest, and most transparent of Cape diamonds, some of which approach, or even equal, the beautiful blue-white Brazilian and Indian stones which are so highly prized.   The abundance of white stones in this mine is sometimes thought to be connected with the complete absence of iron-pyrites, which is found everywhere else and has been supposed to be the cause of the yellow colour of Cape diamonds.   The beauty of these white stones is unfortunately, however, often impaired by the presence of spots and blemishes of various kinds ; moreover, in addition to regularly and symmetrically developed crystals, irregular intergrowths are not infrequently met with, so that a considerable proportion of stones from this mine are unsuitable for cutting and have to be discarded.   The stones found here which are free from faults are of singular beauty ; they are comparable to the diamonds of Bagagem in Brazil and command the very highest price.

It has not hitherto been mentioned that the Kimberley and De Beer's diamonds are supposed to be less hard than stones from Du Toit's Pan and Jagersfontein mines and from the river diggings.

The whole of the South African **diamond trade** centres round Kimberley.   The stones usually change hands in large lots and are often placed on the market directly they come from the washing machinery.   In other cases they are first sorted into parcels containing various qualities; this process offers great scope to the skill and discretion of the diamond merchant, for the amount obtained for a lot of stones depends largely on the arrangement of the stones of different size, quality, colour, &c., into parcels.

Various trade names for different kinds and qualities of diamonds have been evolved side by side with the development of the traffic in these stones.   Only about four such terms were originally in use ; a much greater number are at present in existence, and the significance of some of the most important will now be given below.

Crystals or *Glassies* are white, or nearly white, perfect octahedra.

Round stones are crystals with curved faces ; these are sub-divided according to colour into Cape white, first by-water, and second by-water.

Yellow clean stones is a term which includes all yellow stones, these being grouped according to their shade of colour into off-coloured (the lightest shade), light yellow, yellow, and dark yellow.

Mêlé is a term applied to crystals varying from white to yellow (by-water) and often also to brown, weighing on an average not more than $1\frac{1}{2}$ to $1\frac{3}{4}$ carats.   The term " small mêlé " is applied to similar stones as small as $\frac{1}{20}$ carat.   All stones characterised by this term are round or glassies, never broken fragments.

Cleavage is the term applied to crystals containing spots, to twinned crystals, and to others which need to be cleaved before they can be cut ; thus the term " black cleavage " is applied to stones which, in the rough condition, appear much speckled, but which, after cleaving, give fine stones.   Large blackish diamonds are referred to as "speculative stones"; their value depends on their size and on the probability of obtaining good cleavage fragments from them.   For stones of this sort (cleavage) weighing less than ¾ carat the trade name is *Chips*.

A collection containing black cleavage, stones of a brown or poor yellow colour and bort, forms a "parcel inferior," the contents of which are unsuitable for cutting and are pulverised for grinding powder or applied to other technical purposes.

The London jeweller, Mr. Edwin W. Streeter, in his book "Precious Stones and Gems," gives the following list of trade names for the various kinds of rough Cape diamonds; this differs somewhat from that given above, but will be easily comprehended by the reader of the foregoing pages :

White Clear Crystals	Bright Brown
Bright Black Cleavage	Deep Brown
Cape White	Bort
Light Bywater	Yellows
Light White Cleavage	Large Yellows and Large Bywaters
Picked Mêlé	Fine Quality River Stones
Common and Ordinary Mêlé	Jagersfontein Stones
Bultfontein Mêlé	Splints
Large White Chips	Emden
Small White Chips	Fine Fancy Stones
Mackel or Macle (flat, for roses).	

These different sorts of stones naturally differ widely in **value**; moreover, prices which were current before the discovery of the Cape diamond-fields have been somewhat modified in consequence of the enormous increase in the production due to this discovery. Thus, stones which were rare elsewhere, but abundant at the Cape, have fallen in value, while those which are rare also at the Cape have retained their former value.

In this way the price of perfectly colourless stones of the first water has not fallen in consequence of the discovery of the Cape diamond-fields, but is as high as ever it was.   The price of yellow or yellowish stones of 10 to 150 carats in weight is, on the contrary, much lower.   To such stones Tavernier's old rule, according to which the value varied with the square of the weight, is inapplicable, the price of such stones now varies directly with the weight; sometimes, however, it is in a still lower proportion, so that when the weight of a stone is doubled the price is not always necessarily doubled.

Naturally some considerable time must elapse before prices adjust themselves to new and unknown conditions.   Thus, for the first large stones found at the Cape a price in consonance with Tavernier's rule was demanded; such prices were seen to be too high and were soon regulated to accord with the changed conditions.   As early as 1876 rough Cape white stones of good quality, and up to 6 carats in weight, had fallen in value from 30 to 50 per cent., the largest and smallest stones having suffered the greatest depreciation.   It should be borne in mind, however, that these Cape whites were not quite equal in quality to the white Brazilian stones.   The discovery of the Cape diamond-fields caused a still greater depreciation in the value of bort, the price of this material having in 1873 fallen 85 per cent. ; in 1876, however, the depreciation in the price of bort was only 70 per cent., and it continued to rise in value owing to its increasing application for technical purposes.

The price of diamonds at the Cape, as elsewhere, depends not only on the quality of the stones but also on the conditions of supply and demand, and varies from day to day. According to the statements of E. Cohen, the price of bort varied from 1875 to 1880 between 1s. 9d. and 5s. 8d. per carat ; for Cape whites, of 2 to 6 carats in weight, between £3 15s. and £7 10s. per carat, and for fragments of 1 to 2 carats, between 8s. and 24s. per carat.

According to the estimate of Anton Petersen, as quoted by E. Cohen, the following prices for rough stones were in 1882 paid at the mines :

		£ s.	£ s.	
First water (highest quality) . .	4 carat stones	15 0 to	18 0	per carat.
Best Cape white	1 „ „	1 10 to	1 15	„ „
„ „	6 „ „	4 0 to	5 0	„ „
Light yellow (off-coloured) . .	1 „ „	15 to	1 0	„ „
„ „ „ .	6 „ „	2 0 to	2 10	„ „
„ „ „ .	20–40 „ „	2 5 to	3 0	„ „
„ „ „ . .	100 „ „	3 15 to	6 0	„ „
Bort		6 to	8	„ „

These prices which are exceptionally low, were current only on the Cape diamond-fields and not in the European markets, one market being perhaps influenced by circumstances which do not affect the other.

The prices per carat, stated by Boutan to have been current in Kimberley on July 31, 1883, are tabulated below:

Varieties.	Weight of stones in carats.	Price per carat.
Crystals or Glassies ⎫	1 average	55s.
„ „ ⎪	2 „	75s. to 80s.
„ „ ⎬ Cape White or White	3 „	95s. to 100s.
„ „ ⎪	4 „	120s.
„ „ ⎭	5 to 8 and more	According to size
Cape White Round Stones	1 to 2 „ „	40s. to 45s.
„ „	3 to 4 „ „	47s. 6d. to 52s. 6d.
„ „	5 to 8 „ „	55s. to 60s.
First By-water Round Stones . . .	1 to 2 „ „	⎫
„ „ „ . .	3 to 4 „ „	⎬ 10 per cent. less than the Cape White
„ „ „ . .	5 to 8 „ „	⎭
Second By-water Round Stones . . .	1 to 2 „ „	⎫
„ „ „ . .	3 to 4 „ „	⎬ 5 per cent. less than the First By-water
„ „ „ . .	5 to 8 „ „	⎭
Yellow Clean Stones	1 to 3 „ „	23s. 6d. to 28s. 6d.
„ „	4 to 10 „ „	30s. to 40s.
„ „	up to 40	42s. 6d. to 47s. 6d.
Dark Yellow Clean Stones	1 to 3 and more	22s. 6d. to 27s. 6d.
„ „	4 to 10 „ „	28s. 6d. to 37s. 6d.
„ „	up to 40	40s. to 45s.
Mélé	¼ average	27s. 6d.
„	⅓ „	31s. 6d.
„	½ „	35s.
„	¾ „	40s.
„	1 „	46s.
Cleavage	¾ „	14s. 6d.
„	1 „	17s. 6d.
„	2 „	24s.
„	3 „	28s. 6d.
„	4 to 5 „	32s. 6d.
„	Large Stones	32s. 6d.
Good White Square Chips	½ average	12s. 6d.
„ „	¼ „	8s. 6d.
Small White Square Chips	—	6s. 0d.
Common White Square Chips . . .	—	5s. 6d.
Common Cleavage and Chips . . .	—	5s. 0d.
Bort	—	4s. 6d.

Boutan also remarks on the lowness of these prices which were consequent on a commercial crisis; they reached their lowest level in 1885, having fallen 20 per cent., and after this date began to rise again.   In the following table is given the average value per carat, calculated from the weight and value of the total export during the years 1883 to 1891:

	s.	d.			s.	d.
1883	23	7	1888		19	11
1884	22	9	1889		29	4
1885	19	3	1890		30	7
1886	21	1	1891		25	2
1887	21	8				

It should be remarked here that during the period in which the Cape deposits have been worked, the average quality of the diamonds has remained practically the same, so that the above numbers represent approximately the mean market value for each year.

It will be readily understood that with objects like diamonds, so costly and, yet at the same time, so easily hidden, there are possibilities for very considerable **illicit trade**. Those engaged in the mining, washing, and sorting of diamonds, especially the Kaffirs, constantly find opportunities for secreting stones, in spite of the strict supervision to which they are subjected.   Although the mining employees each time they leave work have to undergo a rigorous personal search, yet diamonds are continually being smuggled through and placed on the market by illicit diamond buyers (I.D.B.).   It is estimated that 30 per cent. of the total output is thus diverted into illegitimate channels.

The strictest of laws and regulations have from time to time been devised and rigidly enforced with the object of suppressing theft of, and illicit trade in, diamonds.   Thus a man convicted of diamond stealing or illicit diamond buying was sentenced to several years of penal servitude ; under no circumstances were natives allowed to sell stones, and white men were obliged to procure a written licence before engaging in the trade of buying and selling diamonds, and to submit for the inspection of the authorities a properly kept register of all transactions.   The difficulty of obtaining witnesses, and therefore of convicting a person of an illicit transaction, made the enforcement of these and similar regulations somewhat of a dead letter ; moreover, the profit attending an illicit transaction successfully carried out was so large that the risk of conviction failed to act as a deterrent.   Since March 1, 1883, still more stringent regulations have been in force; a person suspected of, and charged with, the illicit possession of diamonds, must defend himself against the charge by furnishing a satisfactory explanation of the circumstances leading to his arrest.   Moreover, search-warrants are now granted in the case of white men as well as of natives.

These regulations are in force not only in the diamond-fields but in the whole of Cape Colony, and they were also adopted in the Orange Free State.   The illicit trade has, therefore, been checked, but not altogether stamped out.   The cunning and ingenuity shown by Kaffirs in concealing and disposing of stolen stones is unexampled.   As an illustration we may quote the case of a native who, in 1888, was suspected of being in the unlawful possession of diamonds.   On the approach of his pursuers he shot one of his oxen with a rifle loaded with the stolen stones, and after the police had made an unsuccessful search he extracted the diamonds from the dead body of the ox.   In the same year another native, who died in a mysterious manner, was discovered to have swallowed a 60 carat diamond, which proved itself too much for the constitution even of a Kaffir.

The comparatively recent introduction of the compound system has resulted in making the robbery of diamonds by natives almost an impossibility.   The native workers in the

De Beer's and Kimberley mines, among whom representatives of almost every South African tribe are to be found, live in what is known as a compound, and are debarred from all intercourse with the outside world. This compound is a rigidly guarded enclosure, several acres in extent, in which all the necessaries of life can be purchased as well as other objects specially attractive to the native taste. Water, wood, and medical attendance are supplied to the workers gratuitously, and no effort is spared to make their enforced stay as little irksome as possible. On entering the employment of the company the native contracts to stay for at least three months, during which time he sees no one but the officers of the company ; at the end of this time he may renew his contract or terminate the engagement. Before leaving the compound, however, an exhaustive search of his person and belongings is made, and he has further to submit to the administering of a strong purgative. In spite of the restrictions by which life in the compound is hedged in, and the absolute prohibition laid on the sale of intoxicating liquors, the workers are by no means averse to the system, and often renew their contracts again and again. It has been found, moreover, that the system reduces the possibility of fraud to a minimum.

4. BORNEO.

The information concerning the diamond-fields of Borneo given below is derived, for the greater part, from the investigations of R. D. M. Verbeek, the director of the

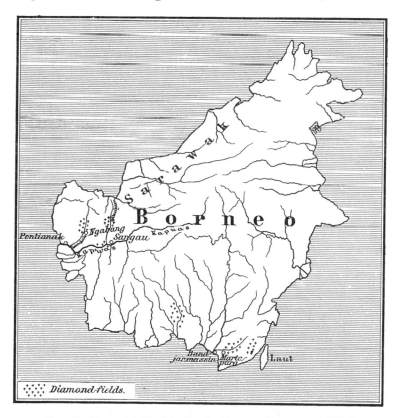

FIG. 42. Diamond-fields of the Island of Borneo. (Scale, 1 : 15,000,000.)

Geological Survey of the Dutch East Indies, as set forth in E. Boutan's book " Le Diamant " (Paris, 1886).

The diamond-fields of Borneo fall into two well-defined groups, one in the west of the island, in the district of the River Kapuas, the mouth of which lies a little below the town of Pontianak, the other in the south-east of the island, not far from the town Bandjarmassin, and nearly opposite the island of Laut (Fig. 42). The three portions into which the western group may be divided are situated on as many different rivers, one being on the river Kapuas, a little below its confluence with the Sikajam, and the other two respectively on the rivers Landak and Sikajam, both tributaries of the Kapuas. The Landak deposits seem to have been known since the time the Malays settled in the island, and were mentioned by the Dutch mariners who first visited the coast ; indeed, from the beginning the Dutch regarded the trade in diamonds in Borneo as their special monopoly.

In the west of the island diamonds occur in beds of alluvium, in masses of débris at the foot of mountains, and in the beds of streams and rivers flowing through diamantiferous districts. The alluvial deposits consist of gravel, sand, and more or less ferruginous clay, more rarely of conglomerate and sandstone. They are distinctly bedded, and vary in thickness from 2 to 12 metres, the diamonds being confined to the lowest bed, which consists of gravel.

These ancient gravels, which themselves show little or no signs of bedding, contain diamonds throughout their whole mass and are formed of more or less rounded rock-fragments. They are essentially river deposits, and occur in isolated patches of small area at the foot of the mountains or in the valleys, but always above the existing high-water level. The rock-fragments of which these gravels are composed differ widely in kind ; white, yellow, or rose quartz predominates, but there are also present hard and compact grey and black quartzites, quartz-schists, clay-slates, quartz-sandstones, hornstones, hornblende, blue and violet corundum, and, in sparing amount, fragments of igneous rocks, so decomposed, however, that it is difficult to determine their original nature. In addition to these there are also to be seen scales of white mica, grains of magnetite, a few particles of cinnabar, and usually a little gold.

It is from these gravels that the diamonds now found in the beds of streams and rivers have been washed out. Both sedimentary and igneous rocks are found *in situ* in the neighbourhood ; among the former are clay-slates and quartz-schists with quartzites of Devonian age, conglomerates and clayey sandstones of much later date, probably belonging to the lowest Tertiary, that is, to the Eocene age. The igneous rocks include granite, diabase, gabbro, andesite, and melaphyre.

Diamonds are only found in places where the beds of Eocene conglomerate and clayey sandstone crop out at the surface, and it has been thought by C. van Schelle, a mining engineer in Borneo, that it is from these beds that the diamond has been derived. In any case the Devonian beds need not be considered, for no diamond has ever been found in alluvial débris derived from, or resting on, Devonian strata, in spite of the fact that such material has been carefully worked over for the gold it contains. The original mother-rock and the mode of origin of the diamond are therefore here as much a mystery as elsewhere, for no single crystal has ever been found in anything but what is obviously a secondary situation.

The working of the diamond-fields is in the hands of Chinese and Malays ; the former work the deposits lying above high-water level, while the latter apply themselves to the alluvium in the present-day water-courses, extracting the diamantiferous gravel by excavating small deep pits reaching down to the solid rock, and washing the gravel in baskets. The methods in use in both cases are very primitive and inadequate, and no thorough investigation of the deposit has yet been made. An improvement in the system might probably be easily made if the diamond-seekers, who are for obvious reasons very uncommunicative, could be persuaded to volunteer the necessary information.

The diamonds of Borneo are, on the average, of poor quality; the proportion of faulty and unpleasingly coloured stones being sometimes stated to be greater here than in Brazil. The diamonds are almost invariably either more or less water-worn or fragmentary and irregular. The predominating crystalline forms are the octahedron and the rhombic dodecahedron; regular octahedra, which are not infrequent, are known to the Malays as "perfect stones," since according to their ideas such stones require little or no cutting. Cubes are rare, but twinned crystals very frequent.

Borneo diamonds exhibit a fair range of colour; the majority though colourless are disfigured by faults or blemishes of some kind or another. A few of the highly prized "blue-white" stones are found, and are of such singular beauty that their equal is nowhere to be found. After the colourless stones, those with a faint blue or yellow tinge are most abundant; more or less darkly coloured stones (bort), as well as those of a grey colour (carbonado), are fairly common. Stones in which a grey or black kernel is enclosed in a colourless and well-crystallised shell are sometimes met with; such a stone is known to the Malays as "soul of the diamond," and is considered to augur a poor deposit. Although the stone itself is regarded as a talisman, and worn round the neck in the belief that it will bring luck to its owner, yet at the spot at which it was found work is immediately abandoned and a fresh place chosen. On the other hand, the presence of the blue corundum is considered to be a good sign by the diamond seekers. Diamond crystals of a deep, black colour, quite distinct from carbonado, are occasionally found; when cut, such stones, though giving no play of prismatic colour, display a magnificent lustre, and are in great request for use in mourning jewellery.

With regard to the size of stones from this locality, it may be asserted that 95 per cent. of the whole output is constituted by stones which weigh less than 1 carat. Next in abundance come stones between 1 and 5 carats in size, while those exceeding this size are very rare. Several diamonds of large size were found in the district belonging to the Malay Prince of Landak, and are now in his possession; owing to their massive silver setting they cannot be weighed, but several have been estimated by C. van Schelle at over 100 carats. In the possession of the Rajah of Mattan is a supposed diamond the size of a pigeon's egg, and weighing 367 carats; this stone, which is probably only rock-crystal, will be again mentioned in the section devoted to large diamonds. The same prince is in possession also of two large and undoubtedly genuine stones, the "Segima" of 70 carats and another of 54 carats, both said to have been found in the island.

While in 1880 the mines on the Sikajam river were worked by about forty Chinese only, those in Landak gave employment to about 350 workers. The alluvial deposits on the Kapuas river are no longer systematically worked; single pits may be sunk here and there, but the production is quite insignificant.

The diamond-bearing deposits in the south of the island, namely, in the districts of Tanah-Laut, Martapura, and Riam are of recent formation, and overlie Eocene strata in the same way as those described above. These Tertiary strata, which in places include thin beds of coal, rest on ancient crystalline rocks, such as mica-schist, chlorite-schist, talc-schist, and hornblende-schist, and like these are inclined and faulted. Interbedded with the Tertiary strata, and specially towards their base, are sheets of recent eruptive rocks (andesites). The diamond-bearing deposits form a broad band round the seaward slopes of the Tertiary hills, while the gold-sands of the region, which contain no diamonds, rest on ancient schists. The actual diamantiferous stratum is constituted of more or less rounded pebbles of various minerals and of sand, either loose or bound together by clay. The mineral most abundantly present is quartz of various colours, after which come fragments of andesite and micaceous sandstone. A blue mineral, formerly thought to be quartz but which has now been proved to be corundum (sapphire), though of no value as a gem has yet a certain importance as an indication of the nature of the deposit. It occurs in the same manner as at Landak, and its presence is regarded by miners as indicating the existence of diamonds in the deposit; it is only after they are satisfied as to the presence of the blue mineral that they apply themselves to an exhaustive search for diamonds.

As a rule, the diamonds are found lying singly and loose in the gravel; sometimes, however, they are cemented by limonite to a pebble or rock-fragment. They are often accompanied by scales of gold and platinum and by grains of chromite and magnetite. The thickness of the diamond-bearing bed varies between 20 centimetres and 2 metres, beds of the greater thickness being usually found filling up depressions in the surface of the ground. The bed rests on a blue clay and is overlain by 1 to 6 metres of gravel and sand, and sometimes, as in the neighbourhood of Bentok, by a layer of nodules of limonite. This diamond-bearing stratum is found mainly in the neighbourhood of rivers and in surface depressions, which in the rainy season become filled with water.

The mining methods adopted here by the Malays appeared to be the same as those practised at Landak; the workings were mainly in the neighbourhood of the village of Tjampaka, in the district of Martapura, where in the year 1868 stones to the value of £1250 had been found; also near Banju-Irang, Bentok, and Liang Angang, all in the district of Tanah-Laut. Thousands of small mines are still to be seen in these districts; the majority of them, however, are now abandoned, for since the great fall in the price of diamonds in 1878, the miners can easily obtain more lucrative employment in the gold mines, tea plantations, &c. A Franco-Dutch Company in 1882 obtained the concession of a stretch of country of 2000 hectares, that is about eight square miles in area, for a period of twenty-five years, with the purpose of diamond mining. On this area, which lies between Tjampaka and Banju-Irang, machinery was set up, but in the very next year, 1883, the work was abandoned, and up to the present has not been recommenced. From the above accounts it is clear that the production of the whole southern group is quite insignificant.

Diamonds are also found in the Kusan district, which lies between the rivers Danau and Wauwan in the State of Pegattan, this latter being a Dutch dependency in East Borneo. The stones are of good quality, but here also the yield is poor.

It is impossible to arrive at anything more than an approximate estimate of the production of diamonds in Borneo. From an early date the Malay princes assumed the right of appropriating at a fixed price all stones exceeding 5 carats in weight found in their own dominions. Thus these stones never leave the country, and no record of their occurrence is kept. In the table given below will be found a few returns published by the Dutch Government in carats and Dutch florins (one florin = 1s. 8d.) of diamonds imported into Java.

Year.	Carats.	Value (in florins).	Year.	Carats.	Value (in florins).
1836	5473	110,601	1843	1315	23,900
1837	5245	97,140	1844	—	46,450
1838	5947	117,750	1845	—	68,825
1839	3884	92,552	1846	—	128,450
1840	1891	62,410	1847	—	96,210
1841	2122	56,520	1848	—	67,200
1842	3980	80,875			

It will be seen from the above table that the practice of recording the weight in carats of the imported diamonds ceased in 1844; moreover, the Customs Register, from which the above table has been compiled, was discontinued after 1848, so that there is no record of succeeding years. The number of stones imported into Java in each year, as set forth in the above table, represents very approximately the yearly production of Borneo, for it was at this period that the old Dutch East India Company was in its most flourishing condition, and the general prosperity created a demand for diamonds which drew almost the whole of the production of Borneo to Batavia; but all this ceased with the abolition of the company. In 1823 and 1831 the Dutch, seeing the demand for diamonds which existed in European markets, sought, unsuccessfully however, to increase the production by organising systematic working of the deposits.

A few estimates of the production of recent times have been made by the merchants at Ngabang, the capital of Landak; they are as follows:

Year.	Carats.		Year.	Carats.
1876	4062		1879	6673
1877	5271		1880	3013
1878	6359			

In the previous century the yield appears to have been much richer; while the deposits have been gradually exhausted, no new ones comparable in richness to the old have been discovered, and the abundance and comparative cheapness of Cape stones has rendered impracticable the exploitation of the poorer deposits. It is stated that in 1738 diamonds to the value of eight to twelve million Dutch florins were mined in Borneo, and even as late as the beginning of last century the value of the annual yield is said to have been as much as a million florins. The estimated annual yield in more recent times has been already given; at the present day it is supposed to be about 5000 carats.

The majority of the stones are roughly cut by the Malays in the island at Ngabang and Pontianak; there are diamond-cutting works also at Martapura, and the natives have been acquainted with the art of gem-cutting for centuries. At the present time scarcely any diamonds are exported, and it has even become customary to import Cape diamonds. The stones which are yielded by the country circulate almost exclusively in Oriental countries, very few finding their way to Europe.

5. AUSTRALIA.

In the year 1851 diamonds were discovered in one or two of the Australian goldfields, and later on in a few of the stanniferous gravels of the same continent. They are present in not altogether insignificant numbers, and up to the year 1890 a total of 50,000 diamonds had been found. New South Wales has up to the present time furnished the greater part of the yield, but a few stones have been found in Victoria and Queensland as well as in South Australia and Western Australia.

Australian diamonds are decidedly small, the largest stone yet found, which came from New South Wales, was an octahedron, and weighed $5\frac{5}{8}$ carats; an octahedral crystal from South Australia weighed $5\frac{5}{16}$ carats. The average weight of diamonds from New South Wales, compared with which the yield from the other States is negligible, is only $\frac{1}{4}$ carat; the great majority of stones vary in weight between $\frac{1}{8}$ and $1\frac{1}{2}$ carats. According to the statements of diamond-cutters, Australian diamonds are harder than the majority of stones from other parts of the world, and can only be cut with their own powder; they have a peculiarly strong surface lustre, and in spite of their extra hardness are usually much water-worn.

In **New South Wales** there are two principal diamond districts (Fig. 43).   One is a stretch of country extending to the north-west of Sydney, as far as the Cudgegong river, and to the west of Sydney, as far as the Lachlan river.   The other diamond

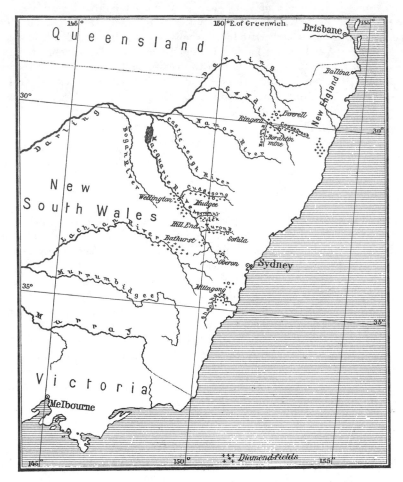

FIG. 43.   Diamond-fields of New South Wales.   (Scale, 1 : 10,000,000.)

district is in the north-east corner of the State, in the district of the Gwydir river, a tributary of the Darling; it embraces the neighbourhood of Inverell and Bingera, and extends to the east of these townships into New England.   In these districts the diamond occurs in sands together with gold and tin-stone (cassiterite), and with one possible exception it has never been found in the solid rock; it is therefore impossible to make any suggestion as to the nature of the rock in which the diamond was formed.

In the southern diamond districts the diamond-bearing débris is mainly confined to ancient water-courses of Pliocene (a subdivision of Tertiary) age.   When the precious stone is found in the beds of recent rivers and streams, it is always associated with material derived from these older deposits, which has been redeposited by natural agencies or during the process of gold-washing, &c.   In this district the diamond is invariably accompanied by gold and it was in the gold-washings that the first discovery of diamonds was made. The diamantiferous gravels and sands of these ancient river deposits, which are always above, and often far above, the present water-courses, are very frequently overlain by a sheet of compact basalt, which must be penetrated before the diamond and gold-bearing stratum can be reached. Re-deposited masses of material, containing both gold and diamonds, often lie on the basalt, having been washed down from the upper part of the valleys.

The first discovery of diamonds in Australia was made in this State in 1851 ; the stones were found in Reedy Creek, a tributary of the Macquarie river, sixteen miles from Bathurst ; a few were found in the same district in 1852, in Calabash Creek. In 1859 a few stones, having the form of triakis-octahedra, were found in the Macquarie river, near Suttor's Bar, and at Burrandong ; in the same year a hexakis-octahedron, weighing $5\frac{1}{8}$ carats, was found in Pyramul Creek. These places are all situated in the same district, and at none of them were more than a few stones found.

In 1867, however, diamonds in greater number were met with near Warburton, or Two Mile Flat, on the Cudgegong river, nineteen miles north-west of Mudgee ; and in 1869 the systematic working of an area of about 500 acres in this district was commenced. The working, which was not very profitable, was carried on at Rocky Ridge, Jordan's Hill, Horse Shoe Bend, and Hassalt Hill, as well as at the places already mentioned. The ancient river-deposits in which the diamonds are found lie under a capping of columnar basalt, and occur at isolated spots along the course of the Cudgegong river, more or less distant from the present river course, and at heights up to 40 feet above the present high-water level. They rest on the eroded edges of perpendicular sedimentary strata, which are interbedded with compact greenstones, and probably belong to the period of Upper Silurian deposits. The diamond-bearing débris consists of coarse sand and mud intermingled with pebbles of quartzite, sandstone, clay-slates, and quartz-slates, accompanied by waterworn grains and crystals of quartz, jasper, agate, silicified wood (this in large amount), and other siliceous minerals, also cassiterite (the "wood-tin" variety), topaz, common corundum (sometimes of a lavender-blue colour), sapphire, ruby, a peculiar variety of corundum called barklyite, zircon, garnet, ruby-spinel, brookite, magnetite, ilmenite, tourmaline, magnesite, nodules of limonite, grains of iridosmine, and, of special importance, gold. The quartz pebbles are frequently encrusted with oxides of iron and manganese. The whole mass of diamantiferous débris is in some places loose and incoherent, and in others bound together to form a solid conglomerate, the cementing material being a green, white, or grey siliceous substance, or a brown or black ferruginous or manganiferous substance. The deposit in places attains a thickness of 70 feet ; the diamonds, which are of small size, are scattered through it so sparingly and irregularly that the working of it cannot be profitably prolonged for any length of time.

In spite of the poor character of the deposit, 2500 stones were found during the first five months of work. All were small, the largest being the octahedron of $5\frac{5}{8}$ carats mentioned above, which, when cut, formed a beautiful colourless brilliant, weighing $3\frac{5}{16}$ carats. The stones average in weight about $\frac{1}{4}$ carat, and vary considerably in colour, passing from perfectly water-clear through various shades of yellow, pale green, and brown to almost black ; a twinned octahedron of a beautiful dark-green colour was once found. The commonest crystalline forms are the octahedron, which occurs both as simple and twinned crystals, the rhombic dodecahedron, triakis-octahedron, and hexakis-octahedron ; one crystal with the form of a deltoid dodecahedron has been found. The crystals are, as a rule, much water-worn ; when this is the case their surfaces are sometimes smooth and bright, at other times rough and dull.   No spheres of bort, such as are found in Brazil and South Africa, appear to occur in Australia.

Solitary specimens of diamond have been found at many other places in this district. At Bald Hill, near Hill End, on the Turon river, a stone of $5\frac{1}{8}$ carats was found, and a number of diamonds, which though of small size were of excellent quality, were met with in the old gold-mines of Mittagong.   Again, near Bathurst, a black diamond, the size of a pea, and having the form of an almost spherical hexakis-octahedron, was found.   Diamonds have also been collected from the gravels underlying the basalt at Monkey Hill and Sally's Flat, in Co. Wellington, just as they occur at Mudgee.   Uralla, Oberon, and Turnkey are other localities at which more than solitary specimens of diamonds have been found.

The occurrences mentioned above were all in ancient river gravels; among existing water-courses in which diamonds have been found may be mentioned the Abercrombie, Cudgegong, Macquarie, Brook's Creek, Shoalhaven, and Lachlan rivers.   The stones found in existing streams are much worn, and many are broken ; from this, and also from the fact that the minerals forming the gravels of these water-courses are identical with those of the ancient river deposits found underlying the basalt, we may conclude that the gravel of the present rivers is redeposited material derived from the ancient river-beds.

The mode of occurrence of the diamond in the north of New South Wales, especially in the district of the Gwydir river, in the neighbourhood of Bingera and Inverell, is of some importance.   As pointed out by Professor A. Liversidge, diamonds are found in the valley of the Horton or Big river, seven or eight miles from Bingera, under just the same conditions as at Mudgee.   The diamond-bearing deposit is 2 to 3 feet thick, and occurs in isolated patches, the material which originally lay between having been denuded away.   These patches of diamantiferous material are scattered over an area measuring four by three miles, in a valley which is opened towards the north, but enclosed otherwise by the Drummond Range.   The deposit consists of sandy and clayey material, and has probably been deposited in former times by the Horton river.   The rocks of the locality are clay-slates of Devonian or Carboniferous age, and the sheet of basalt which occurs in the neighbourhood appears to overlie deposits similar to those now being considered.   Here also the diamantiferous material is in places cemented together to form a solid conglomerate ; it includes boulders and fragments of the underlying clay-slates, and, when clayey, contains crystals of gypsum. The associated minerals are practically the same as at Mudgee, barklyite is, however, here absent.   Black tourmaline is regarded as a specially characteristic associate of the diamond, and its appearance is hailed with joy by the miners.

Diamonds occur here rather more plentifully ; they are colourless or straw-yellow and small, the largest weighing only $2\frac{6}{8}$ carats.   On an average only twenty stones are found in each ton of material, and a stock of 1680 stones weighed no more than about 140 carats.

More recently diamonds have been discovered in the tin-gravels of the neighbourhood of Inverell, and their occurrence here appears to be sufficiently abundant to justify systematic working.   Cassiterite (tin-stone), rock-crystal, sapphire, topaz, tourmaline, monazite, &c., are here associated with diamond ; gold, however, is apparently absent.   Several companies have been formed, and many thousands of stones, averaging $\frac{1}{4}$ to $\frac{1}{3}$ carat in weight, have been obtained from the different mines; the largest of these stones weighed $3\frac{5}{8}$ carats. From the Borah tin-washings, situated at the junction of Cope's Creek with the Gwydir river, 200 stones were obtained in a few months, the largest of which weighed almost $1\frac{1}{2}$ carats ; while in the Bengonover tin-washings, only a few miles away, a stone weighing nearly 2 carats was found.   Diamonds have also been found in most of the alluvial tin-workings on Cope's, Newstead, Vegetable, and Middle Creeks, in the Stanifer, Ruby, and Britannia tin-washings, and elsewhere in the same district.

All the occurrences of diamond in New South Wales, described above, are in secondary deposits of alluvial origin; recently (1901), however, Mr. E. F. Pittman, the Government Geologist of New South Wales, has described its occurrence in what may perhaps be the mother-rock. At Ruby Hill on Bingera Creek, twelve miles to the south of Bingera, diamond has been found in the breccia filling a volcanic pipe. This breccia consists of angular fragments of clay-stone, felsite, basalt, eclogite, &c., with calcite, garnet, zircon, chrome-diopside, and other minerals; it bears a very striking resemblance to the diamantiferous material which fills the Kimberley pipes, the principal difference being that it shows no sign of serpentinisation.

Finally, we must mention the peculiar occurrence of diamond at Ballina, in New England, where solitary specimens have been found in the sands of the sea-shore. The diamond-bearing deposit is here exposed to the action of the waves of the sea, and the solitary specimens found in the shore sands have probably been washed out of the deposit by the waves.

The output of diamonds in New South Wales, as published in the official returns of the Department of Mines, amounted in 1899 to 25,874 carats, valued at £10,349 12s.; in 1900, owing to lack of water, and the reconstruction of the mining company, there was a smaller output of 9828½ carats, valued at £5663 1s. Although a fairly considerable number of diamonds have been met with in New South Wales, yet the other Australian States are very poor in this respect; it is probable, however, that there will be important finds in the future.

In **Queensland,** conglomerates have been observed at Wallerawang and on the Mary river, which are remarkably like the diamond-bearing deposits of Mudgee and Bingera; no diamonds have, however, as yet been discovered in them. At other places in Queensland, namely, on the Palmer river and the Gilbert river, the precious stone has been found.

In **South Australia** about 100 stones have been found in the alluvial gold-washings of Echunga, twenty miles to the south-east of Adelaide; it was here that the octahedron of $5\frac{5}{16}$ carats, previously mentioned, was found. In **Victoria** a few diamonds were met with in 1862 in the Owens and in the Arena goldfields; a larger number of stones were found in the Beechworth district of the same State, upwards of sixty crystals, none however exceeding 1 carat in weight, having been collected. Diamonds have also been stated to occur in the neighbourhood of Melbourne, in association with ruby, sapphire, zircon, and topaz.

Finally, from **Western Australia** also, a certain number of diamonds have come, small crystals rich in faces having been found near Freemantle, in a sand containing zircon, ilmenite, rock-crystal, red, yellow, and white topaz, and apatite. More recently, in 1895, it was reported that diamonds had been found in the north-west of the State, at Nullagine, in the Pilbarra gold-field; many leases have been taken up, but so far no important finds have been made. From 230 tons of auriferous ore treated in 1900 only twenty-five small diamonds were obtained.

Tasmania has recently been added to the list of diamond-producing countries. According to newspaper reports, a large number of stones were found at the end of the year 1894 in Corinna, one of the richest goldfields of the island. The reported occurrence caused a rush of thousands of diamond-seekers into Tasmania from the Australian mainland; many companies for the exploitation of the deposits sprung up, but apparently with no marked results.

6. NORTH AMERICA.

The occurrence of the diamond in the United States of North America is so sparing that it has no effect whatever upon the diamond markets of the world. American stones are, however, greatly prized in the States, both for patriotic reasons and also on account of their scientific interest.

There are two principal diamond-producing regions, one in the extreme east and the other in the extreme west of the country; the former occupies the eastern slope of the southern part of the Appalachians, while the latter lies at the western foot of the Sierra Nevada and the Cascade Range. Though the two regions are so widely severed, yet in each the stones occur under very similar conditions, being found in loose detrital materials, in gravel and sand, and everywhere in association with the same minerals, namely, garnet, zircon, magnetite, anatase, monazite, and specially with gold, in the search for which diamonds are frequently found. This agreement in the mode of occurrence of the diamond in the two regions, must be attributed to the fact that in both the detrital material has been derived by the weathering and denudation of the crystalline silicate rocks of which the surrounding mountains are built, and that these rocks are essentially identical in character in both regions, although those of the east are older than those in the west.

A third diamantiferous district has been recently discovered, namely, the region of the Great Lakes; here the stones have been found in glacial deposits mainly in the State of Wisconsin, but also in Michigan and Ohio.

According to Mr. George F. Kunz, the well-known American expert in precious stones, a considerable number of diamonds have hitherto been found in the eastern region, in the States of Georgia and North Carolina, very few in South Carolina, and still fewer in Kentucky and the southern part of Virginia. From Virginia came the " Dewey " diamond, the largest ever met with in the United States. It was found by a labourer in 1855 in an excavation in the village of Manchester, and had the form of an octahedron with rounded edges, weighing in the rough $23\frac{3}{4}$ carats and when cut $11\frac{11}{16}$ carats. This stone is not of the purest water and its beauty is impaired by other flaws, so that it would not be actually worth more than 300 to 400 dollars (about £62 to £83); in spite of this, however, it has been sold for about ten times this amount. It was the only specimen found at this spot.

In North Carolina diamonds have been found, usually associated with gold, in the district about the east foot of the Blue Ridge. The mountains of the district are built up of crystalline schists; itacolumite, the flexible sandstone which occupies so important a position in the Brazilian deposits, is also found here, though at present no diamonds have been discovered in it. The stones hitherto met with are for the most part octahedra; the largest yet found, weighing $4\frac{1}{2}$ carats, was discovered in 1886. A stone valued at 400 dollars is said to have been found in South Carolina; in this State, as in North Carolina, diamonds are found associated with gold and in the vicinity of itacolumite; the largest, an octahedron of $4\frac{1}{12}$ carats, was found in 1887. About the year 1886 a single small stone was found in a river sand in Kentucky.

During a search for gold in Wisconsin, at the end of the eighties, a few small diamonds were found associated with grains of quartz, magnetite, ilmenite, and with grains and crystals of the varieties of garnet known as almandine and hessonite (or perhaps spessartite). A few large diamonds have been found in this State; thus in 1893 an almost colourless stone, weighing $3\frac{14}{16}$ carats, and having the form of a rhombic dodecahedron, was found in clay at the town of Oregon, Dane County. Previous to this, in 1876, a yellow diamond of the same form and weighing 16 carats was found with several others near Eagle in Waukesha

County; this stone was not recognised as a diamond until 1883. It had been bought by a Milwaukee clockmaker, who did not know what it was, for a dollar. A second still larger stone, stated to have been found about the same time in the same district, was mislaid before its real nature was known. In the year 1886 a pale yellow diamond, with the form of an irregular rhombic dodecahedron, measuring 20 by 13 by 10 millimetres and weighing $21\frac{1}{4}$ carats, was found at Kohlsville in Washington County. Another stone of $6\frac{13}{32}$ carats, found at Saukville in Ozaukee County in 1880, was not determined to be diamond until 1896. Altogether, since 1883, seventeen diamonds, varying in weight from $\frac{1}{2}$ to $21\frac{1}{4}$ carats, have been found in glacial moraines in the region of the Great Lakes of North America, principally in the State of Wisconsin, but also in Michigan and Ohio. The predominating crystalline forms are the rhombic dodecahedron and a hexakis-octahedron; the stones are colourless, or tinted with green or yellow. By plotting the diamond localities and the directions of the glacial striæ on a map, Professor W. H. Hobbs has recently (1899) shown that the diamond probably came from somewhere near James Bay on Hudson Bay; a more detailed study of the glacial geology of the region will, however, be needed before the home of the diamond can be exactly located.

In the western region a certain number of diamonds have been met with in the States of California and Oregon; in the former the stones occur in gold-bearing detritus belonging to two different periods of formation. Here, in Tertiary and even earlier times, the Sierra Nevada and Cascade Range were drained by immense rivers, whose beds, which have been traced for great distances, became filled with auriferous débris derived from these mountains. Later on mighty volcanic eruptions took place, and the surface of northern California and of Oregon was flooded with thick sheets of lava partly filling up the river valleys. The streams and rivers of the present day, rising in the same mountains, cut out for themselves new courses in these volcanic rocks, and have deposited in their beds auriferous débris derived from the same mountains, and therefore of the same nature as the débris laid down in the ancient river-beds. It was from these latter masses of débris that the enormous amount of gold yielded by the Californian goldfields in the early days of their discovery, namely, about the year 1848, was mainly derived. These later gold-sands are now exhausted, and mining is at the present day confined to the earlier deposits of Tertiary and Pre-Tertiary alluvium, which are overlain by lavas and which contain diamonds as well as gold. The pebbles and diamonds of this alluvium are often cemented together by limonite to form a solid conglomerate, which is very similar in appearance to the "tapanhoacanga" of the Brazilian diamond-fields.

The first find of diamonds took place in 1850; since then single stones, usually only of small size, have been met with every year, the largest weighing $7\frac{1}{4}$ carats. The diamonds actually present probably far exceed those actually found in number, for, during the process of winning the precious metal, the solid, auriferous conglomerate is stamped to a fine powder, and thus any diamonds present in it are also crushed and destroyed. As a matter of fact, the presence of splinters of diamond in this crushed material has been frequently observed. Moreover, it is possible that many stones have been lost on account of the practice of ignorant miners, prevalent both here and in other countries, of testing the genuineness of a supposed diamond by means of the hammer and anvil; should such a stone be genuine the test, owing to the brittleness and perfect cleavage of the diamond, results in its complete fragmentation.

The possible occurrence of diamonds in those parts of North America, for example Indiana, which are geologically very similar to other diamond-bearing districts, has often been suggested. In many such places, however, the search has been unsuccessful, and stones reported to have been found in other places have in many cases been placed there with the object of swindling a credulous public.

As an example of the almost incredible extent to which the public have been swindled by false reports as to the occurrence of diamonds, an account of a supposed discovery in the State of Arizona may be given. In the year 1870, a fabulously rich occurrence of various precious stones, including the diamond, was reported to have been discovered somewhere in the West. 80,000 carats of rubies and a single diamond weighing 108 carats, said to form part of the find, were exhibited in a San Francisco bank. Shortly after this a smaller lot of stones, said to be the result of a second search, were on view in this city, and the attention of capitalists was soon enlisted. On May 10, 1872, a Bill in the interests of the diamond-miners was passed by Congress, and an expedition for investigating the locality was fitted out. After some trouble the locality was found, and in the course of a week the members of the expedition had collected 1000 carats of diamonds and 6000 to 7000 carats of rubies, and returned well pleased with their success. Another expedition failed to find the place, and an investigation by the officers of the United States Geological Survey proved the supposed discovery to be a gigantic swindle, the locality having been " salted." The supposed rubies were in reality ordinary garnets, and although the smaller diamonds were genuine the 108-carat stone was nothing but rock-crystal. It was ascertained that a speculative American had imported a large number of rough Cape diamonds, and had scattered them about the neighbourhood so plentifully that stray stones were found there for some years afterwards. The initial outlay of the speculative American was amply repaid by the three-quarters of a million dollars subscribed by Californian and other capitalists.

Diamonds are sought for very diligently in America; it is not unusual for gold miners and labourers engaged in the work of excavation to wear rings set with small, rough diamonds for the purpose of familiarising themselves with the appearance of this stone in its rough condition. In spite, however, of the alertness of persons engaged in such occupations, only a small number of stones continue to be found, though it is always possible that future discoveries of importance may be made.

7. BRITISH GUIANA.

In recent years diamonds have been found in another part of South America, about 2000 miles north-west of the famous Brazilian localities, namely, in the gold-washings on the upper course of the River Mazaruni in British Guiana. The discovery was made accidentally in 1890 by Edward Gilkes, who, while prospecting for gold along the Putareng creek, a tributary of the Mazaruni, found a few diamonds in the batea he was accustomed to use for gold-washing. The locality is situated in latitude 6° 14′ N. and longitude 60° 18′ W., about 150 miles above the town of Bartica on the confluence of the Mazaruni and Essequibo rivers. It is reached after a twelve to twenty days' journey, according to the state of the river, which has many falls and rapids, from Georgetown. The exact spot is situated about four miles from the Mazaruni, and is reached by a narrow trail across swampy land and through tropical jungle, everything having to be carried on the heads of Indians.

The rocks of the Mazaruni valley are largely gneisses and granite traversed by dykes of diabase and other similar rocks. The diamond-workings at present under consideration are situated on the side of a hill, and penetrate (1) 18 inches of pure, white quartz-sand, (2) 18 inches of yellowish sandy clay, with fragments of quartz and portions of sand and gravel cemented together by iron oxide, in which small diamonds are occasionally found, (3) 7 feet⟶

the present limit of working—of clay, which becomes more and more gravelly and the constituent fragments larger and more frequently cemented together with iron oxide, as greater depths are reached. Some of the pebbles are much rounded, and have sand and smaller pebbles attached to them by a felspathic cement, while others are sharp and angular. Some consist of felsite and concretionary iron-stone, but most are of quartz. Associated with these pebbles are grains of ilmenite and small rounded pebbles of black tourmaline and pleonaste, and occasionally of topaz and corundum. When dug out, the gravel is carried in wooden dishes to a little creek hard by, where it is washed in sieves of one-sixteenth inch mesh, the residue being picked over while wet. The diamonds, originally found by Mr. Gilkes in 1890, were obtained at the foot of the hill in the bottom of the valley. The diamantiferous gravel here contains many crystals of quartz, and rests upon a bed-rock of kaolin, differing in both these respects from that which lies on the hill-side.

Up to 1900 diamonds had been found only over an area of country measuring 200 yards in length by about 100 in breadth, but it is probable that the diamantiferous gravels are much more widely distributed. The mode of occurrence of the diamond in these gravels and the minerals with which it is associated are very much the same as in Brazil. The diamond has not yet been found here in its mother-rock; if this should at any time be discovered, British Guiana may become an important diamond-producing country.

As yet there have been found only some few thousands of small diamonds, which have the form of octahedra, and are exceptionally white and brilliant. The smallest are of very small size and the largest about $1\frac{1}{2}$ carats, but there are very few exceeding a carat in weight. During the ten years between 1890 and 1900 between 2000 and 3000 diamonds were found, while, according to custom-house returns, the total export of diamonds up to January 28, 1902, amounted to £10,000. A parcel of 282 stones sent to London during the year 1900 was valued at £2 8s. per carat. During a period of six weeks in the following year, a New York company obtained 8227 small diamonds with an aggregate weight of 767 carats, which were valued at £1920 or £2 10s. per carat. A dozen companies have since been organised and are now at work, and fresh ground is constantly being opened up, so that the diamond-mining industry of British Guiana is likely to develop rapidly.

We are indebted for the above account of the occurrence of diamonds on the upper Mazaruni river to Mr. G. F. Kunz's Annual Reports on Precious Stones, in which is brought together much information from various sources.

Another occurrence of diamond in British Guiana is reported by Professor J. B. Harrison, the Government Geologist of that colony. This is in the gold-washing claims of the Omai creek, this stream being a small tributary of the Essequibo river, which it joins at a spot about 130 miles above the mouth of the latter. From a part of the bed of Gilt creek, one of the tributaries of this stream, measuring about 500 feet in length and 50 in breadth, some 60,000 ounces of gold and some hundreds of small diamonds have been recovered by the somewhat crude methods hitherto in use. The auriferous gravels of this stream consist of fragments of more or less decomposed diabase, pebbles of concretionary iron-stone and angular quartz. They yielded at one time hundreds of very small diamonds, the majority of which were perfectly clear and colourless octahedra, the remainder being of various shades of pink, green, and clear yellow.

It is stated by Mr. G. F. Kunz that in **Dutch Guiana** also diamonds have been found for years past in the tailings of the gold-washings. They have been for the most part small and have attracted but little attention, the gold being the chief object sought for. One fine stone, however, is reported to have been found about the year 1890 and to have been cut in the United States.

8. URAL MOUNTAINS.

The discovery of diamonds in the Urals resulted from the famous expedition made in 1829 to this region by Alexander von Humboldt, with Gustav Rose and Ehrenberg, at the desire of Czar Nicholas. In 1823, in his *Essai géognostique sur le gisement des roches*, Humboldt had expressed an opinion, based on the similarity between the Brazilian and Uralian gold- and platinum-bearing deposits, that diamonds very probably existed in the Uralian deposits just as they were known to do in Brazil. This conclusion, which was supported by the fact that the minerals associated with gold and platinum in the Urals and in Brazil were practically identical, had been previously and independently expressed by Professor Moritz von Engelhardt of Dorpat, who later investigated and reported on the first diamond occurrence in the Urals. Humboldt was so convinced of the truth of his opinion, that on his departure he assured the Czarina that he would not again appear before her Majesty unless he had some Russian diamonds to show.

Throughout their whole journey the explorers spared no pains in their search for the precious stones; every gold-bearing sand they met with was subjected to microscopical examination in order to detect the presence of diamond; their efforts, however, were not crowned with the success they deserved.

Better fortune fell to the lot of Count Polier, who accompanied the expedition part of the way and to whom Humboldt had communicated some of his own enthusiasm on the subject. On leaving Humboldt's party, therefore, the Count set himself seriously to work in the mining district of Bissersk (about latitude $58\frac{1}{2}°$ N.), in the gold-washings on the estates of his wife, the Princess Shachovskoi. Here the first Uralian diamond, indeed the first European diamond, was found on July 5, 1829. The exact locality was the small gold-washing of Adolphskoi, near the larger washing of Krestovosvidshenskoi, twenty-five versts (seventeen miles) to the north-east of Bissersk, and four versts (about three miles) from the mountain ridge on the western or European slope of the Urals. It is situated in a side stream of the Paludenka, one of the head-streams of the Koiva, which flows into the Chussovaya, itself a tributary of the Kama river.

The muddy-looking gold-sand of the Adolphskoi washings contains, besides diamonds and gold, a few grains of platinum, also quartz, limonite, magnetite, much iron-pyrites (either bright yellow and unaltered or altered on the surface to limonite), chalcedony, anatase, &c., and fragments of the neighbouring rocks. All these minerals and rock-fragments have been derived from the mountain ridge which overhangs the stream, and which is principally composed of a quartzose chloritic talc-schist, which has been suggested to be identical with the Brazilian itacolumite, but which, according to the investigation of G. Rose, in no way agrees with this. This chlorite-talc-schist contains subordinate beds of hæmatite, grey limestone, and especially dolomite, coloured black by carbonaceous matter. This dolomite immediately underlies the gold-sands of the Adolphskoi washings, and was considered by von Engelhardt to be the original mother-rock of the diamond. Other observers regard the chlorite-talc-schist as the mother-rock; since diamonds have as yet only been found loose in the sands and never embedded in rock, the point remains a disputed one.

Only about 150 diamonds have hitherto been found in the Adolphskoi gold-washings, so that the discovery of this locality has not in any way affected the diamond market. The stones are colourless to yellowish, perfectly transparent and very brilliant; a few crystals, however, show dark brown or black enclosures; their crystalline form is almost invariably the rhombic dodecahedron, the faces of which are curved and nicked along the short diagonal (Fig. 31 c).   The largest weighed $2\frac{17}{32}$ carats, the five coming next in order of size weighed respectively $1\frac{1}{4}$, $1\frac{1}{8}$, $1\frac{1}{16}$, $1\frac{1}{32}$, and 1 carat, while the remainder were small, the smallest weighing only $\frac{1}{8}$ carat; the twenty-eight stones first found had an aggregate weight of $17\frac{9}{16}$ carats.   A recent discovery of five stones at this locality led to the institution of a systematic search, which, however, has been unattended with any marked success.

The occurrence of diamonds in the Urals is not confined to the Adolphskoi washings, for in 1831 two small diamonds, one weighing $\frac{5}{8}$ carat, were found in the Medsher gold-washings, fourteen versts (nine miles) east of Ekaterinburg.   Again in 1838, a small stone weighing $\frac{7}{16}$ carat was found in the Kushaisk mine, which lies in the sands of the Goroblagodatsk mining district, twenty-five versts (seventeen miles) from the smelting furnaces of Kushvinsk and east of Bissersk.   In the next year a crystal weighing $\frac{7}{8}$ carat was found in the Uspenskoi mine in the sands of the district of Verchne-Uralsk in the government of Orenburg. Solitary stones have been found also at other places, for example, recently in the Charitono-Companeiski sands on the Serebrianaya river in the district of Kungur, government of Perm ; among these was a twin-crystal showing several hexakis-tetrahedra in combination, and having a thickness of 5 millimetres.   Among these finds were a small colourless stone from the Kamenka mine, in the Troitzk district, government of Orenburg, and also the first diamond discovered in the southern Urals; this latter is a perfectly transparent, yellow hexakis-octahedron of $\frac{3}{5}$ carat, and was found in 1893 in the gold-washings of Katshkar. This latter discovery is of interest, for it shows that the diamond is more widely distributed in the Urals than was formerly thought to be the case ; moreover, the minerals with which it is associated in the southern Urals are the same as in Brazil, from which we may conclude that the diamond originated in a similar manner in the two countries.

The occurrence of diamonds in the Urals is so rare and the stones themselves are so small that there are persons who doubt the genuineness of the occurrence, and express the opinion that the stones which have been met with found their way to this locality in order to fulfil Humboldt's prophecy.   No definite proof of fraud has, however, been brought forward, and Russian mineralogists, who have closely studied the question, are satisfied that the occurrence is genuine, their opinion being supported by recent discoveries of stones having been made.   A number of rough diamonds acquired from other countries, have been distributed by the Russian Government among the managers of mines, in order that the miners shall become familiar with the appearance of rough stones, and the possibility of diamonds being overlooked avoided.

9. LAPLAND.

In a far western corner of the Russian empire, namely, in Russian Lapland, a few small diamonds have been recently found.   At the time of the expedition of C. Rabot in the second half of the eighties, a few stones were found in the valley of the Pasevig river which flows into the Varanger Fjord, an arm of the Arctic Ocean, and forms the Russo-Norwegian frontier in longitude 30° E of Greenwich.   This river flows over gneiss which is penetrated by numerous veins of granite and pegmatite, by the weathering of which the diamond-bearing sand has originated.

According to the investigations of C. Vélain, these sands contain the following minerals, named in the order of their frequency of occurrence: Garnet (almandine) in rose-red rounded grains, forming half the total bulk of the sand, zircon, brown and green hornblende, glaucophane, kyanite, green augite, quartz, corundum, rutile, magnetite, staurolite, andalusite, tourmaline, epidote, felspar (oligoclase), and, lastly, the rarest constituent, diamond. The diamonds occur as small, angular, rarely water-worn, water-clear grains or broken fragments. These rarely exceed 0·25 mm. in diameter, though one crystal has been found measuring 1·5 mm. across. The transparency of the stones is greatly impaired by the presence of numerous enclosures, some being cavities containing gas, while others are microscopic crystals of an unknown substance. That these small grains are indubitable diamonds is demonstrated both by their hardness and by the fact that they are combustible in oxygen, the products of combustion being pure carbon dioxide.

The minerals associated with the diamond in Lapland are practically the same as in India and Brazil; epidote, however, while present in India is absent in Brazil, and the hydrous chloro-phosphates, so abundant in Brazil, are here absent. Vélain has expressed the opinion that the diamond here originated in the pegmatite; the mother-rock and the precious stone having been formed concurrently, as is also supposed by Chaper to be the case with the pegmatite at Wajra Karur, near Bellary in southern India, described above. In any case the Lapland diamonds must have been derived from one of the ancient crystalline rocks mentioned above, since no other type of rock is present in the whole of this region. No recent accounts of this locality have been published, and further investigation is much to be desired, since it would probably shed fresh light on the problem as to the nature of the original mother-rock of the diamond, and on the doubtful occurrence of diamond in pegmatite at Wajra Karur. Though of no commercial or economic importance, the discovery of Rabot and Vélain is of great scientific interest and should be assiduously followed up.

10. DIAMONDS IN METEORITES.

In recent years our knowledge of the distribution of diamonds has been extended in an interesting direction by the observation that this mineral occurs in a number of meteorites in the form of small, usually microscopic, grains of a grey or black colour, closely resembling carbonado. Diamond is thus a substance which is not confined to the earth, but is present in extra-terrestrial bodies, fragments of which, from time to time, fall upon the earth's surface. From an æsthetic point of view, meteoric diamonds are, of course, valueless; their interest and importance in connection with the natural history of the mineral is, on the contrary, very considerable, and it is fitting that this particular occurrence of the diamond should receive appropriate mention, especially as on it various theories as to the mode of origin of the diamond in the earth's crust have been based.

The meteorite which fell in Russia on the morning of September 22, 1886, in a field three miles from the village of Novo-Urei, on the right bank of the Alatyr, a river in the Krasnoslobodsk district of the government of Penza, was the first in which diamonds were observed. It was examined by Messrs. Jerofejeff and Latshinoff, and was found to consist of the minerals olivine and augite, with interspersed carbonaceous matter and metallic nickeliferous iron. The carbonaceous substance contained small greyish grains, the hardness, specific gravity, chemical composition, and appearance under the microscope of which, proved them to be undoubtedly diamond (carbonado). The results arrived at by these investigators have been fully confirmed by others. In this stone diamond was present in the proportion of 1 per cent., so that, since the whole meteorite weighed 1762·3 grams (rather less than 4lbs.), the total amount of diamond present was 17·62 grams or 85·43 carats.

Black grains of diamond have also been observed in the meteoric stone of Carcote, in the desert of Atacama, in Chile. Diamonds have subsequently been found, usually in meteoric irons, for example, in that which fell at Arva in Hungary, and at Cañon Diablo in Arizona.

Many meteoric irons contain small cubes of graphite, having the same crystalline form as cubes of diamond; this cubic form of graphitic carbon, to which the name cliftonite has been given, is present in the meteoric irons which fell at Arva in Hungary, at Toluca in Mexico, at Youndegin in Western Australia, in Cocke County and at Smithville in Tennessee, and perhaps also in a few others. It is, however, highly probable that these cubes were originally diamond, and were changed into graphite by the agency of heat, the possibility of such a change being effected artificially by heating diamonds away from contact with air, having been previously mentioned above. If this be so, the occurrence of diamond in meteorites is much more general than has been supposed, and now that attention has been drawn to the matter, it is probable that closer examination will demonstrate the presence of diamond in many other meteorites in which its absence has not been definitely proved.

C. ORIGIN AND ARTIFICIAL PRODUCTION OF THE DIAMOND.

The problem as to the origin of the diamond in nature is one which has received no small amount of attention. Many and various are the suggestions which have been from time to time put forward, but few are based on scientific considerations, the majority being purely imaginative, and therefore valueless as working hypotheses.

Before formulating any useful hypothesis as to the mode of origin of the diamond in nature, it is necessary to learn as much as possible of the conditions under which diamonds occur in all parts of the globe and in extra-terrestrial matter, and of the manner in which they may be artificially produced. Of neither of these subjects is our knowledge much in advance of that of former years. The artificial production of diamonds is an art yet in its infancy, and for the elucidation of the problem as to their natural origin, experiments in the artificial production under conditions corresponding as closely as possible to natural conditions, are required. A close study of the conditions under which diamonds occur in nature is desirable therefore, not only as an aid in making such experiments, but also with a view to the collection of detailed information respecting the minerals associated with the diamond, a knowledge of the mode of origin of which would be a substantial help in the solution of the problem.

Almost every mode of origin, possible and impossible, for the diamond has, at one time or another, been brought forward: thus some have conceived the formation of diamond to have been brought about by the vital processes of plants; others have derived the precious stone from organic remains: and yet others from inorganic substances. Some assume a high temperature to have been an essential of the process; while others, on the ground that diamond subjected to high temperatures is converted into graphite, consider such a condition an impossibility. A few of the many theories which have been brought forward are set forth below:

The earliest is that of the celebrated physicist, Sir David Brewster, who unreservedly expressed the opinion that the formation of diamonds was due to the vital processes of plants, the process of formation having been similar to that of the formation of resin, and that the diamond at one time was viscous like a resin. The view that the diamond had been separated out from the sap of some plant in much the same way as a form of silica, the so-called tabasheer, is separated out and deposited in the knotty stems of the bamboo, was accepted also by the Scotch mineralogist, Jameson, and similar views were held by Petzholdt.

D'Orbigny regarded the diamond as a decomposition product of extinct plants; the same view was held by Wöhler, who assumed the alteration to have taken place at a low temperature, and vigorously disputed the theory which demanded a high temperature for the formation of the diamond. J. D. Dana, on the other hand, regarded a high temperature if not as an essential, at least as a probable condition, and considered that the diamond might have been derived from organic substances by the same processes which effect the metamorphism of rocks. Göppert entertained similar views, owing to his belief that in the course of his detailed investigation of the enclosures of diamond he had detected plant remains. By theorists of this school it was supposed that the decomposition products of decaying vegetable gradually escaped, leaving behind a substance which contained an ever-increasing proportion of carbon ; this substance, having been at length transformed into pure, amorphous carbon, was supposed to be capable of taking on the crystalline form of diamond. Very similar views were held by G. Wilson, who supposed that hydrogen and oxygen gradually escaped from woody matter and left behind a substance resembling anthracite, which, by further alteration, he conceived might be transformed into diamond. These processes were assumed to take place at a low temperature, since at higher temperatures graphite, and not diamond, would be formed.

The opposite opinion was held by Parrot, namely, that the alteration of woody matter took place at high temperatures; he suggested that the transformation into diamond of small particles of carbonaceous matter, strongly heated by volcanic agencies, was effected by a sudden cooling. Carvill Lewis has expressed similar views with special reference to the origin of South African diamonds. He considered the diamond to be formed in the kimberlite at the time this was erupted into the pipes as a molten igneous rock. Numerous fragments of the carbonaceous and bituminous shales, through which the igneous rock forced a passage, were caught up by the igneous mass, and Carvill Lewis supposed the heat of this mass to have converted the carbon contained in the fragments of shale into diamond. According to the views of other investigators the igneous rock itself contained carbon, which crystallised out of the molten mass as diamond, before it was erupted into the volcanic pipes.

C. C. von Leonhard also invoked the aid of volcanic heat, but supposed the carbon to be volatilised, and diamonds to be formed by the crystallisation of the sublimed carbon. G. Bischof, while he does not combat the view that the diamond has been formed from vegetable matter, makes no definite statement with regard to its mode of origin other than the opinion that a high temperature is an impossible condition.

Liebig supposed that by some kind of decomposition a product growing ever richer in carbon was separated out from a fluid hydrocarbon, and that this product, when it finally became pure carbon, crystallised in the form of diamond ; he pointed out that such a mode of origin could only conceivably take place at low temperatures, the combustibility of the substance rendering the process impossible under the conditions of a high temperature and the presence of oxygen. Berthelot, though not referring specially to diamond, has asserted that a separation of carbon from such a liquid could take place only under the influence of heat, a statement directly opposed to the assumption made by Liebig. Chancourtois brought forward a theory to the effect that diamond has been formed by the slow oxidation of emanations of a gaseous hydrocarbon, the hydrogen forming water, part of the carbon forming carbon dioxide, and the remainder crystallising as diamond, the whole process being analogous to the formation of sulphur from hydrogen sulphide, the oxidation of the hydrogen being accompanied by the separation of crystals of sulphur.

A few authorities, including the mineralogist, J. N. Fuchs, suppose that large amounts of carbon dioxide are present in many places in the interior of the earth, and that this has played an essential part in the formation of the diamond.

Göbel suggested that at high temperatures carbon dioxide might be reduced by certain metals, such as aluminium, magnesium, calcium, iron, or by silicon, &c., the carbon crystallising out during the reduction as diamond. In connection with the existence of drops of liquid carbon dioxide enclosed in cavities in certain diamonds, Simmler stated the opinion that liquid carbon dioxide at a high temperature and pressure is capable of dissolving carbon, and that from this solution carbon may crystallise out as diamond. These assumptions were not, however, supported by the investigations of Gore and of Dölter, who failed to establish the solubility of carbon either in liquid carbon dioxide or in gaseous carbon dioxide above the critical temperature and under great pressure.

The compounds of carbon with chlorine are sometimes assumed to be the original source of the carbon of the diamond. A. Favre, and later H. St. Claire Deville, admitted the possibility of the formation of diamond from such compounds, the former having been induced to entertain this idea from the fact that certain minerals associated with the diamond in Brazil can be artificially produced from chlorine compounds. Goreeix, who was well acquainted with Brazilian deposits, considered such an origin for diamonds from this region to be quite within the range of possibility, the carbon being supplied by chlorine or fluorine compounds. Without committing himself to any details of the process, Damour also considers it possible that Brazilian diamonds may have originated by the interaction of a variety of suitable compounds. Finally, we may mention the hypothesis for which Gannal is responsible, namely, that the diamond may have been formed by the decomposition of carbon bisulphide. Other possible modes of formation will be mentioned when the methods employed in the artificial production of diamonds are under consideration.

A consideration of the character of the various diamantiferous deposits, which are scattered over the face of the globe, and often widely separated, leads inevitably to the conclusion that there is no single mode of origin common to all diamonds, but that in different deposits the diamond has originated in different ways. If the diamond really occurs in India and Lapland embedded as an original constituent in granite-veins penetrating gneiss, and in South Africa in an olivine-rock associated with gneiss, it almost certainly follows that in these localities the diamond has been formed in the same way as the igneous rock in which it occurs.

Unfortunately, however, this brings us no nearer the truth, for the exact mode of formation of such rocks is still one of the obscure questions of geology. It is probable that such granite-veins (pegmatite) are due to the solidification, presumably at not very high temperatures, of masses of fused silicates saturated with water. Such a mode of origin is, however, not likely in the case of the gneisses and rocks interfoliated with them, such as the olivine-rock in which the diamonds of the Cape were originally contained, and in the alteration product of which, namely, the serpentine-breccia, they are now found. If, on the other hand, this original olivine-rock is not interfoliated with gneiss at some depth below the earth's surface, but is an eruptive rock which has penetrated the gneiss, as Carvill Lewis has urged, then its mode of origin may not differ essentially from that of pegmatite, and the South African diamonds may have been formed in the same way as those of India and Lapland. If, then, in these countries the diamond is, or has been, a constituent of an igneous rock, we must conclude that it was during the cooling and solidifying of the fused mass of rock that the diamond crystallised out, the carbon of its substance, if not present as a normal constituent of the rock, being derived from bituminous foreign matter. The existence of enclosures of liquid carbon dioxide in the minerals of such rocks, especially in the quartz of granite and gneiss, but also in olivine and even in diamond itself, has a special significance in this connection. Luzi has demonstrated that in the presence of water and of compounds of fluorine, carbon is soluble in a fused silicate. Though in the experiments of this investigator the dissolved carbon separated out in the form not of diamond but of graphite, it is not inconceivable that under other conditions, such as the high pressure and temperature of the earth's interior and the presence of other substances, diamond itself might be formed. The more recent experiments of I. Friedländer (1898) have demonstrated that graphite is soluble in fused olivine, and that on cooling diamond separates out, facts which have an important bearing on the origin of South African diamonds.

The diamonds which, together with crystals of quartz and other minerals, are found in crevices in the itacolumite of Brazil must have originated in quite a different way. There is not the slightest doubt that the crystals of quartz and of the accompanying minerals have been deposited from aqueous solution, perhaps even at ordinary temperatures. If this mode of occurrence of the diamond is really a fact, we can only assume that the diamond originated in the same way as the minerals with which it is associated, but as to the nature of a solution capable of depositing diamonds only negative statements can be made. Gorceix, who first expressed the opinion that Brazilian diamonds originated in the same way as the minerals with which they are associated, considered that this common origin must be sought not in the direction of deposition by solution, but rather in the interaction of gases, such as compounds of chlorine and fluorine and water-vapour rising up from the interior of the earth.

The diamonds of the Adolphskoi gold-washings in the Urals have been supposed to have been originally embedded in a bituminous dolomite. Engelhardt, who first suggested the probable identity of the bituminous dolomite with the mother-rock of the diamonds of this district, supposed the diamonds to have originated by the transformation of the bituminous material, but made no suggestion as to the means whereby such a transformation might be effected.

The occurrence of diamond in the meteoric stone of Novo-Urei, on account of its association with olivine and augite, is comparable with its occurrence in South Africa, and the mode of origin of the diamond in South Africa and in extra-terrestrial matter of this type is probably one and the same. The origin of diamond in meteoric iron is, however, of a different kind. Under ordinary conditions the excess of carbon taken up by molten iron crystallises out on cooling in the form of graphite, and crystals of this modification of carbon are frequently to be seen both in ordinary cast-iron and in most meteoric irons. Moissan, however, has recently demonstrated that when a mass of molten iron solidifies under great pressure, the carbon separates out as diamond. The existence of diamond in meteoric iron may conceivably be due to causes other than high pressure during its solidification, such, for example, as the presence of nickel, phosphorus, and other chemical elements.

It is obvious from what has been said that the origin of the diamond in nature is still to a great extent shrouded in mystery. For the complete elucidation of the problem further study of the various modes of occurrence of the precious stone, and of the minerals associated with it, together with more extended experiments in its artificial production, are desirable and necessary.

Very few facts shedding light on the problem as to the origin of diamond have hitherto been gleaned from experiments in the **artificial production** of the precious stone, the observation and study of diamantiferous deposits having been a much more fruitful method of attacking the problem. Although various experimenters have from time to time asserted their success in producing artificial diamonds, it is only in quite recent years that absolutely unquestionable results have been attained. The methods employed by various workers have, as a rule, differed from each other ; thus in one case it has been sought to obtain diamonds by fusing or volatilising carbon in the intense heat of the electric furnace ; in another by the separation of carbon from one of its liquid compounds at ordinary temperatures, or at a high temperature combined with great pressure. Some of the most important investigations have been made by Despretz and Hannay, and of the more recent experimenters Moissan has been most successful. Even the crystals obtained by Moissan scarcely exceed microscropic size, so that though their scientific interest is great, they are inapplicable as gems.

In the experiments of Despretz (1853), electric sparks were continually passed through a vacuum for a period exceeding a month ; the terminals made use of were respectively a carbon cylinder and a platinum wire. At the conclusion of the experiment it was found that the latter had received a coating of particles of carbon, which, under a magnification of 30 diameters, appeared as small octahedra, and were said to scratch corundum. In other experiments in which the terminals were a carbon point and a platinum wire, the electric current was passed through acidulated water, and similar, though less marked, results were stated to have been obtained. In neither case, however, were the particles of carbon proved beyond question to be diamond.

In 1880 it was demonstrated by J. B. Hannay, a Scotch chemist, that metallic sodium, and still more metallic lithium, is capable at a high temperature of separating carbon from a hydrocarbon ; the same worker also claimed to have established the fact that, at a high temperature and pressure, carbon could be separated in the same way from nitrogenous organic substances. The investigation was carried out by placing lithium and paraffin (the latter to play the part of a hydrocarbon) and a little sperm-oil in a very strong, sealed, wrought-iron cylinder, which was then exposed to a very high temperature ; the interaction of these substances thus took place under great pressure. It was hoped that the carbon separated from the paraffin by the lithium would, at the moment of its separation, dissolve in the sperm-oil, and that on cooling diamond would crystallise out from this solution. A crystalline mass, containing 97 per cent. of carbon was indeed obtained, but its identity with diamond is, as before, doubtful.

Since 1893 more successful experiments having the same object in view have been devised and performed by Moissan, the celebrated French chemist. He adopted the method of causing carbon to dissolve in iron at the very high temperature of the electric furnace, and then subjecting the molten mass to rapid cooling by immersing the crucible in which it was contained in cold water, or by pouring the molten substance into a mould of iron filings, or by other means. The object of the rapid cooling was to produce an exterior shell of solid metal enclosing liquid material under tremendous pressure ; it was hoped that the carbon crystallising out from the liquid under this great pressure would assume the form of diamond, instead of graphite as under ordinary conditions. At the close of the experiment, the mass of iron was dissolved in acid, and some black grains and small, water-clear crystals were obtained, which possessed all the properties of diamond and were completely combustible in oxygen, yielding carbon dioxide as the sole product of combustion. The largest crystals obtained by this method are about $\frac{1}{2}$ millimetre ($\frac{1}{50}$ inch) in diameter ; identical results are obtained when the experiment is performed with molten silver instead of iron. There can no longer be any doubt that by this method it is possible to produce genuine diamonds.

Another successful method of producing artificial diamonds has been recently (1898) discovered by I. Friedländer. A small piece of olivine was fused in the gas-blowpipe and the molten mass stirred with a rod of graphite. After cooling, those parts of the silicate which had been in contact with the carbon were found to contain vast numbers of brown crystals of microscopic size (0·001 mm. in diameter). These crystals were found to be octahedral or tetrahedral in form, to be unattacked by hydrofluoric and sulphuric acids, to have a high refractive index, to sink slowly in methylene iodide, to burn away when heated in a current of oxygen, and to be unaltered by heating in a current of carbon dioxide. All these characters point to one conclusion, namely, that the crystals so produced are diamond, and still further proof is furnished by the fact that the stony mass containing them is capable of scratching corundum.

When we essay to draw conclusions as to the origin of diamonds in nature from experiments in their artificial production, we should be careful that the conditions of the artificial production and of the natural occurrence are parallel. For example, the experiments of Moissan are most helpful in explaining the origin of diamond in meteoric iron, while those of Friedländer suggest that the diamonds of South Africa and of meteoric stones have been formed by the action of a molten silicate magma upon graphite or some carbonaceous material. Friedländer himself suggests that, in the case of the South African deposits, the carbon may have been derived from the carbonaceous shales which are penetrated by the pipe of diamond-bearing " blue ground "; there are, however, objections to this view, as has been shown already.

Further experiments in the artificial production of diamonds are desirable, especially with a view to the discovery of a liquid capable of depositing carbon in the crystalline form of diamond ; when more knowledge on this point has been obtained, it will, perhaps, be possible to explain the origin of diamond in pegmatite veins and in itacolumite, as well as in olivine-rock and in eclogite (p. 197).

D. APPLICATIONS OF DIAMOND.

The natural beauty of the diamond renders it primarily an object of æsthetic value; only faulty, opaque, or unpleasingly coloured specimens are applied to technical purposes for which a material of great hardness is essential.

1. APPLICATION IN JEWELLERY.

The beauty of a diamond depends but rarely upon the tint of the pigment it may contain, but is rather associated with the marvellous lustre and play of prismatic colours so characteristic of this stone. This play of colour differs in degree in different stones, and depends to a large extent on the method of cutting which has been adopted ; uncut stones, with their rough, and often somewhat irregular faces, either do not show any play of colours, or display it to only a very limited extent.

How far the ancients were acquainted with the art of diamond-cutting, or even with the polishing of the natural crystal-faces of the stone, is not certainly known, but that they were not entirely ignorant of the process is apparent from their writings.

In India, the ancient home of the diamond, the art of polishing the faces of natural crystals was practised in the remotest times ; when or how the device of faceting stones was discovered or introduced in this country is not known, but it was practised in the seventeenth century, at the time of Tavernier's visit (1665). According to native ideas in

India, the only object of cutting a stone is the removal of faulty portions, and a natural octahedron of which the faces have been polished is preferred. In spite of this, various forms of cutting are more or less in vogue; the most general forms are the thick-stones (Plate IV., Fig. 15, *a*, *b*), table-stones, and thin-stones; the first named form, on account of its generality in India, is often referred to as the "Indian cut." Such forms of cutting, together with others in which the facets are more numerous, are admired and met with, not only in India, but elsewhere in the Orient, namely in Persia, Arabia, Bagdad, &c. The Oriental diamond-cutter follows the outlines of the rough stone as closely as may be, striving to reduce the loss of material to a minimum. The European diamond-cutter, on the contrary, aims at developing to their fullest extent the optical properties of the stone, and makes economy of material only a secondary consideration. In many cases an irregular gem, "lumpy stone," or "pebble," cut by an Oriental, has passed into the hands of a European, and has been re-cut, the greatly enhanced beauty of the European cut stone compensating for the loss of material involved in a second cutting process. Such was the fate of the famous "Koh-i-noor," among other stones; Plate X., Fig. 4, *a*, *b*, shows the form of the Indian cut of this stone, and Fig. 5, *a*, *b*, *c*, its form after being re-cut.

Diamond-cutting in India was not, however, entirely in the hands of native lapidaries, for Tavernier states that the "Great Mogul," the large diamond named after the ruler of Delhi, its possessor, was cut rather unsuccessfully by Hortensio Borgis, a Venetian cutter. The diamond-cutting industry has flourished in Europe since the end of mediæval times, and European ideas have had a certain influence upon the development of the art in India.

In all Western countries in the middle ages, diamonds were used in the rough condition, with the natural faces polished, or in the form of point-stones, thick-stones, table-stones, &c., which were the forms of cutting usual in India at that time. At this period gems were used not so much as personal ornaments by women, but more often for enriching robes of state, such, for example, as the coronation mantle of Charles the Great, and for ornamenting reliquary shrines, sceptres, crowns, scabbards, &c. Practically nothing is known of the diamond-cutting industry in Europe until the beginning of the fifteenth century, when a skilled artisan named Hermann appeared in Paris, and did much to develop the art. As early as 1373 the existence of diamond-polishers at Nürnberg is mentioned, but nothing is known as to the methods practised by these workers.

The gradual development of the art of gem-cutting and the spread of the knowledge of this art was accompanied by the gradual growth of the custom of wearing diamonds as personal ornaments by women. The custom was introduced subsequently to the year 1431, in the time of Charles VII., in the French court by Agnes Sorel. The taste for this form of personal ornament had grown amongst the ladies of the court of Francis I. to such an extent, that edicts, levelled against the excessive use of gems as personal ornaments, were issued both by Charles IX. and Henry IV., without, however, having any effect. From the French court the custom gradually spread over the whole of Europe.

The large demand for diamonds thus caused gave a new impulse to the diamond-cutting industry, and during the course of the fifteenth century the art made its greatest strides. The advancement was due, to a certain extent, to the influence of the Dutch lapidary of Bruges, Ludwig van Berquen, who invented his particular process in the year 1476. Although some credence has been given to the statement of Robert van Berquen, that his grandfather was the originator of the modern method of diamond-cutting, namely, the use of diamond powder, yet the device had probably long before been known in Europe. What L. van Berquen probably did was so to improve the technique of the art that greater precision in the arrangement of the facets could be attained. Owing to irregularity in cutting, the play of prismatic colours in table- and point-stones was often scarcely observable at all, and the stones cut after Berquen's method offered a great contrast in this particular.

Among the first and most famous diamonds cut by L. van Berquen are said to have been those in the possession of Charles the Bold, Duke of Burgundy, some of which were lost at the time of his defeat by the Swiss. Certain of these stones are probably still at the present day in the form in which they left Berquen's hands, and they bear witness to the high degree of perfection to which this artist had attained. According to Schrauf, the cutting of the "Florentine" (Plate XI., Fig. 10, a, b), and of the "Sancy" diamonds (Fig. 11, a, b), was the work of van Berquen ; they are cut in the briolette or pendeloque form, a form which van Berquen was the first to adopt.

This particular form of cutting was not often copied, at the present time it is quite unused and is never shown by recently cut stones. The different varieties of the rose or rosette form of cutting (Plate IV., Figs. 1–7), which were introduced in the sixteenth century (about 1520), on the contrary grew quickly into favour, and at the present day are in general use and frequently seen. This form of cutting is most advantageous for thin and flat stones, since it involves but little waste of material and at the same time permits the full display of the lustre of the stone. It is, however, inferior to certain other forms of cutting in that it fails to develop to the fullest extent possible the beautiful play of colours so characteristic of a cleverly cut diamond.

The rose form was in vogue for about a century, and was then largely superseded by the most perfect form of cutting yet devised, namely by the brilliant form. This form was invented in the middle of the seventeenth century, and the idea is said to have originated with Cardinal Mazarin. The stones first cut in this form were double-cut brilliants (Plate II., Fig. 1, a, b, c), and had on the upper part sixteen facets besides the table. At the end of the seventeenth century the triple-cut brilliant (Fig. 3, a, b, c,), with thirty-two facets on the upper part, was introduced by Vincent Peruzzi of Venice. This form of cutting, which is still more favourable for the display of the optical properties of the stone, is in use at the present day, with no alteration except that the size of the facets is more equalised, as shown in Fig. 4, a, b, c, of the same plate. The various forms of cutting which have subsequently come into use do not differ in any essential respect from those already described. The star-cut, for example, of M. Caire (Plate III., Fig. 1, a, b, c), by means of a slight modification, combines an economy in the rough material, with no inferiority in the display of the optical qualities of the stone.

The loss of material involved in the cutting of a stone in the brilliant form is very considerable, sometimes amounting to one-third, one-half, or to an even greater proportion of the rough stone. The "Regent," for example, which is the most perfect brilliant known, weighed in the rough 410 carats, while its weight after cutting was only $136\frac{7}{8}$ carats, or only one-third as much. The "Koh-i-noor," as cut in India, weighed $186\frac{1}{16}$ carats, and after re-cutting in England, $106\frac{1}{16}$ carats; again, the "Star of the South" weighed in the rough $254\frac{1}{2}$ carats, and as a brilliant 125 carats. The beautiful play of colours obtained by the use of this form of cutting more than compensates for the waste of rough material; in no other form is this character brought out so prominently, and a good brilliant with a fine play of colours, though small, is more highly prized than a larger stone cut in an inferior form, and consequently with a less fine play of colours.

The brilliant is always so mounted in its setting that the broad facet or table is turned towards the observer. Only rarely, in cases in which the stone has faults to be concealed,

Actual Sizes of Brilliants (Diamonds) of ¼ to 100 carats.   Plate IX.

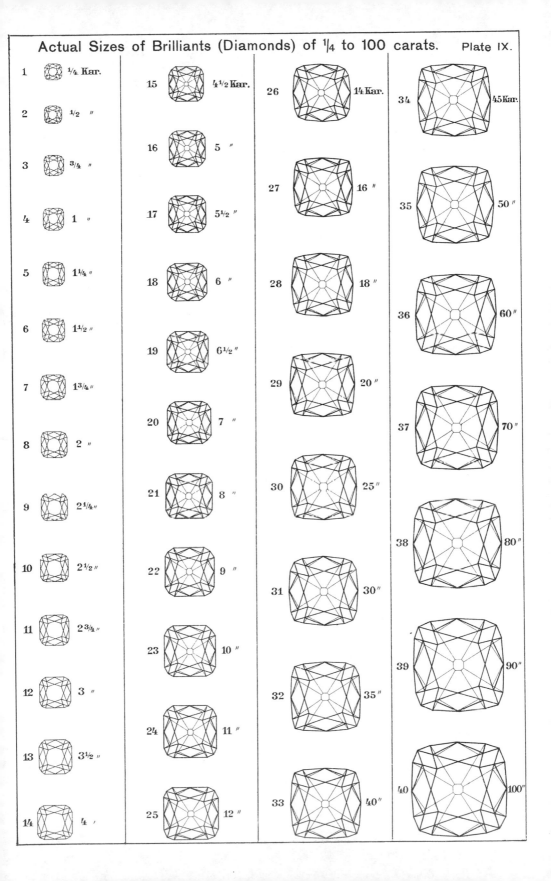

1 ¼ Kar.
2 ½ "
3 ¾ "
4 1 "
5 1¼ "
6 1½ "
7 1¾ "
8 2 "
9 2¼ "
10 2½ "
11 2¾ "
12 3 "
13 3½ "
14 4 "

15 4½ Kar.
16 5 "
17 5½ "
18 6 "
19 6½ "
20 7 "
21 8 "
22 9 "
23 10 "
24 11 "
25 12 "

26 14 Kar.
27 16 "
28 18 "
29 20 "
30 25 "
31 30 "
32 35 "
33 40 "

34 45 Kar.
35 50 "
36 60 "
37 70 "
38 80 "
39 90 "
40 100 "

is the position reversed and the so-called Indian setting adopted. The dazzling appearance of the brilliant is due to the way in which the rays of light entering the stone are reflected from the facets at the back, pass out by the front facets and reach the eye of the observer in the manner already described. As compared with a brilliant, other diamonds appear dull and lifeless, for the arrangement of their facets does not admit of the passage of light rays in such a way as to produce the most favourable effect, either as to brilliancy or play of colours. It is essential, however, in order to produce the greatest effect, that even the brilliant form should conform rigidly to the rules of proportion which have been already laid down. The "Regent," for example, the proportions of which are strictly accurate, is a far more dazzling and brilliant stone than is the "Koh-i-noor," which is cut with too small a depth.

The brilliancy and fire of a cut diamond is enhanced by suitable illumination; the source of light should not be too large, otherwise the separation of the differently coloured refracted rays is masked, and the light reflected appears to be white; neither should the source of light be surrounded by an opal shade. A brilliant appears at its best when illuminated by a number of small flames; and the effect can be still further increased by attaching the stone in its setting to a thin metal rod or wire, when the quivering motion imparted to the stone by every movement produces rapidly changing flashes of colour.

Although each of the different forms of cutting which have been described may under various circumstances be made use of for the diamond, yet there are but two forms in general use, the first and most important being the brilliant, the other, the rose or rosette. The diamond is the only gem which is so invariably cut in these forms, and hence is often loosely referred to as a brilliant or a rosette.

Whenever the form of a rough stone permits, it is always cut as a brilliant, no matter what its size may be. Plate IX. shows in actual size a series of diamonds cut in the brilliant form, viewed from above, and ranging in weight from $\frac{1}{4}$ to 100 carats; from a study of this series some idea of the actual sizes of stones of different weights may be gained. As a rule only small diamonds of little thickness, fragments cleaved off in the fashioning of a brilliant from a large stone, and large diamonds which have not sufficient depth to be cut as brilliants, are cut as rosettes. Sometimes, by preference, a large thin stone is made into several small brilliants and not into a single large rosette.

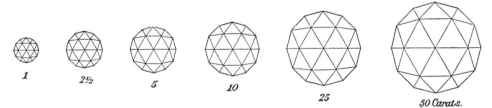

FIG. 44.   Actual sizes of rose diamonds of 1 to 50 carats.

As in Plate IX., a series of brilliants in actual size is shown, so Fig. 44 shows the actual sizes of a series of rose diamonds ranging in weight from 1 to 50 carats. Stones of surprisingly small size are often to be met with cut as roses, with regularly arranged facets; they may be so small that 1500 or even more weigh no more than a carat. Stones of very small size are not generally cut as brilliants. Small roses, of which 100 or 160 are required to make up the weight of a carat, are known as piece-roses; the Dutch lapidaries are specially skilled in the art of cutting extremely small stones. Very minute splinters of diamond are often furnished with a few irregular facets, when they are known as senaille, and, like the smallest roses, are used to form a setting round a large gem (carmoizing). Very thin laminæ of diamond are ground down, polished on both sides, and the edges furnished with small facets ; they are then used to cover small miniatures set in rings, &c., producing a very striking effect, and are now known as portrait-stones or brilliant-glass.

A few trade expressions made use of by jewellers may be explained here. Very small diamonds are referred to as salt-grains, very large ones as solitaires, nonpareils or paragons. All cut diamonds exceeding 50 carats in weight were formerly known as solitaires, while those exceeding 100 carats have been referred to as majestic diamonds. Stones weighing less than a carat are known as carat-goods, those of 1, 2, &c., carats as carat-stones, two-carat-stones, &c.

2. DIAMOND-CUTTING.

The general principles of the art of gem-cutting have been already considered ; the cutting of the diamond, however, on account of the perfect cleavage and enormous hardness of the stone, requires special methods, which must be separately considered.

As we have seen, diamonds are most frequently cut in the brilliant form, which form is comparable to that of an octahedron of which two opposite corners have been truncated. A truncated octahedron, therefore, can be transformed into a brilliant by simply adding the necessary facets, hence the form of crystal which can be most conveniently cut as a brilliant is the octahedron. Crystals of the form shown in Fig. 31, n and o, are therefore specially suited for the fashioning of brilliants. The rhombic dodecahedron and the hexakis-octahedron (Fig. 31, c and d) are also suitable ; but stones, whose form differs widely from that of the regular octahedron, for example Fig. 31, e and f, cannot be so easily transformed into brilliants. Before such a stone is faceted it is reduced by cleavage to the octahedral form, in order to avoid the tedious process of grinding away portions which need to be removed. The property of cleavage then is very useful to the diamond-cutter, for not only is he spared much labour in grinding, but the fragments removed by cleavage can be utilised in the fashioning of smaller gems ; moreover, the property can be made use of for the purpose of removing the faulty portions of a stone or of dividing a large stone of unsuitable form into several smaller ones. The operation is, however, one which demands the greatest care ; the worker should be capable not only of detecting the direction of cleavage from the outward form of the rough diamond, but also of recognising the existence of twinning in a crystal. Any attempt either to cleave a twinned crystal or to cleave an ordinary stone in a wrong direction, will probably be attended with more or less complete fracture of the stone.

The operation of **cleaving** a diamond, which is entrusted to trained workmen, is performed in a manner now to be described. The stone to be cleaved is fixed to the end of a rod with some kind of cement, such as a mixture of shellac, turpentine, and the finest brick-dust, and in such a position that the direction of cleavage is parallel to the length of the rod. A second diamond with a projecting edge is fixed to a similar rod with the edge uppermost. By grinding the sharp edge of the second diamond against the first, a nick in the direction in which the stone is to be cleaved is cut to a sufficient depth. The rod supporting this diamond is set on a firm elastic base, a sharp, strong chisel is placed in the proper direction in the nick, and the cleavage effected by dealing the chisel a single sharp blow with a hammer. The cement may be loosened by heating, and the stone placed in another position if it is desired to cleave it in another direction. The powder produced when the nick is made is caught in a small box provided with a sieve, and is utilised in the process of grinding.

It has been stated by Tavernier that the custom of cleaving diamonds has been practised in India since very early times ; in Europe, however, the art was acquired much more recently, and is said to have originated with the English chemist and physicist Wollaston (1766–1828), of whom it is related that by cleaving away the faulty exterior portions of a large diamond he was enabled to dispose of the stone at a considerable profit.

Stones which, either by nature or by the hand of man, have been given the shape of an octahedron, have already the ground-form of the brilliant and can be at once faceted. The work of grinding the facets is facilitated by a preliminary operation, which is entrusted to special workers and is not performed in the process of cutting any other gem. This operation, by which the shape and position of the facets are roughly marked out, is known as **bruting**, rubbing, or greying the stone. The stone to be bruted is fixed to a handle, and, with the exception of the area on which the facet is to be made, is embedded in cement or in a fusible alloy of lead and tin. This projecting portion is rubbed with a strong pressure upon the projecting portion of another stone similarly mounted and prepared, and thus a facet, in approximately the correct position and with a fairly even but rough surface, is developed upon each of the stones. The powder abraded from the two stones during the operation is carefully preserved for use in grinding. During the operation of greying, which, by the way, derives its name from the grey metallic appearance of the facets so made, any over-heating of the stone by friction must be carefully avoided, since it leads to the development of " icy flakes " in the interior of the diamond. The operation is attended by a peculiar grating sound, which is said to be so characteristic that by this alone the possessor of a practised ear can determine whether the two stones which are rubbed together are diamonds or some less hard gems.

At the completion of the first stage of the operation the stone is removed by warming the cement or alloy, placed in another position, and the remaining larger facets successively marked out in the same way. The smaller facets are not so marked out by a preliminary operation, but by the subsequent process of grinding. When ready for the grinding process the stones are bounded by a number of fairly even, rough faces, with a grey, somewhat metallic lustre ; they have no longer the appearance of diamonds, but resemble dull grey metallic bodies with the general contours of the form of cutting the stone is finally to take.

In the combined process of **grinding and polishing**, the preliminary disposition of the facets, which may be slightly incorrect, is rendered strictly accurate by the completion of the grinding, their rough surfaces are rendered smooth and shining, and the smaller facets are added. The grinding process is the same as in the case of other gems as already described, the diamond being imbedded in the fusible alloy of a dop and placed on the grinding disc. Since in the grinding of diamonds the disc must be charged with diamond powder, which of course has the same hardness as the stone itself, the operations of grinding and polishing take place simultaneously, and any separate polishing process is superfluous. Any dirt or foreign matter which may adhere to the stone after the process of grinding is removed by treatment with fine bone-ash or tripolite.

In the process of grinding it is by no means immaterial in which direction the grinding disc moves across the facet which is being worked. Owing to the fact that the diamond, as well as other precious stones, has a different degree of hardness in different directions, the grinding of its facets can be accomplished with comparative ease in some directions, while in others the process is extremely tedious. To avoid injury both to the stone and to the grinding disc, the diamond must be ground " with the grain " ; and the operator ought to make himself familiar with these directions of least resistance, otherwise his work will be unnecessarily prolonged. When, for example, the table of a brilliant is to be developed upon an octahedron, the grinding disc should move from centre to centre of two opposite octahedral faces; if allowed to move from edge to edge, the facet can only be developed with the greatest difficulty, for in this direction the hardness of the diamond is much greater

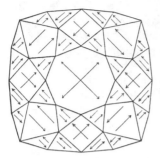

than in the other. The directions of least resistance to grinding on each of the facets of a brilliant are indicated in Fig. 45 by arrows. The large four-sided facets above and below and to the right and left of the table, are the faces of the octahedron.

We have thus seen that the three operations in the process of diamond-cutting are entrusted to as many classes of skilled workmen, namely cleavers, bruters, and grinders or polishers.

The order in which the facets of a brilliant are ground has also a certain importance. Starting from the octahedral ground-form, the table and culet are in every case first developed. The correct proportions of a brilliant are

FIG. 45. Directions of least hardness on the facets of a brilliant.

attained by grinding away five-ninths of the upper half for the table, and one-ninth of the lower half of the stone for the culet, the upper portion of a perfect brilliant being one-third, and the lower being two-thirds of the whole thickness of the stone from table to culet.

Some rough stones are of such a shape that they cannot be cut into the usual brilliant form, but are given an oval or triangular outline; in this case, the method of procedure

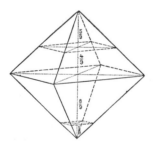

described above requires a slight modification, as also when the stone is to be cut as a rose or in some form other than the brilliant. In the latter cases the cleavage of the stone does not play so important a part, but otherwise the mode of procedure is much the same.

In past times the diamond-cutting industry has centred now round this town, now round that. The important invention ascribed to Ludwig van Berquen was made in 1476 at Bruges; in the fifteenth and sixteenth centuries, however, most of the work was done in Antwerp, where the workers of L. van Berquen had settled. Later on Amsterdam became the centre of the industry, and through many vicissitudes this city has retained its supremacy up to the present day. In Amsterdam there are now seventy diamond-cutting establish-

FIG. 46. First stage in the development of a brilliant from an octahedron, showing the position of the table and culet.

ments, large and small, fitted up with all the modern technical appliances, and using steam as the motive-power. The industry gives employment to more than 12,000 people, all of whom are Jews; in one establishment alone there are 450 grinding machines and 1000 employees, and in the whole of Amsterdam there are said to be a total of about 7000 grinding machines (skaifs) at work. Amsterdam, however, no longer monopolises the industry, for skilful cutters are now to be found in Antwerp, Ghent, Paris, St. Claude in the French Jura, London, and specially in Germany, more particularly at Hanau; diamond-cutting works are also to be found at Berlin and at Oberstein on the Nahe, a town which has long been well known as the centre for the working of agate and other varieties of quartz. In America the diamond-cutting industry has been introduced at Boston. But Amsterdam still holds the lead, and the largest and most valuable stones are always entrusted to the cutters there, who are held to be the most skilful and who best understand how to provide quite small stones with regularly arranged facets.

While the softer precious stones are often submitted to the operation of **engraving**, the diamond is very rarely so treated. Whether diamonds were engraved at all by the ancients is a matter open to grave doubt; only a few engraved diamonds belonging to more recent times are known; one of these bears the portrait of Don Carlos, and another was engraved with the Spanish arms for Charles V. The engraving of the diamond is never attempted at the present day, since the result of such attempts is in no way commensurate with the labour expended. That the art was not entirely unknown in the Orient is evidenced by the existence of a beautiful Indian octahedron of 30 carats, described by Boutan, one face of which was engraved with a devotional motto. According to G. Rose's account, some of the faces of the irregularly-formed diamond known as the " Shah," now in the Russian crown jewels, were engraved in Persian characters with the names of Persian kings. Two of the faces of the " Akbar Shah," another large diamond, were engraved with Arabic inscriptions, but these, like the engravings on the " Shah," had to be sacrificed when the stone was re-cut.

Only rarely have diamonds been bored and threaded as beads; the art is said to be practised at the present day at Ghent and Venice, in the latter city being the last remains of an old diamond-cutting industry which once flourished there. The boring is effected by the use of diamond powder, the perforation being started with a diamond point and continued with a steel point charged with diamond dust.

3. TECHNICAL APPLICATIONS.

The value of the diamond in its technical applications depends, as a rule, upon its exceptional hardness, and only in a few cases on its high refractive index.

The high index of refraction of the diamond led to an attempt to utilise this substance for the construction of microscope objectives. A diamond lens having only a slight curvature will give the same magnification as a lens of much greater curvature constructed of some less strongly refracting substance, such as glass. The disadvantages connected with the use of strongly curved lenses would be thus avoided, and, moreover, diamond lenses on account of their hardness would not be liable to be scratched by particles of dust and other matter. Experiments in this direction were made mainly by Pritchard, under the direction of Dr. Goring, between 1824 and 1826; although Pritchard was successful in preparing a few suitable lenses their use has never been adopted, probably on account of the difficulties connected with their construction and the prohibitive cost of the material.

Very extensive use is made of the diamond for the purpose of cutting glass, every glazier being in the possession of a tool known as a **glazier's diamond.** The stone set in this tool must be bounded by two rounded crystal-faces meeting in the curved cutting edge, which should not be too obtuse. This cutting edge is drawn with a slight pressure over the surface of the glass to be cut, and produces a fine scratch, not more than $\frac{1}{200}$ inch in depth, but which is sufficient to cause the glass to be easily broken in the direction of the scratch. The cutting edge, when properly used, cuts into the surface layer of glass like a wedge; when drawn across the glass by an unpractised hand, it simply tears instead of cutting the surface, and this also happens when a sharp pointed splinter of diamond is used instead of the curved cutting edge. In the direction of the jagged furrow produced by such means glass will not break as it does along a clean cut. Wollaston investigated this matter in detail, and found that the edge formed by the intersection of two curved natural faces of the diamond is specially adapted for cutting glass, and that artificially prepared edges of similar construction, made of softer stones, were almost equally efficacious. He constructed such edges of ruby, sapphire, quartz, &c., and while these were well adapted for cutting glass, the straight edge formed by the intersection of two plane faces of a diamond was found to be quite unsuitable for the purpose.

For a glazier's diamond, a small natural crystal of suitable form, such, for example, as the curved rhombic dodecahedron, or other crystal with curved faces (Fig. 31, *c*, *d*, &c.), is used. This is fixed in a metal setting by a fusible alloy, with the cutting edge projecting, and the whole furnished with a wooden handle. The use of the tool requires a little practice, which needs to be acquired for each particular diamond; a slight deviation from the correct method of handling the tool militates against its effectiveness. The majority of the stones used in the construction of glazier's diamonds are said to come from Borneo and Bahia.

Sharp splinters of diamond, such as are often obtained in cleaving diamonds preparatory to the process of grinding, are often mounted in a similar manner and used for the purpose of drawing and writing on glass and other hard substances. Such diamond-points are also useful in boring holes in glass, porcelain, precious stones, &c., and are employed now, as in ancient times, in the engraving of precious stones, such as the ruby and sapphire. In recent times, however, the diamond-point for engraving has been replaced by a small, rapidly rotating metal disc or point charged with diamond powder and olive-oil.

There are many purposes for which the extreme hardness of the diamond renders it specially valuable. Thus it is used as a cutting tool in the lathe for turning the edges of watch-glasses and pivots of specially hard steel, such as are used in instruments of precision of all kinds; for the boring of cannon, as, for example, in the works of Krupp at Essen; in the manufacture of the finer mechanical tools; for boring small holes in large stones through which fine gold and silver wire is drawn; in the turning and working of hard stones, such as granite, gneiss, porphyry, &c.; for the pivot supports of the most accurate chronometers and other delicate instruments, and for many other purposes.

The diamond, however, at the present time, has a wider application in a direction which has not hitherto been indicated, namely, in the construction of **rock-drills**. Since its introduction in 1860, the diamond rock-drill has grown steadily in importance, and is now widely used in the many boring operations connected with mining, tunnelling, and the sinking of artesian wells, prospecting bores, &c. The rotating, boring crown of the drill is studded with carbonado, and, compared with other boring appliances, penetrates the rock with extraordinary rapidity. The fashioning of sharp-edged furrows on the grinding surfaces of millstones, for which in recent years special machines have been devised, is also effected by means of the diamond; for these, and all other technical purposes, small diamonds of poor quality, bort, and carbonado are used.

Diamond powder is now used very largely, not only in the grinding of the diamond itself, but also in the grinding and cutting of other precious stones, even such as are soft enough to be cut with emery. The economy in time and labour, and the superiority of the results attending its use, more than compensate for the increased cost of the material. The same substance is also used for the purpose of slicing through hard stones, the cutting being effected by a rotating disc of soft iron, in the edge of which are embedded fine splinters of diamond.

4. LARGE AND FAMOUS DIAMONDS.

There are a comparatively small number of diamonds in existence which, either on account of their size, beauty, or historic and ancient associations, possess a special interest. While the origin and early history of many stones in existence at the present day is a complete blank, there are others, of which reliable accounts and drawings are given in ancient writings, whose present whereabouts is entirely unknown; the latter may have been destroyed or lost, or, on the other hand, they may lie hidden in the treasure-houses of some Oriental princes, whose predecessors possessed a taste for the collection of gems.

All the older famous diamonds of large size and enormous value, which are known by special names, come from India; only in comparatively recent times, namely, about the middle of the eighteenth century, have stones of remarkable size been found in Brazil, while the discovery of the South African diamonds was still later. The South African deposits have already yielded more large stones than are comprised in the aggregate yield of India and Brazil during hundreds of years; these stones are usually, however, of a yellowish tinge, and are, in consequence, less highly valued than are the blue-white diamonds of India and Brazil. Only a few of the many large stones which have been found in South Africa have, in consequence, received distinctive names. The value of these rare stones is naturally enormous, and they usually find a place amongst the crown jewels of different countries, rarely entering the possession of private individuals except in the case of wealthy collectors, especially in eastern countries.

The subject of famous diamonds is specially dealt with in a book entitled, *The Great Diamonds of the World* (London, 1882), by Mr. E. W. Streeter; also in *Le Diamant* (Paris, 1886), by Mons. E. Boutan, who has made a careful study of the tangled history of each stone. Much of the information given in the account which follows has been derived from these sources, as well as from older works. Most of the well-known famous diamonds are figured in their cut condition and actual size in Plates X. and XI.; the form of cutting most general is that of the brilliant, but examples of other forms will be found.

A few of the large stones, of which accounts have been given, are very probably not diamond at all, but some one of the minerals with which diamond is often confused. Among these is, in all probability, the "Braganza," which, if genuine, would rank as the largest of known diamonds. This stone, which is the size of a hen's egg, and weighs 1680 carats, came from Brazil, but the exact locality is unknown. It is preserved with the Portuguese crown jewels, and is not available for detailed examination; should it be proved to be topaz, which is very probably what it is, its value, formerly placed at £224,000,000, would at once sink to a comparatively insignificant amount.

Another large diamond, the genuineness of which is open to question, is a stone belonging to the Rajah of Mattan, in Borneo; it is known there as the "Danau Rajah," but is generally referred to as the "Mattan." It weighs 367 carats, and, if a genuine diamond, is by far the largest ever found in Borneo. It is pear-shaped, and about the size of a pigeon's egg, and is said to have been found in 1787, in the district of Landak, in western Borneo; the name "Danau Rajah," however, suggests the neighbourhood of the River Danau, in the south-east of the island, as a more probable locality. The stone is said to have been examined at Pontianak, in Borneo, in 1868, when it was declared to be rock-crystal; this decision is generally accepted, although it has been stated that an imitation, and not the real stone, was submitted for examination.

The genuineness of the diamonds now to be described is unquestionable; of these the Indian stones will be first considered, and afterwards the Brazilian and the South African.

The large Indian diamonds are often supposed to be of very ancient discovery, the majority, however, probably do not date back to very early times. No definite information can be gleaned from ancient writings, but it is a well-established fact that the diamonds in the possession of the Romans were all of small size.

Probably the largest of Indian diamonds is the **Great Mogul,** the history of which is very obscure. This was seen in the treasury of the Great Mogul, Aurungzebe, in 1665, by Tavernier, who both drew and described the stone in detail. This diamond had then the form of a very high and round rosette (Plate X., Fig. 2), and was of good water. It weighed $319\frac{1}{2}$ ratis, which Tavernier calculated to be equivalent to 280 carats, assuming 1 rati = $\frac{7}{8}$ carat. By authorities, who consider this value of the rati too high, the equivalent is given as 188 carats. The rough stone is supposed to have been found between 1630 and 1650, in the mines at Kollur, and to have originally weighed $787\frac{1}{2}$ carats, a weight which would make it unquestionably the largest of Indian diamonds.

The considerable disparity between the weight of the rough stone and its weight when cut, has been attributed to the unskilful manner in which it was cut by Hortensio Borgis, the Venetian diamond-cutter, who at that time was domiciled in India. The subsequent history of the " Great Mogul " is a complete blank ; it has been variously supposed to have been lost or destroyed, to be in existence under another name, such as the " Orloff" diamond, or the " Koh-i-noor," to be in the possession of the Shah of Persia, or to be lying forgotten among the jewels of some Indian prince.

Another large diamond of the same weight, namely, 320 ratis, is described in the memoirs of Baber, the founder of the Mogul dynasty. According to this account the stone had long been famous in India, and had formed part of the spoils of war of many an Indian prince, finally passing into the possession of Baber in 1556. This stone is regarded by Professor Story-Maskelyne as being identical with the diamond seen at Delhi, and described as the " Great Mogul " by Tavernier, and identical with the stone at present known as the " Koh-i-noor "; this view is very generally accepted.

The **Koh-i-noor** was appropriated in 1739 by Nadir Shah, the Persian conqueror of the Mogul Empire ; in 1813 it passed into the possession of the Rajah of Lahore, and after the British annexation of the Punjab, became the property of the East India Company, which in 1850 presented it to Queen Victoria. The stone had then the form of an irregular rosette (Plate X., Figs. 4a, 4b), with numerous facets above, below a broad cleavage surface, and on the side a second smaller cleavage surface. The weight of the Indian-cut stone was $186\frac{1}{16}$ carats, which agrees closely with the weight of 320 ratis, recorded long before as the weight of the stone described by Baber. In order to improve its form, which was very far from perfect, it was re-cut in England in 1852 by the diamond-cutter, Voorsanger, of the Amsterdam firm of Coster ; the work of re-cutting occupying thirty-eight days, of twelve hours each.

The " Koh-i-noor " is now a stone of considerable beauty, weighing $106\frac{1}{16}$ carats ; its new form (Plate X., Figs. 5a, 5b, 5c,) is, however, too thin for a perfect brilliant ; moreover, it is not of the purest water, and the colour is slightly greyish. In spite of these blemishes it is valued at £100,000. The question as to the identity of the " Great Mogul " with the " Koh-i-noor " can scarcely now be decided. Tennant regarded them as identical, and suggested that the " Koh-i-noor " and the " Orloff" are both parts of the rough stone of $787\frac{1}{2}$ carats, mentioned by Tavernier, and that the third and remaining portion of it is the plate of diamond weighing 132 carats, often mentioned as having been taken by Abbas Mirza with other jewels from Reeza Kuli Khan at the capture of Coocha, in Khorassan. Tennant constructed models of these separate portions in fluor-spar, a mineral which has the

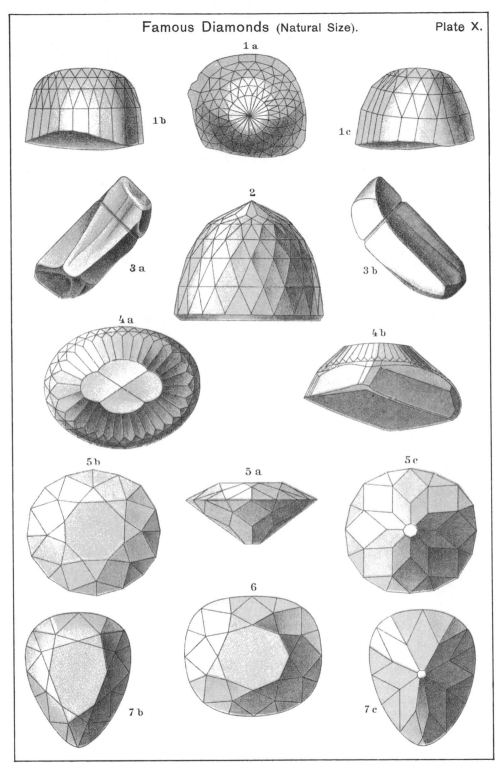

Famous Diamonds (Natural Size).   Plate X.

1a, b, c. Orloff.   2. Great Mogul.   3a, b. Shah.   4a, b. Koh-i-noor, Indian cut.
5a, b, c. Koh-i-noor, new form.   6. Stewart (from South Africa).
7b, c. Mr. Dresden's (from Brazil).

same octahedral cleavage as diamond, and by piecing the portions together arrived at the conclusion that the rough stone had the size of a hen's egg, the form of a rhombic dodecahedron, and a weight of about 793⅝ carats, which agrees closely with Tavernier's account.

Opinions differ also as to the derivation of the name " Koh-i-noor," which is sometimes said to signify " Mountain of Light," and is supposed to have been given to the stone by Nadir Shah. It has also been supposed to be a corruption of Kollur, the locality at which it was found, and the name by which it is said to have been formerly known in India.

The **Orloff** is the largest of the diamonds comprised in the Russian crown jewels, and usually forms the termination of the imperial sceptre; it is a stone of the finest water, perfectly pure and with a brilliant lustre. In form (Plate X., Figs. 1a, 1b, 1c,) it is very similar to that of Tavernier's drawing of the " Great Mogul," being an almost hemispherical rosette bounded on the lower side by a cleavage surface, as was the case with the Indian-cut " Koh-i-noor." Its height is 10 lines, its greatest diameter 15½ lines, and its weight 194¾ carats. This stone has had a chequered career; it is said at one time to have formed one of the eyes of an idol in the Brahmin temple on the island of Sheringham, in the Cauvery river near Trichinopoly. It was stolen from here, at the beginning of the eighteenth century, by a French soldier, passed into the hands of an English ship's captain, and so found its way into Europe, and in 1791 was bought in Amsterdam (being on this account sometimes known as the " Amsterdam " diamond) by Prince Orloff for the Empress Catharine II. of Russia for the sum of 1,400,000 Dutch florins.

There is a story to the effect that this stone came into the possession of the Russian crown through an Armenian, named Schafras; this story probably, however, applies not to the " Orloff" but to another large diamond in the Russian crown jewels, namely, the **Moon of the Mountains**. This diamond, which weighs 120 carats, became the booty of Nadir Shah, who used it for the adornment of his throne. At his assassination, it was stolen with other jewels by an Afghan soldier, from whom it passed into the possession of the Armenian Schafras. The latter sold it in 1775 to Catharine II. for 450,000 roubles, an annuity of 4000 roubles, and letters of nobility.

The **Polar Star**, a beautiful brilliant of 40 carats (Plate XI., Fig. 15), also belongs to the Russian crown, as does the peculiarly shaped stone known as the **Shah**. This latter stone was presented in 1829 to the Czar Nicholas by the Persian prince, Chosroes, the younger son of Abbas Mirza. It is of the purest water and in form a very irregular prism (Plate X., Figs. 3a, 3b), 1 inch 5½ lines long and 8 lines wide in the thickest part. The boundaries of the stone are partly cleavage faces and partly artificially cut facets; on three of the latter the names of three Persian kings are engraved, so that the " Shah " is one of the few examples of engraved diamonds. Professor Gustav Rose, who saw the stone soon after it was brought to St. Petersburg, gave the weight as 88 carats, but this does not agree with a subsequent statement to the effect that the stone has been re-cut and its weight reduced from 95 to 86 carats, the interesting inscriptions being lost in the process.

Another engraved diamond is the **Akbar Shah**, so called from its first possessor, the Great Mogul, Akbar; when in the possession of Jehan, Akbar's successor, Arabic inscriptions were engraved on two of its faces. It subsequently disappeared for a long period, reappearing again in Turkey, under the name of the " Shepherd's Stone," comparatively recently, and still recognisable as the " Akbar Shah " by its Arabic inscriptions. It at first weighed 116 carats, but after re-cutting in 1866 its weight was reduced to 71 or 72 carats and the inscriptions were lost in the process. In 1867 the stone was sold to the Gaikwar of Baroda for £35,000.

One of the largest of Indian diamonds is the **Nizam**, a stone of 277 carats, which has been known only since 1835, and which is supposed to have been picked up by a child on the ground in the neighbourhood of Golconda. This, however, is not the only version of the discovery of this stone, and its original weight has been placed at 440 carats; it is supposed to be at present in the possession of the Nizam of Haidarabad.

The **Great Table**, of Tavernier, was seen in 1642 at Golconda by this traveller, who states that it weighed $242\frac{3}{16}$ carats, and that it was the largest diamond he had seen in

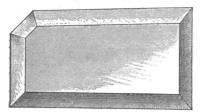

India in the hands of dealers. His offer of 400,000 rupees for the stone was rejected and, as in the case of the " Great Mogul," its subsequent history is obscure.

The Shah of Persia is in the possession of two large diamonds of which also very little is known. One of these, the **Darya-i-noor** (Sea of Light), weighs 186 carats, and the other, **Taj-e-mah** (Crown of the Moon), weighs 146 carats. Both are of the purest water and cut as rosettes; they were formerly set in a pair of armlets which were valued at one million sterling.

FIG. 47.   The "Great Table," a large Indian diamond mentioned by Tavernier.

A large diamond of singular beauty, perhaps the most perfect of all, is the **Regent** or **Pitt**, at present preserved with the French crown jewels. In its rough condition it was the largest of all Indian diamonds, the genuineness of which is unquestionable. It was found in 1701 in the Partial mines on the river Kistna in southern India (or according to another account in the Malay Peninsula), and was bought for £20,400 by Governor Pitt of Fort St. George, Madras. In 1717 it was acquired in its rough state by the Duke of Orleans, then Regent of France, for 2,000,000 francs (£80,000). The operation of cutting was performed in London; it occupied two years and cost £5000; the weight of the stone was reduced from 410 to $136\frac{11}{16}$ carats, and the portions detached in the cuttings remained the property of the former owner. The stone when cut (Plate XI., Figs. 8a, 8b, 8c), was a brilliant of the most perfect form; its colour, however, does not reach the same high standard of perfection. In the valuation of the French crown jewels, made in 1791, this diamond was stated to be worth 12,000,000 francs (£480,000). In 1792 it was stolen in company with many other crown jewels, but was subsequently recovered, and after being pledged at the time of the Revolution was redeemed by Napoleon. Being an object of general interest, it was not disposed of with the other crown jewels, but has remained up to the present time one of the most beautiful and valuable of the jewels belonging to the French nation.

The **Florentine**, " Grand Duke of Tuscany " or the " Austrian," is a large diamond now in the treasury of the imperial palace at Vienna. It has the form of a briolette (Plate XI., Figs. 10a, 10b), with the facets arranged in nine groups radiating from the centre. Its weight is $133\frac{1}{5}$ Vienna carats (27·454 grams), the weight of $139\frac{1}{2}$ carats, which is sometimes given, being in the smaller Florentine carats. The stone though distinctly yellow in colour, is beautifully clear and shows a fine fire. According to the usual but disputed account, this stone was cut by L. van Berquen for Charles the Bold, who lost it on the battlefield of Granson, where it was found by a Swiss soldier. After frequently changing hands, it passed into the possession of the Grand Duke Francis Stephen of Tuscany, who brought it to Vienna, its present resting-place.

The **Sancy** diamond, a stone of $53\frac{1}{2}$ carats, though much smaller is very similar in form to the " Florentine," and is also stated to have been cut by L. van Berquen for Charles the Bold. At the death of the latter, at the battle of Nancy in 1477, the stone is supposed to have been taken by a soldier to Portugal, where it was acquired by the French nobleman de Sancy, who about 1600 sold it to Queen Elizabeth of England. It was carried back to France by Henrietta Maria, the Queen of Charles I., and passed into the possession of Cardinal Mazarin as a pledge. Together with seventeen other large diamonds it was left by the latter to Louis XIV., and in the inventory, made in 1791, of the French crown jewels was valued at 1,000,000 francs. At the time of the Revolution it was stolen in company with the "Regent" ("Pitt"), but unlike the latter was not recovered. It reappeared ten years later as the property of the Spanish crown; from 1828 to 1865 it was in the possession of Prince Demidoff, by whom it was sold for £20,000. It is now said to be the property of the Maharaja of Patiala, and so, after many vicissitudes, to have returned to the land of its origin. So many stories are related of this stone that it seems not improbable that the history of other large diamonds has been confused with that of the "Sancy." It was exhibited at the Paris Exhibition of 1867, and is figured in Plate XI., Figs. 11a, 11b.

The **Nassak** diamond derives its name from its long sojourn in the temple to Siva at Nassak on the upper Godavari river. From the possession of the last independent Prince of Peshawar, it passed in 1818 into the hands of the East India Company. At that time it was of an unsymmetrical cut-form and weighed 89½ carats; the form in which it was re-cut, namely, that of a triangular brilliant, is shown in Plate XI., Figs. 13a, 13b, 13c. In 1831 it was bought for £7200 by Emanuel, a London jeweller, who soon afterwards disposed of it to the Duke of Westminister, in whose family it still remains.

The **Empress Eugénie** diamond is a beautiful brilliant of unknown origin, weighing 51 carats. It was given by Catharine II. of Russia to her favourite, Potemkin, in whose family it remained until it was acquired by Napoleon III. for a wedding-gift to his bride Eugénie. After the dethronement of the latter it came into the possession of the Gaikwar of Baroda in India.

The **Pigott** is a brilliant brought by Lord Pigott from India to England about the year 1775, and afterwards disposed of to Ali Pasha, the Viceroy of Egypt. All trace of this stone has since been lost, and, according to report, it has been destroyed. Its weight is given by Mawe, who saw the stone shortly before it was sold to Ali Pasha, as 49 carats, but other values up to 81½ carats have been given at various times.

The **White Saxon Brilliant** is one of the most beautiful of known diamonds; it is square in outline with an edge measuring $1\frac{1}{12}$ inches in length, and weighs 48¾ carats. For this stone August the Strong is said to have paid 1,000,000 thalers.

The **Pasha of Egypt** is a fine eight-sided brilliant of 40 carats, purchased by the Viceroy Ibrahim of Egypt for £28,000.

The comparatively small diamond known as the **Star of Este** surpasses in beauty many of those already mentioned. Its intrinsic beauty is absolutely flawless, and the brilliant form in which it is cut is as perfect. Its weight is $25\frac{13}{32}$ Vienna carats (5232 milligrams), only about half the weight, that is to say, of the "Empress Eugénie" or the "Sancy" diamond. Compared with these stones, however, it does not appear sensibly smaller, so perfect are its proportions and so regular the cutting It is at present in the possession of the Archduke Franz Ferdinand of Austrian-Este, eldest son of the Archduke Karl Ludwig. In 1876 it was valued at 64,000 Austrian florins, a former valuation having been 200,000 to 250,000 francs.

Excluding the yellow South African diamonds, stones which combine large size with beauty of colour are rare and are all of Indian origin. Of these the following are most famous:

The **Hope Blue** diamond is characterised not only by the possession of a beautiful sapphire-blue colour—an extremely rare tint in diamonds—but also by a brilliant lustre and fine play of colours. Its existence has been known of since 1830, and it at one time formed part of the famous collection of precious stones of Henry Philip Hope, who bought it for £18,000. It is a perfect brilliant weighing $44\frac{1}{4}$ carats.

A beautiful blue, triangular brilliant of $67\frac{2}{16}$ carats, and valued in 1791 at 3,000,000 francs, was preserved among the French crown jewels up to the year 1792, when it was stolen, together with the "Regent" and others. It had been cut from a rough stone, weighing $112\frac{3}{16}$ carats, brought from India by Tavernier for Louis XIV. There are substantial grounds for the suggestion that when this brilliant was stolen it was divided, and the portions re-cut and placed on the market about 1830 in a new form. It is very possible that the "Hope Blue" diamond is one of these portions; another being a stone of $13\frac{3}{4}$ carats of the same blue colour, and formerly in the possession of Duke Karl of Brunswick, who sold it in 1874 in Geneva for 17,000 francs; the third portion may be identical with a stone of $1\frac{1}{4}$ carats of the same colour, once bought for £300 and now in the possession of an English family.

The **Dresden Green** diamond, preserved in the "Green Vaults" of Dresden, is the most famous representative of stones of this colour. It is of a very fine clear apple-green, intermediate between the colour of emerald and chrysoprase, perfectly transparent and faultless in every way. It is almond-shaped in form, being $1\frac{1}{12}$ inches long and $\frac{5}{8}$ inch thick, and weighs 40 carats, not, as is sometimes stated, $31\frac{1}{4}$ or 48 carats. Since 1743 it has been the property of the Saxon crown, and 60,000 thalers is said to have been paid for it by August the Strong.

Very few diamonds famous for their size have come from Brazil, the only important exceptions being two stones found, in the 'fifties of the nineteenth century, in the district of Bagagem, in the western part of Minas Geraes, both of which were acquired by the Gaikwar of Baroda, a purchase which would seem to indicate that India can no longer satisfy the taste of her native princes for gorgeous jewels.

The **Star of the South,** found at the end of July 1853, is one of these two famous Brazilian diamonds. The rough stone, which was examined by the French mineralogist,

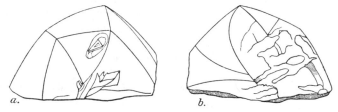

FIG. 48. "Star of the South." Two views of the rough stone. (Natural size.)

Dufrénoy, was described as being an irregular rhombic dodecahedron with convex faces (Fig. 48), and as weighing $254\frac{1}{2}$ carats. The stone showed in a few places small octahedral impressions of other diamonds, as if the larger diamond had once formed one of a group of crystals; in other places the octahedral cleavage was discernible. A few small black plates enclosed in the stone have been considered to be ilmenite (titaniferous iron ore), since this mineral has been shown to occur as an enclosure in diamond. The rough stone fetched 430 contos de reis, about £40,000. It was cut in Amsterdam, and produced a beautiful pure brilliant of 125 carats (Plate XI., Figs. 9a, 9b, 9c), which was bought by the Gaikwar of Baroda for £80,000.

Mr. E. Dresden's diamond was found at the same place as the last-mentioned

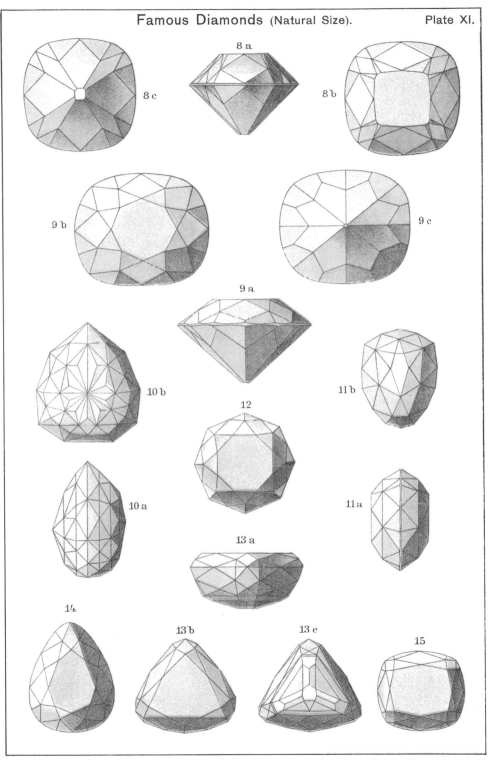

8a, b, c. Regent. 9a, b, c. Star of the South (from Brazil). 10a, b. Florentine. 11a, b, c. Sancy. 12. Pasha of Egypt. 13a, b, c. Nassak. 14. Star of South Africa. 15. Polar Star.

stone, and almost at the same time. It weighed, in the rough, 119½ carats, and was therefore smaller than the " Star of the South," and its appearance suggested that it might be a fragment of a larger crystal. It was transformed into an egg-shaped brilliant (Plate X., Figs. 7b, 7c) of 76½ carats, the process of cutting not involving in this case a very large loss of material.

It has already been mentioned that the supposed large diamond, the " Braganza," came from Brazil. Some other large Brazilian diamonds have been mentioned above under the description of Brazilian deposits, one of these being the large stone found at the beginning of the nineteenth century on the Rio Abaété, in Minas Geraes, as to the history of which nothing is known.

Only a few of the large diamonds which have been found in South Africa are distinguished by special names. Some of these were discovered and named before the comparative abundance of large stones in these deposits was known; others, however, so far surpass other large diamonds in size and beauty that it is only fitting that they should receive distinctive names. Some of these diamonds have already been mentioned under the description of the South African deposits.

The first large diamond found in this country was discovered in 1869 in the river diggings, and is known as the **Star of South Africa.** It weighed, in the rough, 83½ carats, and formed, when cut, an oval, three-sided brilliant (Plate XI., Fig. 14) of 46½ carats of the purest water, comparable with the best Indian and Brazilian stones. It was sold to the Countess of Dudley for nearly £25,000, and is therefore sometimes referred to as the " Dudley" diamond.

The **Stewart,** a much larger stone, was found in 1872 in the river diggings, known as Waldeck's Plant, on the Vaal. It weighed, in the rough, 288⅜ carats, and for many years remained the largest of Cape diamonds. The rough stone was first disposed of for £6000, but on again changing hands made £9000; it gave a slightly yellowish brilliant of 120 carats (Plate X., Fig. 6).

The **Porter Rhodes** diamond was found at Kimberley on February 12, 1880. Its weight in the rough has been variously given at 150 and 160 carats. It is a perfectly colourless blue-white stone, and, on the whole, may be considered to surpass all other South African diamonds in beauty. It was valued by its owner at £200,000.

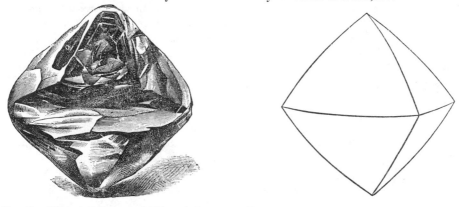

FIG. 49. " Victoria" diamond of 457½ carats from South Africa. (Actual size.)   FIG. 50. Outline of a diamond of 428½ carats from South Africa. (Actual size.)

A stone of 457½ carats, from the South African deposits, reached Europe in 1884, of which nothing as to its exact origin is known. The rough stone, which had the form of an irregular octahedron, is shown in its actual size in Fig. 49.   A very beautiful colourless brilliant of 180 carats was cut from it, which is variously known as the **Victoria,**

"Imperial," or "Great White," and is valued at £200,000.

The largest brilliant, the genuineness of which is unquestionable, is one of 288½ carats, which was cut from a stone of 428½ carats found on March 28, 1880, in the De Beers mine.   This was yellowish in colour and had the form of a fairly regular octahedron, the outline of which is shown in Fig. 50 in its actual size. In the direction of its longest axis it measured 1⅞ inches. Another large diamond, which weighed in the rough 655 carats, was found in the Jagersfontein mine at the end of the year 1895.

The largest of all known diamonds is the **Excelsior,** afterwards called the "Jubilee," in honour of the celebration of the sixtieth anniversary of the accession of Queen Victoria.   The rough stone is represented in its actual size and form in Fig. 51.   It came from the Jagersfontein mine in Orange River Colony, and weighed 971¼ carats, measuring 2½ inches in length,

FIG. 51.   Largest known diamond, the "Excelsior," weight 971¾ carats.   From the Jagersfontein mine in South Africa. (Actual size.)

2 inches in breadth, and 1 inch in thickness, thus surpassing in size even the "Great Mogul," which in its rough condition is supposed to have weighed 787½ carats.   It was found on June 30, 1893, by a Kaffir, who received as a reward £500 in money and a horse equipped with saddle and bridle.   It is said that an agreement existed between the mine-owners and certain diamond merchants by which the latter were to purchase every stone found in the mine during a certain period at a uniform price per carat.   This period ended on June 30, and the "Excelsior" was one of the last stones to be found on that day, so that the mine-owners instead of the merchants came very near to profiting by this lucky find.   The stone is of a beautiful blue-white colour and of the purest water, and has been valued by different experts at amounts which vary between £50,000 and £1,000,000; the latter value, however, seems somewhat prohibitive.   The rough stone, though of such perfection of colour, lustre, and water, had a black spot near the centre of its mass which had to be removed by cleaving the stone in two.   From the larger portion was cut an absolutely perfect brilliant weighing 239 international carats of 205 milligrams, and measuring 1⅝ inches in length, 1⅜ in breadth, and 1 inch in depth.

FIG. 52.   The "Tiffany Brilliant," 125½ carats.   (Actual size.)

The orange-yellow **Tiffany Brilliant,** now in the possession of the Tiffany Company of New York, is also a Cape diamond.   It is one of the finest of yellow diamonds, and at the present time is the largest brilliant in America, weighing 125½ carats.   The form of the stone can be seen in Fig. 52 in its actual size.

5. VALUE OF DIAMONDS.

The valuation of a diamond, involving as it does a nice appreciation of the defects and of the good points of the stone, and the striking of a just balance between the two, is a matter of no little difficulty, and can only be performed with accuracy and rapidity by an expert. In this section we shall confine ourselves to a consideration of the value of diamonds which are to be used as gems, neglecting those to be applied to technical purposes, the value of which depends on the weight and the current market price.

Of all the characters which help to determine the value of a diamond there is perhaps none more potent than that of size. Other things being equal, the larger the diamond the greater its value, and, moreover, the ratio of progression in price is greater than that of progression in weight, owing to the comparative rarity of large stones. Since the discovery of the South African deposits, however, this disparity has been less marked, and the value of stones not exceeding a certain size and which are of frequent occurrence, is influenced to a large extent by the exigencies of the trade. Exceptionally large and beautiful stones, the so-called solitaires, paragons, or nonpareils, have, corresponding to the rarity of their occurrence, an exceptional value, which is subject to no rules and is governed solely by the special circumstances of the case.

The value of a diamond depends very largely upon the form in which it is cut. Although during the process of cutting the weight of a rough stone is reduced by one half or even more, yet its intrinsic value is greater than before, on account of the almost immeasurable improvement in its appearance effected by the faceting. The brilliant is by far the most effective form of cutting, and at the same time is the form which involves the greatest expenditure of skill in the cutting, hence a brilliant-cut diamond commands a higher price than a rose or indeed any other form. Among brilliants themselves different degrees are recognisable, a stone which is correctly proportioned and which bears a large number of facets having a greater value than one less admirable in these respects. A brilliant which possesses no cross facets, the large facets being produced until they meet in the girdle, is described as being " once formed " ; while the terms " twice formed " and " thrice formed " are applied respectively to stones which bear cross facets only below the girdle, and to those which possess these facets both above and below the girdle. The value of a brilliant, therefore, is the greater the more complex is its form of cutting, and in the same way the value of stones cut in any of the other forms varies with the symmetry and completeness of that form. A perfect brilliant of one carat has at least four times the value of a rough stone of the same weight and quality, and five-fourths the value of a rose of this size and quality.

The value of a rough stone also is influenced to a certain extent by its form, for, as we have seen, stones whose form in the rough approximates most nearly to that of the cut stone are most favourable for cutting. Thus octahedral and rhombic dodecahedral crystals can be fashioned into brilliants with less labour and loss of material than is the case with irregularly shaped stones, which often need considerable preliminary shaping, if not actual division into portions suitable for cutting. Among such stones must be included flat specimens, like the twinned crystals shown in Fig. 31, *g* and *h*, which cannot be cut as brilliants and are suitable only for cutting as roses. Another property which greatly facilitates the process of cutting is that of cleavage ; a simple crystal, from which the cleavage octahedron can be readily developed, is therefore far more desirable than a twinned crystal, such as is shown in Fig. 31 *i*, or an irregular crystal group which, as often as not, can be utilised only as bort.

The value of a diamond depends most of all, however, on the degree of its transparency, clearness, and purity, the colour it possesses, and its freedom from flaws.   Of these qualities transparency and clearness stand first in importance, and the possession of these qualities in perfection renders a diamond extremely valuable.

Those faults which impair the transparency and lustre of a stone diminish its value very considerably.ʹ   Large enclosures of black, brown, or of some other colour are frequently seen, as are also enclosures of " sand " and " ash," and yellow spots technically known as " straw."   A fine surface polish over certain areas of a stone is often made impossible by the presence of white, grey, or brown " clouds " or by " icy flakes " of no definite colour, which are developed when the stone is allowed to become over-heated during the process of grinding.   The existence of internal cracks following the direction of cleavage, and known as " feathers," not only impairs the transparency of the stone, but also renders it liable to fracture during the process of grinding or when in use as an ornament.   All these faults, even if insignificant in extent, become very obvious in the cut stone, numerous images of them being reflected into the eye of the observer from the various facets of the stone.   Should they be present in large numbers the stone is not worth cutting, but is regarded as bort.

With regard to the colour of diamonds, stones which are perfectly colourless and water-clear are, as a rule, most highly prized, the so-called blue-white quality, which is more rare in stones from the Cape than in those from India or Brazil, being specially admired. Even a trace of colour, so small as to be indistinguishable to an unpractised eye, lowers the value of a stone very considerably, the diminution in value being still greater when the colour is more perceptible.   Of coloured diamonds, those displaying tones of blue, grey, red, and yellow are preferred to those which are coloured brown or black.   A coloured diamond which is lacking in transparency is of very much less value than one of the same colour which is clear and transparent.

Those diamonds which, in addition to perfect transparency and clearness, possess a pronounced and beautiful colour, are on account both of their rarity and beauty very highly esteemed, and always command a much higher price than the most perfect of colourless specimens.   Among these so-called " fancy stones," red, blue, green, and yellow specimens are included, the last-named, however, since the discovery of the Cape deposits, are by far the most common.   Compared with colourless diamonds, coloured specimens exist in quite insignificant numbers.

Diamonds showing different degrees of transparency and clearness and freedom from faults are usually classified as stones of the first, second, and third water, and are valued accordingly.   Stones of the *first water* (1st quality) are perfectly colourless, transparent, and water-clear ; they are free from any fault or blemish or tinge of colour and stand first in point of value.   Colourless stones showing insignificant faults, or stones which are free from faults, but tinged with colour, are placed in the second division and referred to as stones of the *second water ;* while stones of the *third water* display very obvious faults or a colour of undesirable depth.   A further division of the stones of the latter description is sometimes made, and in this class are placed the smallest diamonds which can be used as gems.   It is by no means easy, however, in every case to place any given stone without hesitation in one or other of these three or four classes, and it may often be observed that a stone referred to as being of the second water by one jeweller will be placed in the first class by another.   Generally speaking, it may be said that a brilliant of the second water has only about two-thirds of the value of a similar gem of the first water ; while the values of two roses of the first and second qualities are in the ratio of four to three.

Taking the value of a brilliant of the first water as unity, that of a similar brilliant of the second water will be $\frac{2}{3}$, while the values of roses of the first and second water will be expressed by the fractions $\frac{4}{5}$ and $\frac{3}{5}$. It may be remarked here that it is almost impossible to classify rough stones in this way, since the qualities on which the classification depends are not sufficiently obvious until the stone has been cut.

It appears from the writings of Pliny, that among the ancients the diamond was regarded as the most costly of precious stones, and indeed of all personal possessions. Such, however, is not the case at the present time, for the price of a colourless diamond of good size is always exceeded by that of a ruby of the same size, and generally also by that of an emerald, or even of a blue sapphire if of special beauty. This, of course, does not apply to the few diamonds which possess a fine colour in addition to their other beautiful qualities, the price of such stones being more or less prohibitive.

While the relative value of diamonds of different qualities changes but little, the absolute prices paid depend on a variety of conditions and are subject to considerable fluctuation.

The earliest record in existence of the price of a diamond is that made by the Arabian Teifaschius, who, in the twelfth century, valued a 1 carat diamond at 2 dinar (about £6). In the year 1550, Benvenuto Cellini placed the value of a beautiful stone of the same weight at 100 golden scudi, a sum which is stated by Schrauf to be equivalent to 200 Austrian florins (£20), and by Boutan to 1100 francs (£44). This latter value is abnormally high, and is probably based on an incorrect estimate of the value of the scudi. In 1609 Boetius de Boot gave the value of the carat-stone at 130 ducats (about £22), while the price mentioned in the anonymous work, *The History of Jewels*, published in London in 1672, is from 40 to 60 crowns (£8 to £12). This large fall in the value of the diamond is probably to be attributed to the effects of the Thirty Years War. According to Tavernier, the price of a carat-stone in 1676 was £8, and this statement is confirmed by contemporary writers both in Holland and at Hamburg. The price of rough diamonds had sunk in 1733 to £1 per carat, but this fall was due to the panic which followed the discovery of diamonds in Brazil. In the next year the price of the carat-stone had risen to £1 10s., at which it stood for several years subsequently. In 1750 the famous London jeweller, David Jeffries, the author of a *Treatise on Diamonds and Pearls*, records the value of a fine one-carat cut stone at £8, which is the same as the value given by Tavernier in 1676. In a work on precious stones, entitled *Der aufrichtige Jubelier*, published at Frankfurt-on-the-Main in 1772, the high price of 120 thalers (£18) is mentioned for a stone of the same description.

At the time of the French Revolution prices fell very considerably, and as far as can be ascertained from the valuation of the French crown jewels and from the prices fetched by the many less valuable stones which changed hands at this time, it would seem that in 1791 a one-carat cut stone would fetch no more on an average than £6. When more settled times came, however, and Napoleon's luxurious court was established, the price again rose, and in 1832 £9 could be obtained for a one-carat brilliant, and rough stones of a quality suitable for cutting fetched 42s. to 48s. or even £3 per carat. Later on still, in the year 1859, rough stones of the same description were worth from £4 to £5 5s. per carat, while in 1860 and 1865 £13 to £18 was paid for a one-carat cut stone.

In the year 1869, shortly before the Cape diamonds came on the market, the following prices, according to Schrauf, were current: rough stones suitable for cutting, and similar to those which come in large parcels from the countries in which they are mined, cost £5 per carat; parcels of stones, the larger proportion of which could be used only as bort, made £1 to £2 per carat; while parcels containing nothing but bort were sold for 4s. to 6s. per carat. The prices recorded for cut stones show the importance which was attached not only to the quality of a stone but also to the form and manner in which it had been cut.

A one-carat brilliant of the first water was worth £20 to £25, one of the second water £15, while one-carat roses of the first water were worth only £15 to £18; a brilliant of $\frac{1}{2}$ carat would fetch £6, one of $\frac{3}{4}$ carat £12, and one of $\frac{1}{10}$ carat £1; for small roses, of which 50 go to the carat, £15 per carat was paid; very small roses of about 1000 to the carat cost about 6d. each.   Only at most prosperous times, in the sixteenth and at the beginning of the seventeenth centuries, were such high prices paid as were current for diamonds in 1869.   In the following table, compiled by L. Dieulafait, may be seen the prices in francs (25 francs = £1) which were paid for brilliants of 1 to 5 carats in the years 1606, 1750, 1865, and 1867.   The prices current in the year 1878, which are given further on, are incorporated in this table in order to show the fall which took place in consequence of the discovery of the South African diamond-fields, and which followed a steady rise in the years 1867 to 1869.

Brilliant of	1606.	1750.	1865.	1867.	1878.
1 carat	545	202	453	529	110
2 ,,	2182	807	1639	2017	350
3 ,,	4916	1815	3151	3529	625
4 ,,	6554	2470	—	—	975
5 ,,	8753	5042	8067	8823	1375

The prices current for brilliants of ordinary size at the end of the 'seventies is best seen from the following table, which was compiled by Vanderheym, on behalf of the syndicate of Parisian jewellers, for the Paris Exhibition of 1878.   Two brilliants of weights from $\frac{1}{2}$ to 12 carats and of four qualities were exhibited, and the prices in francs given in the table are for the pair of stones :

Weight in carats.	1st quality.	2nd quality.	3rd quality.	4th quality.
1	220	180	150	120
1$\frac{1}{2}$	400	300	250	200
2	700	600	480	400
2$\frac{1}{2}$	950	800	625	525
3	1250	1020	780	660
3$\frac{1}{2}$	1600	1225	945	720
4	1950	1440	1120	960
4$\frac{1}{2}$	2350	1642	1305	1080
5	2750	1900	1500	1250
5$\frac{1}{2}$	3250	2117	1705	1430
6	3700	2340	1920	1620
6$\frac{1}{2}$	4250	2567	2112	1820
7	5000	2765	2310	1995
7$\frac{1}{2}$	5800	3000	2550	2175
8	6700	3240	2800	2360
8$\frac{1}{2}$	7600	3485	3060	2550
9	8500	3735	3330	2700
9$\frac{1}{2}$	9400	3990	3562	2897
10	10300	4250	3800	3050
10$\frac{1}{2}$	11400	4515	4042	3255
11	12500	4840	4290	3465
11$\frac{1}{2}$	13700	5175	4600	3737
12	15000	5400	4800	3900

The prices given in the above table of course apply only to the time at which it was compiled. A striking feature of the table is the difference which exists between the prices of stones of the same weight but of different qualities, especially in the case of stones of the first and second waters. The difference between the value of a 1-carat stone of the first water and one of the second water is much greater than between stones of the second and third waters, and in larger stones the difference is still greater. Thus a 12-carat stone of the first water is worth almost three times as much as a stone of equal weight of the second water, the values of stones of this size of the second and third quality being in the ratio of nine to eight. The explanation of the apparent anomaly lies in the fact that in the Cape deposits large diamonds of the first water are rare, while stones of large size but inferior quality are abundant.

A consideration of the table will also show to what a small extent the values of diamonds at the present day are in agreement with the so-called Tavernier's rule, according to which the value of a stone is proportional to the square of its weight. While the value of a 12-carat stone of the first quality would be, according to Tavernier's rule, $110 \times 12 \times 12 = 15,840$ francs, its actual value in 1878, according to the table, was 7500 francs, or not quite half. The application of the rule to smaller stones results in a calculated value which is still further removed from the actual value; thus the value of a 6-carat diamond of the first water calculated by this rule would be $110 \times 6 \times 6 = 3960$ francs, while it is actually worth but 1850 francs. At the present time, this tendency is even more marked than it was in 1878; the value of stones up to 15 carats is approximately proportional to their weight, so that a 2-carat stone is worth about double, and a 3-carat stone about three times as much as a 1-carat diamond. This holds good, at any rate, for the three inferior qualities of stones, but in the case of diamonds of the first water the increase in value is not proportional to the increase of weight.

The price of a 12-carat stone of the first water calculated by Schrauf's rule, according to which the value of a 1-carat stone is multiplied by the product of half the weight of the stone into its weight plus 2, would be $110 \times 6 \times 14 = 9240$ francs, the tabulated value being 7500 francs; the value thus calculated, although nearer the mark than in the former case, is still considerably too much. As in the case with Tavernier's rule, the values calculated by Schrauf's rule for smaller stones are still further from their actual value, the calculated worth of a 6-carat stone being $110 \times 3 \times 8 = 2640$ francs while it is actually worth but 1850 francs. At the present time the market price of a fine 1-carat brilliant is £15; in exceptional cases, however, £20 to £25 may be given for such a stone.

The price of stones of exceptional size, that is of those weighing anything over 12 carats, is not governed by rule, and depends very much on what a rich person or State is disposed to give for them. Diamonds of exceptional size and of unusual colours are not common articles of commerce, and their price, while always, of course, very high, depends on the number of would-be purchasers which can be found for them.

With regard to the prices current for smaller diamonds, it is impossible to say much more than has been already said, for, after all, the value of stones of ordinary size depends to a very large extent on their quality. The price of cut gems and of rough stones always differs very widely; the latter are not, as a rule, bought and sold singly but come into the markets in large parcels, those from the Cape being carefully sorted and arranged according to quality, while parcels from Brazil consist of unsorted stones of all qualities.

6. IMITATION AND COUNTERFEITING.

Attempts have often been made by unprincipled dealers to pass off stones of little value or worthless imitations as genuine diamonds. The gems of inferior value most frequently used for this purpose are colourless topaz, zircon rendered colourless by heating, white sapphire, spinel, beryl, tourmaline, phenakite, and even rock-crystal and other minerals. In all these stones, however, the beautiful play of prismatic colours so characteristic of the diamond is far less marked, as is also, except, perhaps, in the case of colourless zircon, the peculiarly high lustre of the diamond. No one in the least degree familiar with the appearance of the diamond would for a moment confuse it with any of the stones just mentioned. Among the physical characters by which the diamond may be distinguished from other colourless gems are hardness, specific gravity, and refraction of light, the spinel alone of the minerals mentioned above being singly refracting like the diamond. The diamond is much less frequently confused with coloured than with colourless gems.

In absence of colour, in transparency, lustre, and play of prismatic colours, some kinds of glass, especially strass, resemble the diamond with astonishing closeness. This material is, therefore, largely used in the manufacture of imitation diamonds, and so closely does the appearance of a piece of freshly-cut strass simulate that of a genuine diamond that it is possible even for an expert to be deceived. The genuineness, or otherwise, of such a stone can, however, be easily and conclusively proved by a simple test of the hardness with a steel point or file.

The construction of so-called doublets for the purpose of deception is by no means infrequent. In such cases the upper portion of the brilliant is of diamond, while the lower is of glass or of some colourless stone such as white sapphire. The device by which the yellow tinge of a diamond is temporarily concealed, namely, by applying a thin coating of some blue substance, has been already referred to. The play of prismatic colours, characteristic of the diamond, is imitated with a certain amount of success by painting the under side of the counterfeit stone. Articles of this description, known as irises, have found a ready sale, without any attempt at passing them off for anything other than what they are.

A CATALOG OF SELECTED
DOVER BOOKS
IN ALL FIELDS OF INTEREST

A CATALOG OF SELECTED DOVER
BOOKS IN ALL FIELDS OF INTEREST

CONCERNING THE SPIRITUAL IN ART, Wassily Kandinsky. Pioneering work by father of abstract art. Thoughts on color theory, nature of art. Analysis of earlier masters. 12 illustrations. 80pp. of text. 5⅜ x 8½.   23411-8 Pa. $3.95

ANIMALS: 1,419 Copyright-Free Illustrations of Mammals, Birds, Fish, Insects, etc., Jim Harter (ed.). Clear wood engravings present, in extremely lifelike poses, over 1,000 species of animals. One of the most extensive pictorial sourcebooks of its kind. Captions. Index. 284pp. 9 x 12.   23766-4 Pa. $12.95

CELTIC ART: The Methods of Construction, George Bain. Simple geometric techniques for making Celtic interlacements, spirals, Kells-type initials, animals, humans, etc. Over 500 illustrations. 160pp. 9 x 12. (USO)   22923-8 Pa. $9.95

AN ATLAS OF ANATOMY FOR ARTISTS, Fritz Schider. Most thorough reference work on art anatomy in the world. Hundreds of illustrations, including selections from works by Vesalius, Leonardo, Goya, Ingres, Michelangelo, others. 593 illustrations. 192pp. 7⅛ x 10¼.   20241-0 Pa. $9.95

CELTIC HAND STROKE-BY-STROKE (Irish Half-Uncial from "The Book of Kells"): An Arthur Baker Calligraphy Manual, Arthur Baker. Complete guide to creating each letter of the alphabet in distinctive Celtic manner. Covers hand position, strokes, pens, inks, paper, more. Illustrated. 48pp. 8¼ x 11.   24336-2 Pa. $3.95

EASY ORIGAMI, John Montroll. Charming collection of 32 projects (hat, cup, pelican, piano, swan, many more) specially designed for the novice origami hobbyist. Clearly illustrated easy-to-follow instructions insure that even beginning papercrafters will achieve successful results. 48pp. 8¼ x 11.   27298-2 Pa. $3.50

THE COMPLETE BOOK OF BIRDHOUSE CONSTRUCTION FOR WOOD-WORKERS, Scott D. Campbell. Detailed instructions, illustrations, tables. Also data on bird habitat and instinct patterns. Bibliography. 3 tables. 63 illustrations in 15 figures. 48pp. 5¼ x 8½.   24407-5 Pa. $2.50

BLOOMINGDALE'S ILLUSTRATED 1886 CATALOG: Fashions, Dry Goods and Housewares, Bloomingdale Brothers. Famed merchants' extremely rare catalog depicting about 1,700 products: clothing, housewares, firearms, dry goods, jewelry, more. Invaluable for dating, identifying vintage items. Also, copyright-free graphics for artists, designers. Co-published with Henry Ford Museum & Greenfield Village. 160pp. 8¼ x 11.   25780-0 Pa. $10.95

HISTORIC COSTUME IN PICTURES, Braun & Schneider. Over 1,450 costumed figures in clearly detailed engravings–from dawn of civilization to end of 19th century. Captions. Many folk costumes. 256pp. 8⅜ x 11¾.   23150-X Pa. $12.95

FRANK LLOYD WRIGHT'S HOLLYHOCK HOUSE, Donald Hoffmann. Lavishly illustrated, carefully documented study of one of Wright's most controversial residential designs. Over 120 photographs, floor plans, elevations, etc. Detailed perceptive text by noted Wright scholar. Index. 128pp. 9¼ x 10¾. 27133-1 Pa. $11.95

THE MALE AND FEMALE FIGURE IN MOTION: 60 Classic Photographic Sequences, Eadweard Muybridge. 60 true-action photographs of men and women walking, running, climbing, bending, turning, etc., reproduced from rare 19th-century masterpiece. vi + 121pp. 9 x 12. 24745-7 Pa. $10.95

1001 QUESTIONS ANSWERED ABOUT THE SEASHORE, N. J. Berrill and Jacquelyn Berrill. Queries answered about dolphins, sea snails, sponges, starfish, fishes, shore birds, many others. Covers appearance, breeding, growth, feeding, much more. 305pp. 5¼ x 8¼. 23366-9 Pa. $8.95

GUIDE TO OWL WATCHING IN NORTH AMERICA, Donald S. Heintzelman. Superb guide offers complete data and descriptions of 19 species: barn owl, screech owl, snowy owl, many more. Expert coverage of owl-watching equipment, conservation, migrations and invasions, etc. Guide to observing sites. 84 illustrations. xiii + 193pp. 5⅜ x 8½. 27344-X Pa. $8.95

MEDICINAL AND OTHER USES OF NORTH AMERICAN PLANTS: A Historical Survey with Special Reference to the Eastern Indian Tribes, Charlotte Erichsen-Brown. Chronological historical citations document 500 years of usage of plants, trees, shrubs native to eastern Canada, northeastern U.S. Also complete identifying information. 343 illustrations. 544pp. 6½ x 9¼. 25951-X Pa. $12.95

STORYBOOK MAZES, Dave Phillips. 23 stories and mazes on two-page spreads: Wizard of Oz, Treasure Island, Robin Hood, etc. Solutions. 64pp. 8¼ x 11. 23628-5 Pa. $2.95

NEGRO FOLK MUSIC, U.S.A., Harold Courlander. Noted folklorist's scholarly yet readable analysis of rich and varied musical tradition. Includes authentic versions of over 40 folk songs. Valuable bibliography and discography. xi + 324pp. 5⅜ x 8½. 27350-4 Pa. $9.95

MOVIE-STAR PORTRAITS OF THE FORTIES, John Kobal (ed.). 163 glamor, studio photos of 106 stars of the 1940s: Rita Hayworth, Ava Gardner, Marlon Brando, Clark Gable, many more. 176pp. 8⅜ x 11¼. 23546-7 Pa. $12.95

BENCHLEY LOST AND FOUND, Robert Benchley. Finest humor from early 30s, about pet peeves, child psychologists, post office and others. Mostly unavailable elsewhere. 73 illustrations by Peter Arno and others. 183pp. 5⅜ x 8½. 22410-4 Pa. $6.95

YEKL and THE IMPORTED BRIDEGROOM AND OTHER STORIES OF YIDDISH NEW YORK, Abraham Cahan. Film Hester Street based on Yekl (1896). Novel, other stories among first about Jewish immigrants on N.Y.'s East Side. 240pp. 5⅜ x 8½. 22427-9 Pa. $6.95

SELECTED POEMS, Walt Whitman. Generous sampling from *Leaves of Grass*. Twenty-four poems include "I Hear America Singing," "Song of the Open Road," "I Sing the Body Electric," "When Lilacs Last in the Dooryard Bloom'd," "O Captain! My Captain!"—all reprinted from an authoritative edition. Lists of titles and first lines. 128pp. 5³⁄₁₆ x 8¼. 26878-0 Pa. $1.00

THE BEST TALES OF HOFFMANN, E. T. A. Hoffmann. 10 of Hoffmann's most important stories: "Nutcracker and the King of Mice," "The Golden Flowerpot," etc. 458pp. 5⅜ x 8½. 21793-0 Pa. $9.95

FROM FETISH TO GOD IN ANCIENT EGYPT, E. A. Wallis Budge. Rich detailed survey of Egyptian conception of "God" and gods, magic, cult of animals, Osiris, more. Also, superb English translations of hymns and legends. 240 illustrations. 545pp. 5⅜ x 8½. 25803-3 Pa. $13.95

FRENCH STORIES/CONTES FRANÇAIS: A Dual-Language Book, Wallace Fowlie. Ten stories by French masters, Voltaire to Camus: "Micromegas" by Voltaire; "The Atheist's Mass" by Balzac; "Minuet" by de Maupassant; "The Guest" by Camus, six more. Excellent English translations on facing pages. Also French-English vocabulary list, exercises, more. 352pp. 5⅜ x 8½. 26443-2 Pa. $8.95

CHICAGO AT THE TURN OF THE CENTURY IN PHOTOGRAPHS: 122 Historic Views from the Collections of the Chicago Historical Society, Larry A. Viskochil. Rare large-format prints offer detailed views of City Hall, State Street, the Loop, Hull House, Union Station, many other landmarks, circa 1904-1913. Introduction. Captions. Maps. 144pp. 9⅜ x 12¼. 24656-6 Pa. $12.95

OLD BROOKLYN IN EARLY PHOTOGRAPHS, 1865-1929, William Lee Younger. Luna Park, Gravesend race track, construction of Grand Army Plaza, moving of Hotel Brighton, etc. 157 previously unpublished photographs. 165pp. 8⅞ x 11¾. 23587-4 Pa. $13.95

THE MYTHS OF THE NORTH AMERICAN INDIANS, Lewis Spence. Rich anthology of the myths and legends of the Algonquins, Iroquois, Pawnees and Sioux, prefaced by an extensive historical and ethnological commentary. 36 illustrations. 480pp. 5⅜ x 8½. 25967-6 Pa. $8.95

AN ENCYCLOPEDIA OF BATTLES: Accounts of Over 1,560 Battles from 1479 B.C. to the Present, David Eggenberger. Essential details of every major battle in recorded history from the first battle of Megiddo in 1479 B.C. to Grenada in 1984. List of Battle Maps. New Appendix covering the years 1967-1984. Index. 99 illustrations. 544pp. 6½ x 9¼. 24913-1 Pa. $14.95

SAILING ALONE AROUND THE WORLD, Captain Joshua Slocum. First man to sail around the world, alone, in small boat. One of great feats of seamanship told in delightful manner. 67 illustrations. 294pp. 5⅜ x 8½. 20326-3 Pa. $5.95

ANARCHISM AND OTHER ESSAYS, Emma Goldman. Powerful, penetrating, prophetic essays on direct action, role of minorities, prison reform, puritan hypocrisy, violence, etc. 271pp. 5⅜ x 8½. 22484-8 Pa. $6.95

MYTHS OF THE HINDUS AND BUDDHISTS, Ananda K. Coomaraswamy and Sister Nivedita. Great stories of the epics; deeds of Krishna, Shiva, taken from puranas, Vedas, folk tales; etc. 32 illustrations. 400pp. 5⅜ x 8½. 21759-0 Pa. $10.95

BEYOND PSYCHOLOGY, Otto Rank. Fear of death, desire of immortality, nature of sexuality, social organization, creativity, according to Rankian system. 291pp. 5⅜ x 8½. 20485-5 Pa. $8.95

A THEOLOGICO-POLITICAL TREATISE, Benedict Spinoza. Also contains unfinished Political Treatise. Great classic on religious liberty, theory of government on common consent. R. Elwes translation. Total of 421pp. 5⅜ x 8½. 20249-6 Pa. $9.95

MY BONDAGE AND MY FREEDOM, Frederick Douglass. Born a slave, Douglass became outspoken force in antislavery movement. The best of Douglass' autobiographies. Graphic description of slave life. 464pp. 5⅜ x 8½. 22457-0 Pa. $8.95

FOLLOWING THE EQUATOR: A Journey Around the World, Mark Twain. Fascinating humorous account of 1897 voyage to Hawaii, Australia, India, New Zealand, etc. Ironic, bemused reports on peoples, customs, climate, flora and fauna, politics, much more. 197 illustrations. 720pp. 5⅜ x 8½. 26113-1 Pa. $15.95

THE PEOPLE CALLED SHAKERS, Edward D. Andrews. Definitive study of Shakers: origins, beliefs, practices, dances, social organization, furniture and crafts, etc. 33 illustrations. 351pp. 5⅜ x 8½. 21081-2 Pa. $8.95

THE MYTHS OF GREECE AND ROME, H. A. Guerber. A classic of mythology, generously illustrated, long prized for its simple, graphic, accurate retelling of the principal myths of Greece and Rome, and for its commentary on their origins and significance. With 64 illustrations by Michelangelo, Raphael, Titian, Rubens, Canova, Bernini and others. 480pp. 5⅜ x 8½. 27584-1 Pa. $9.95

PSYCHOLOGY OF MUSIC, Carl E. Seashore. Classic work discusses music as a medium from psychological viewpoint. Clear treatment of physical acoustics, auditory apparatus, sound perception, development of musical skills, nature of musical feeling, host of other topics. 88 figures. 408pp. 5⅜ x 8½. 21851-1 Pa. $10.95

THE PHILOSOPHY OF HISTORY, Georg W. Hegel. Great classic of Western thought develops concept that history is not chance but rational process, the evolution of freedom. 457pp. 5⅜ x 8½. 20112-0 Pa. $9.95

THE BOOK OF TEA, Kakuzo Okakura. Minor classic of the Orient: entertaining, charming explanation, interpretation of traditional Japanese culture in terms of tea ceremony. 94pp. 5⅜ x 8½. 20070-1 Pa. $3.95

LIFE IN ANCIENT EGYPT, Adolf Erman. Fullest, most thorough, detailed older account with much not in more recent books, domestic life, religion, magic, medicine, commerce, much more. Many illustrations reproduce tomb paintings, carvings, hieroglyphs, etc. 597pp. 5⅜ x 8½. 22632-8 Pa. $11.95

SUNDIALS, Their Theory and Construction, Albert Waugh. Far and away the best, most thorough coverage of ideas, mathematics concerned, types, construction, adjusting anywhere. Simple, nontechnical treatment allows even children to build several of these dials. Over 100 illustrations. 230pp. 5⅜ x 8½. 22947-5 Pa. $7.95

DYNAMICS OF FLUIDS IN POROUS MEDIA, Jacob Bear. For advanced students of ground water hydrology, soil mechanics and physics, drainage and irrigation engineering, and more. 335 illustrations. Exercises, with answers. 784pp. 6⅛ x 9¼. 65675-6 Pa. $19.95

SONGS OF EXPERIENCE: Facsimile Reproduction with 26 Plates in Full Color, William Blake. 26 full-color plates from a rare 1826 edition. Includes "The Tyger," "London," "Holy Thursday," and other poems. Printed text of poems. 48pp. 5¼ x 7. 24636-1 Pa. $4.95

OLD-TIME VIGNETTES IN FULL COLOR, Carol Belanger Grafton (ed.). Over 390 charming, often sentimental illustrations, selected from archives of Victorian graphics—pretty women posing, children playing, food, flowers, kittens and puppies, smiling cherubs, birds and butterflies, much more. All copyright-free. 48pp. 9¼ x 12¼. 27269-9 Pa. $7.95

PERSPECTIVE FOR ARTISTS, Rex Vicat Cole. Depth, perspective of sky and sea, shadows, much more, not usually covered. 391 diagrams, 81 reproductions of drawings and paintings. 279pp. 5⅜ x 8½. 22487-2 Pa. $7.95

DRAWING THE LIVING FIGURE, Joseph Sheppard. Innovative approach to artistic anatomy focuses on specifics of surface anatomy, rather than muscles and bones. Over 170 drawings of live models in front, back and side views, and in widely varying poses. Accompanying diagrams. 177 illustrations. Introduction. Index. 144pp. 8⅜ x11¼. 26723-7 Pa. $8.95

GOTHIC AND OLD ENGLISH ALPHABETS: 100 Complete Fonts, Dan X. Solo. Add power, elegance to posters, signs, other graphics with 100 stunning copyright-free alphabets: Blackstone, Dolbey, Germania, 97 more—including many lower-case, numerals, punctuation marks. 104pp. 8⅛ x 11. 24695-7 Pa. $8.95

HOW TO DO BEADWORK, Mary White. Fundamental book on craft from simple projects to five-bead chains and woven works. 106 illustrations. 142pp. 5⅜ x 8. 20697-1 Pa. $4.95

THE BOOK OF WOOD CARVING, Charles Marshall Sayers. Finest book for beginners discusses fundamentals and offers 34 designs. "Absolutely first rate . . . well thought out and well executed."–E. J. Tangerman. 118pp. 7¾ x 10⅝. 23654-4 Pa. $6.95

ILLUSTRATED CATALOG OF CIVIL WAR MILITARY GOODS: Union Army Weapons, Insignia, Uniform Accessories, and Other Equipment, Schuyler, Hartley, and Graham. Rare, profusely illustrated 1846 catalog includes Union Army uniform and dress regulations, arms and ammunition, coats, insignia, flags, swords, rifles, etc. 226 illustrations. 160pp. 9 x 12. 24939-5 Pa. $10.95

WOMEN'S FASHIONS OF THE EARLY 1900s: An Unabridged Republication of "New York Fashions, 1909," National Cloak & Suit Co. Rare catalog of mail-order fashions documents women's and children's clothing styles shortly after the turn of the century. Captions offer full descriptions, prices. Invaluable resource for fashion, costume historians. Approximately 725 illustrations. 128pp. 8⅜ x 11¼. 27276-1 Pa. $11.95

THE 1912 AND 1915 GUSTAV STICKLEY FURNITURE CATALOGS, Gustav Stickley. With over 200 detailed illustrations and descriptions, these two catalogs are essential reading and reference materials and identification guides for Stickley furniture. Captions cite materials, dimensions and prices. 112pp. 6½ x 9¼. 26676-1 Pa. $9.95

EARLY AMERICAN LOCOMOTIVES, John H. White, Jr. Finest locomotive engravings from early 19th century: historical (1804–74), main-line (after 1870), special, foreign, etc. 147 plates. 142pp. 11⅜ x 8¼. 22772-3 Pa. $10.95

THE TALL SHIPS OF TODAY IN PHOTOGRAPHS, Frank O. Braynard. Lavishly illustrated tribute to nearly 100 majestic contemporary sailing vessels: Amerigo Vespucci, Clearwater, Constitution, Eagle, Mayflower, Sea Cloud, Victory, many more. Authoritative captions provide statistics, background on each ship. 190 black-and-white photographs and illustrations. Introduction. 128pp. 8⅞ x 11¾. 27163-3 Pa. $13.95

EARLY NINETEENTH-CENTURY CRAFTS AND TRADES, Peter Stockham (ed.). Extremely rare 1807 volume describes to youngsters the crafts and trades of the day: brickmaker, weaver, dressmaker, bookbinder, ropemaker, saddler, many more. Quaint prose, charming illustrations for each craft. 20 black-and-white line illustrations. 192pp. 4⅝ x 6. 27293-1 Pa. $4.95

VICTORIAN FASHIONS AND COSTUMES FROM HARPER'S BAZAR, 1867–1898, Stella Blum (ed.). Day costumes, evening wear, sports clothes, shoes, hats, other accessories in over 1,000 detailed engravings. 320pp. 9⅜ x 12¼. 22990-4 Pa. $14.95

GUSTAV STICKLEY, THE CRAFTSMAN, Mary Ann Smith. Superb study surveys broad scope of Stickley's achievement, especially in architecture. Design philosophy, rise and fall of the Craftsman empire, descriptions and floor plans for many Craftsman houses, more. 86 black-and-white halftones. 31 line illustrations. Introduction 208pp. 6½ x 9¼. 27210-9 Pa. $9.95

THE LONG ISLAND RAIL ROAD IN EARLY PHOTOGRAPHS, Ron Ziel. Over 220 rare photos, informative text document origin (1844) and development of rail service on Long Island. Vintage views of early trains, locomotives, stations, passengers, crews, much more. Captions. 8⅞ x 11¾. 26301-0 Pa. $13.95

THE BOOK OF OLD SHIPS: From Egyptian Galleys to Clipper Ships, Henry B. Culver. Superb, authoritative history of sailing vessels, with 80 magnificent line illustrations. Galley, bark, caravel, longship, whaler, many more. Detailed, informative text on each vessel by noted naval historian. Introduction. 256pp. 5⅜ x 8½. 27332-6 Pa. $7.95

TEN BOOKS ON ARCHITECTURE, Vitruvius. The most important book ever written on architecture. Early Roman aesthetics, technology, classical orders, site selection, all other aspects. Morgan translation. 331pp. 5⅜ x 8½. 20645-9 Pa. $8.95

THE HUMAN FIGURE IN MOTION, Eadweard Muybridge. More than 4,500 stopped-action photos, in action series, showing undraped men, women, children jumping, lying down, throwing, sitting, wrestling, carrying, etc. 390pp. 7⅞ x 10⅝. 20204-6 Clothbd. $25.95

TREES OF THE EASTERN AND CENTRAL UNITED STATES AND CANADA, William M. Harlow. Best one-volume guide to 140 trees. Full descriptions, woodlore, range, etc. Over 600 illustrations. Handy size. 288pp. 4½ x 6⅜. 20395-6 Pa. $6.95

SONGS OF WESTERN BIRDS, Dr. Donald J. Borror. Complete song and call repertoire of 60 western species, including flycatchers, juncoes, cactus wrens, many more–includes fully illustrated booklet. Cassette and manual 99913-0 $8.95

GROWING AND USING HERBS AND SPICES, Milo Miloradovich. Versatile handbook provides all the information needed for cultivation and use of all the herbs and spices available in North America. 4 illustrations. Index. Glossary. 236pp. 5⅜ x 8½. 25058-X Pa. $6.95

BIG BOOK OF MAZES AND LABYRINTHS, Walter Shepherd. 50 mazes and labyrinths in all–classical, solid, ripple, and more–in one great volume. Perfect inexpensive puzzler for clever youngsters. Full solutions. 112pp. 8⅛ x 11. 22951-3 Pa. $4.95

PIANO TUNING, J. Cree Fischer. Clearest, best book for beginner, amateur. Simple repairs, raising dropped notes, tuning by easy method of flattened fifths. No previous skills needed. 4 illustrations. 201pp. 5⅜ x 8½. 23267-0 Pa. $6.95

A SOURCE BOOK IN THEATRICAL HISTORY, A. M. Nagler. Contemporary observers on acting, directing, make-up, costuming, stage props, machinery, scene design, from Ancient Greece to Chekhov. 611pp. 5⅜ x 8½. 20515-0 Pa. $12.95

THE COMPLETE NONSENSE OF EDWARD LEAR, Edward Lear. All nonsense limericks, zany alphabets, Owl and Pussycat, songs, nonsense botany, etc., illustrated by Lear. Total of 320pp. 5⅜ x 8½. (USO) 20167-8 Pa. $6.95

VICTORIAN PARLOUR POETRY: An Annotated Anthology, Michael R. Turner. 117 gems by Longfellow, Tennyson, Browning, many lesser-known poets. "The Village Blacksmith," "Curfew Must Not Ring Tonight," "Only a Baby Small," dozens more, often difficult to find elsewhere. Index of poets, titles, first lines. xxiii + 325pp. 5⅝ x 8¼. 27044-0 Pa. $8.95

DUBLINERS, James Joyce. Fifteen stories offer vivid, tightly focused observations of the lives of Dublin's poorer classes. At least one, "The Dead," is considered a masterpiece. Reprinted complete and unabridged from standard edition. 160pp. 5³⁄₁₆ x 8¼. 26870-5 Pa. $1.00

THE HAUNTED MONASTERY and THE CHINESE MAZE MURDERS, Robert van Gulik. Two full novels by van Gulik, set in 7th-century China, continue adventures of Judge Dee and his companions. An evil Taoist monastery, seemingly supernatural events; overgrown topiary maze hides strange crimes. 27 illustrations. 328pp. 5⅜ x 8½. 23502-5 Pa. $8.95

THE BOOK OF THE SACRED MAGIC OF ABRAMELIN THE MAGE, translated by S. MacGregor Mathers. Medieval manuscript of ceremonial magic. Basic document in Aleister Crowley, Golden Dawn groups. 268pp. 5⅜ x 8½. 23211-5 Pa. $8.95

NEW RUSSIAN-ENGLISH AND ENGLISH-RUSSIAN DICTIONARY, M. A. O'Brien. This is a remarkably handy Russian dictionary, containing a surprising amount of information, including over 70,000 entries. 366pp. 4½ x 6⅛. 20208-9 Pa. $9.95

HISTORIC HOMES OF THE AMERICAN PRESIDENTS, Second, Revised Edition, Irvin Haas. A traveler's guide to American Presidential homes, most open to the public, depicting and describing homes occupied by every American President from George Washington to George Bush. With visiting hours, admission charges, travel routes. 175 photographs. Index. 160pp. 8¼ x 11. 26751-2 Pa. $11.95

NEW YORK IN THE FORTIES, Andreas Feininger. 162 brilliant photographs by the well-known photographer, formerly with *Life* magazine. Commuters, shoppers, Times Square at night, much else from city at its peak. Captions by John von Hartz. 181pp. 9¼ x 10¾. 23585-8 Pa. $12.95

INDIAN SIGN LANGUAGE, William Tomkins. Over 525 signs developed by Sioux and other tribes. Written instructions and diagrams. Also 290 pictographs. 111pp. 6⅛ x 9¼. 22029-X Pa. $3.95

ANATOMY: A Complete Guide for Artists, Joseph Sheppard. A master of figure drawing shows artists how to render human anatomy convincingly. Over 460 illustrations. 224pp. 8⅜ x 11¼. 27279-6 Pa. $10.95

MEDIEVAL CALLIGRAPHY: Its History and Technique, Marc Drogin. Spirited history, comprehensive instruction manual covers 13 styles (ca. 4th century thru 15th). Excellent photographs; directions for duplicating medieval techniques with modern tools. 224pp. 8⅜ x 11¼. 26142-5 Pa. $12.95

DRIED FLOWERS: How to Prepare Them, Sarah Whitlock and Martha Rankin. Complete instructions on how to use silica gel, meal and borax, perlite aggregate, sand and borax, glycerine and water to create attractive permanent flower arrangements. 12 illustrations. 32pp. 5⅜ x 8½. 21802-3 Pa. $1.00

EASY-TO-MAKE BIRD FEEDERS FOR WOODWORKERS, Scott D. Campbell. Detailed, simple-to-use guide for designing, constructing, caring for and using feeders. Text, illustrations for 12 classic and contemporary designs. 96pp. 5⅜ x 8½. 25847-5 Pa. $2.95

SCOTTISH WONDER TALES FROM MYTH AND LEGEND, Donald A. Mackenzie. 16 lively tales tell of giants rumbling down mountainsides, of a magic wand that turns stone pillars into warriors, of gods and goddesses, evil hags, powerful forces and more. 240pp. 5⅜ x 8½. 29677-6 Pa. $6.95

THE HISTORY OF UNDERCLOTHES, C. Willett Cunnington and Phyllis Cunnington. Fascinating, well-documented survey covering six centuries of English undergarments, enhanced with over 100 illustrations: 12th-century laced-up bodice, footed long drawers (1795), 19th-century bustles, 19th-century corsets for men, Victorian "bust improvers," much more. 272pp. 5⅜ x 8¼. 27124-2 Pa. $9.95

ARTS AND CRAFTS FURNITURE: The Complete Brooks Catalog of 1912, Brooks Manufacturing Co. Photos and detailed descriptions of more than 150 now very collectible furniture designs from the Arts and Crafts movement depict davenports, settees, buffets, desks, tables, chairs, bedsteads, dressers and more, all built of solid, quarter-sawed oak. Invaluable for students and enthusiasts of antiques, Americana and the decorative arts. 80pp. 6½ x 9¼. 27471-3 Pa. $8.95

HOW WE INVENTED THE AIRPLANE: An Illustrated History, Orville Wright. Fascinating firsthand account covers early experiments, construction of planes and motors, first flights, much more. Introduction and commentary by Fred C. Kelly. 76 photographs. 96pp. 8¼ x 11. 25662-6 Pa. $8.95

THE ARTS OF THE SAILOR: Knotting, Splicing and Ropework, Hervey Garrett Smith. Indispensable shipboard reference covers tools, basic knots and useful hitches; handsewing and canvas work, more. Over 100 illustrations. Delightful reading for sea lovers. 256pp. 5⅜ x 8½. 26440-8 Pa. $7.95

FRANK LLOYD WRIGHT'S FALLINGWATER: The House and Its History, Second, Revised Edition, Donald Hoffmann. A total revision—both in text and illustrations—of the standard document on Fallingwater, the boldest, most personal architectural statement of Wright's mature years, updated with valuable new material from the recently opened Frank Lloyd Wright Archives. "Fascinating"—*The New York Times*. 116 illustrations. 128pp. 9¼ x 10¾. 27430-6 Pa. $11.95

PHOTOGRAPHIC SKETCHBOOK OF THE CIVIL WAR, Alexander Gardner. 100 photos taken on field during the Civil War. Famous shots of Manassas Harper's Ferry, Lincoln, Richmond, slave pens, etc. 244pp. 10⅝ x 8¼.   22731-6 Pa. $9.95

FIVE ACRES AND INDEPENDENCE, Maurice G. Kains. Great back-to-the-land classic explains basics of self-sufficient farming. The one book to get. 95 illustrations. 397pp. 5⅜ x 8½.   20974-1 Pa. $7.95

SONGS OF EASTERN BIRDS, Dr. Donald J. Borror. Songs and calls of 60 species most common to eastern U.S.: warblers, woodpeckers, flycatchers, thrushes, larks, many more in high-quality recording.   Cassette and manual 99912-2 $9.95

A MODERN HERBAL, Margaret Grieve. Much the fullest, most exact, most useful compilation of herbal material. Gigantic alphabetical encyclopedia, from aconite to zedoary, gives botanical information, medical properties, folklore, economic uses, much else. Indispensable to serious reader. 161 illustrations. 888pp. 6½ x 9¼. 2-vol. set. (USO)   Vol. I: 22798-7 Pa. $9.95
                                                                             Vol. II: 22799-5 Pa. $9.95

HIDDEN TREASURE MAZE BOOK, Dave Phillips. Solve 34 challenging mazes accompanied by heroic tales of adventure. Evil dragons, people-eating plants, blood-thirsty giants, many more dangerous adversaries lurk at every twist and turn. 34 mazes, stories, solutions. 48pp. 8¼ x 11.   24566-7 Pa. $2.95

LETTERS OF W. A. MOZART, Wolfgang A. Mozart. Remarkable letters show bawdy wit, humor, imagination, musical insights, contemporary musical world; includes some letters from Leopold Mozart. 276pp. 5⅜ x 8½.   22859-2 Pa. $7.95

BASIC PRINCIPLES OF CLASSICAL BALLET, Agrippina Vaganova. Great Russian theoretician, teacher explains methods for teaching classical ballet. 118 illustrations. 175pp. 5⅜ x 8½.   22036-2 Pa. $5.95

THE JUMPING FROG, Mark Twain. Revenge edition. The original story of The Celebrated Jumping Frog of Calaveras County, a hapless French translation, and Twain's hilarious "retranslation" from the French. 12 illustrations. 66pp. 5⅜ x 8½.
                                                                             22686-7 Pa. $3.95

BEST REMEMBERED POEMS, Martin Gardner (ed.). The 126 poems in this superb collection of 19th- and 20th-century British and American verse range from Shelley's "To a Skylark" to the impassioned "Renascence" of Edna St. Vincent Millay and to Edward Lear's whimsical "The Owl and the Pussycat." 224pp. 5⅜ x 8½.
                                                                             27165-X Pa. $4.95

COMPLETE SONNETS, William Shakespeare. Over 150 exquisite poems deal with love, friendship, the tyranny of time, beauty's evanescence, death and other themes in language of remarkable power, precision and beauty. Glossary of archaic terms. 80pp. 5¹⁵⁄₁₆ x 8¼.   26686-9 Pa. $1.00

BODIES IN A BOOKSHOP, R. T. Campbell. Challenging mystery of blackmail and murder with ingenious plot and superbly drawn characters. In the best tradition of British suspense fiction. 192pp. 5⅜ x 8½.   24720-1 Pa. $6.95

THE WIT AND HUMOR OF OSCAR WILDE, Alvin Redman (ed.). More than 1,000 ripostes, paradoxes, wisecracks: Work is the curse of the drinking classes; I can resist everything except temptation; etc. 258pp. 5⅜ x 8½. 20602-5 Pa. $5.95

SHAKESPEARE LEXICON AND QUOTATION DICTIONARY, Alexander Schmidt. Full definitions, locations, shades of meaning in every word in plays and poems. More than 50,000 exact quotations. 1,485pp. 6½ x 9¼. 2-vol. set.
Vol. 1: 22726-X Pa. $16.95
Vol. 2: 22727-8 Pa. $16.95

SELECTED POEMS, Emily Dickinson. Over 100 best-known, best-loved poems by one of America's foremost poets, reprinted from authoritative early editions. No comparable edition at this price. Index of first lines. 64pp. 5³⁄₁₆ x 8¼.
26466-1 Pa. $1.00

CELEBRATED CASES OF JUDGE DEE (DEE GOONG AN), translated by Robert van Gulik. Authentic 18th-century Chinese detective novel; Dee and associates solve three interlocked cases. Led to van Gulik's own stories with same characters. Extensive introduction. 9 illustrations. 237pp. 5⅜ x 8½. 23337-5 Pa. $6.95

THE MALLEUS MALEFICARUM OF KRAMER AND SPRENGER, translated by Montague Summers. Full text of most important witchhunter's "bible," used by both Catholics and Protestants. 278pp. 6⅝ x 10. 22802-9 Pa. $12.95

SPANISH STORIES/CUENTOS ESPAÑOLES: A Dual-Language Book, Angel Flores (ed.). Unique format offers 13 great stories in Spanish by Cervantes, Borges, others. Faithful English translations on facing pages. 352pp. 5⅜ x 8½.
25399-6 Pa. $8.95

THE CHICAGO WORLD'S FAIR OF 1893: A Photographic Record, Stanley Appelbaum (ed.). 128 rare photos show 200 buildings, Beaux-Arts architecture, Midway, original Ferris Wheel, Edison's kinetoscope, more. Architectural emphasis; full text. 116pp. 8¼ x 11. 23990-X Pa. $9.95

OLD QUEENS, N.Y., IN EARLY PHOTOGRAPHS, Vincent F. Seyfried and William Asadorian. Over 160 rare photographs of Maspeth, Jamaica, Jackson Heights, and other areas. Vintage views of DeWitt Clinton mansion, 1939 World's Fair and more. Captions. 192pp. 8⅞ x 11. 26358-4 Pa. $12.95

CAPTURED BY THE INDIANS: 15 Firsthand Accounts, 1750-1870, Frederick Drimmer. Astounding true historical accounts of grisly torture, bloody conflicts, relentless pursuits, miraculous escapes and more, by people who lived to tell the tale. 384pp. 5⅜ x 8½. 24901-8 Pa. $8.95

THE WORLD'S GREAT SPEECHES, Lewis Copeland and Lawrence W. Lamm (eds.). Vast collection of 278 speeches of Greeks to 1970. Powerful and effective models; unique look at history. 842pp. 5⅜ x 8½. 20468-5 Pa. $14.95

THE BOOK OF THE SWORD, Sir Richard F. Burton. Great Victorian scholar/adventurer's eloquent, erudite history of the "queen of weapons"–from prehistory to early Roman Empire. Evolution and development of early swords, variations (sabre, broadsword, cutlass, scimitar, etc.), much more. 336pp. 6⅛ x 9¼.
25434-8 Pa. $9.95

AUTOBIOGRAPHY: The Story of My Experiments with Truth, Mohandas K. Gandhi. Boyhood, legal studies, purification, the growth of the Satyagraha (nonviolent protest) movement. Critical, inspiring work of the man responsible for the freedom of India. 480pp. 5⅜ x 8½. (USO) 24593-4 Pa. $8.95

CELTIC MYTHS AND LEGENDS, T. W. Rolleston. Masterful retelling of Irish and Welsh stories and tales. Cuchulain, King Arthur, Deirdre, the Grail, many more. First paperback edition. 58 full-page illustrations. 512pp. 5⅜ x 8½. 26507-2 Pa. $9.95

THE PRINCIPLES OF PSYCHOLOGY, William James. Famous long course complete, unabridged. Stream of thought, time perception, memory, experimental methods; great work decades ahead of its time. 94 figures. 1,391pp. 5⅜ x 8½. 2-vol. set.
Vol. I: 20381-6 Pa. $12.95
Vol. II: 20382-4 Pa. $12.95

THE WORLD AS WILL AND REPRESENTATION, Arthur Schopenhauer. Definitive English translation of Schopenhauer's life work, correcting more than 1,000 errors, omissions in earlier translations. Translated by E. F. J. Payne. Total of 1,269pp. 5⅜ x 8½. 2-vol. set. Vol. 1: 21761-2 Pa. $11.95
Vol. 2: 21762-0 Pa. $12.95

MAGIC AND MYSTERY IN TIBET, Madame Alexandra David-Neel. Experiences among lamas, magicians, sages, sorcerers, Bonpa wizards. A true psychic discovery. 32 illustrations. 321pp. 5⅜ x 8½. (USO) 22682-4 Pa. $8.95

THE EGYPTIAN BOOK OF THE DEAD, E. A. Wallis Budge. Complete reproduction of Ani's papyrus, finest ever found. Full hieroglyphic text, interlinear transliteration, word-for-word translation, smooth translation. 533pp. 6½ x 9¼.
21866-X Pa. $10.95

MATHEMATICS FOR THE NONMATHEMATICIAN, Morris Kline. Detailed, college-level treatment of mathematics in cultural and historical context, with numerous exercises. Recommended Reading Lists. Tables. Numerous figures. 641pp. 5⅜ x 8½.
24823-2 Pa. $11.95

THEORY OF WING SECTIONS: Including a Summary of Airfoil Data, Ira H. Abbott and A. E. von Doenhoff. Concise compilation of subsonic aerodynamic characteristics of NACA wing sections, plus description of theory. 350pp. of tables. 693pp. 5⅜ x 8½. 60586-8 Pa. $14.95

THE RIME OF THE ANCIENT MARINER, Gustave Doré, S. T. Coleridge. Doré's finest work; 34 plates capture moods, subtleties of poem. Flawless full-size reproductions printed on facing pages with authoritative text of poem. "Beautiful. Simply beautiful."–*Publisher's Weekly.* 77pp. 9¼ x 12. 22305-1 Pa. $6.95

NORTH AMERICAN INDIAN DESIGNS FOR ARTISTS AND CRAFTSPEOPLE, Eva Wilson. Over 360 authentic copyright-free designs adapted from Navajo blankets, Hopi pottery, Sioux buffalo hides, more. Geometrics, symbolic figures, plant and animal motifs, etc. 128pp. 8⅜ x 11. (EUK) 25341-4 Pa. $8.95

SCULPTURE: Principles and Practice, Louis Slobodkin. Step-by-step approach to clay, plaster, metals, stone; classical and modern. 253 drawings, photos. 255pp. 8⅛ x 11.
22960-2 Pa. $11.95

THE INFLUENCE OF SEA POWER UPON HISTORY, 1660–1783, A. T. Mahan. Influential classic of naval history and tactics still used as text in war colleges. First paperback edition. 4 maps. 24 battle plans. 640pp. 5⅜ x 8½.   25509-3 Pa. $12.95

THE STORY OF THE TITANIC AS TOLD BY ITS SURVIVORS, Jack Winocour (ed.). What it was really like. Panic, despair, shocking inefficiency, and a little hero- ism. More thrilling than any fictional account. 26 illustrations. 320pp. 5⅜ x 8½.
20610-6 Pa. $8.95

FAIRY AND FOLK TALES OF THE IRISH PEASANTRY, William Butler Yeats (ed.). Treasury of 64 tales from the twilight world of Celtic myth and legend: "The Soul Cages," "The Kildare Pooka," "King O'Toole and his Goose," many more. Introduction and Notes by W. B. Yeats. 352pp. 5⅜ x 8½.   26941-8 Pa. $8.95

BUDDHIST MAHAYANA TEXTS, E. B. Cowell and Others (eds.). Superb, accu- rate translations of basic documents in Mahayana Buddhism, highly important in his- tory of religions. The Buddha-karita of Asvaghosha, Larger Sukhavativyuha, more. 448pp. 5⅜ x 8½.   25552-2 Pa. $12.95

ONE TWO THREE . . . INFINITY: Facts and Speculations of Science, George Gamow. Great physicist's fascinating, readable overview of contemporary science: number theory, relativity, fourth dimension, entropy, genes, atomic structure, much more. 128 illustrations. Index. 352pp. 5⅜ x 8½.   25664-2 Pa. $8.95

ENGINEERING IN HISTORY, Richard Shelton Kirby, et al. Broad, nontechnical survey of history's major technological advances: birth of Greek science, industrial revolution, electricity and applied science, 20th-century automation, much more. 181 illustrations. ". . . excellent . . ."—*Isis.* Bibliography. vii + 530pp. 5⅜ x 8¼.
26412-2 Pa. $14.95

DALÍ ON MODERN ART: The Cuckolds of Antiquated Modern Art, Salvador Dalí. Influential painter skewers modern art and its practitioners. Outrageous evalu- ations of Picasso, Cézanne, Turner, more. 15 renderings of paintings discussed. 44 calligraphic decorations by Dalí. 96pp. 5⅜ x 8½. (USO)   29220-7 Pa. $4.95

ANTIQUE PLAYING CARDS: A Pictorial History, Henry René D'Allemagne. Over 900 elaborate, decorative images from rare playing cards (14th–20th centuries): Bacchus, death, dancing dogs, hunting scenes, royal coats of arms, players cheating, much more. 96pp. 9¼ x 12¼.   29265-7 Pa. $11.95

MAKING FURNITURE MASTERPIECES: 30 Projects with Measured Drawings, Franklin H. Gottshall. Step-by-step instructions, illustrations for constructing hand- some, useful pieces, among them a Sheraton desk, Chippendale chair, Spanish desk, Queen Anne table and a William and Mary dressing mirror. 224pp. 8⅛ x 11¼.
29338-6 Pa. $13.95

THE FOSSIL BOOK: A Record of Prehistoric Life, Patricia V. Rich et al. Profusely illustrated definitive guide covers everything from single-celled organisms and dinosaurs to birds and mammals and the interplay between climate and man. Over 1,500 illustrations. 760pp. 7½ x 10⅛.   29371-8 Pa. $29.95

Prices subject to change without notice.

Available at your book dealer or write for free catalog to Dept. GI, Dover Publications, Inc., 31 East 2nd St., Mineola, N.Y. 11501. Dover publishes more than 500 books each year on science, elementary and advanced mathematics, biology, music, art, literary history, social sciences and other areas.